D1757542

WITHDRAWN

009957

BIOTECHNOLOGY IN PLANT DISEASE CONTROL

WILEY SERIES IN
ECOLOGICAL AND
APPLIED MICROBIOLOGY

BIOTECHNOLOGY IN PLANT DISEASE CONTROL

Edited by ILAN CHET

**Department of Plant Pathology and Microbiology,
The Hebrew University of Jerusalem, Faculty of Agriculture, Rehovot, Israel**

 **WILEY-LISS**

A JOHN WILEY & SONS, INC., PUBLICATION
New York • Chichester • Brisbane • Toronto • Singapore

Address All Inquiries to the Publisher
Wiley-Liss, Inc., 605 Third Avenue, New York, NY 10158-0012

Copyright © 1993 Wiley-Liss, Inc.

Printed in the United States of America.

Library of Congress Cataloging-in-Publication Data

Biotechnology in plant disease control / edited by Ilan Chet.
        p. cm. — (Wiley series in ecological and applied microbiology)
        Includes bibliographical references and index.
        ISBN 0-471-56084-7
        1. Plants—Disease and pest resistance—Genetic aspects.
2. Phytopathogenic microorganisms—Biological control. 3. Plant diseases. 4. Agricultural biotechnology. I. Chet, Ilan, 1939–   .
II. Series.
SB750.B56 1992
632'.3—dc20   92-32197
                                                  CIP

The text of this book is printed on acid-free paper.

"Imagination is more important than knowledge"

ALBERT EINSTEIN

To
Ruthie
Gal, Guy, Dana, Tal, and Tom

CONTENTS

CONTRIBUTORS   xi

PREFACE   xv
Ilan Chet

I. IMPACT

1. THE IMPACT OF BIOTECHNOLOGY ON PLANT
   BREEDING, OR HOW TO COMBINE INCREASES IN
   AGRICULTURAL PRODUCTIVITY WITH AN IMPROVED
   PROTECTION OF THE ENVIRONMENT   1
   J. Logemann and J. Schell

2. INFLUENCE OF BIOTECHNOLOGY ON BIOCONTROL
   OF TAKE-ALL DISEASE OF WHEAT   15
   Bruce C. Hemming and John M. Houghton

II. MOLECULAR GENETIC RESEARCH IN DISEASE CONTROL

3. PROTEIN SECRETION BY PLANT PATHOGENIC BACTERIA 39
   Sheng Yang He, Magdalen Lindeberg, and Alan Collmer

4. PLANT DISEASE RESISTANCE GENES: INTERACTIONS
   WITH PATHOGENS AND THEIR IMPROVED UTILIZATION
   TO CONTROL PLANT DISEASES   65
   N.T. Keen, Andrew Bent, and Brian Staskawicz

III. INTRODUCTION OF DISEASE RESISTANCE INTO PLANTS

5. VIRUS RESISTANCE THROUGH EXPRESSION OF COAT
   PROTEIN GENES   89
   Roger N. Beachy

6. THE LOCAL LESION RESPONSE TO VIRUSES:
    POSSIBILITIES FOR ENGINEERING RESISTANT PLANTS   **105**
    G. Loebenstein and A. Gera

7. TRANSGENIC PLANTS RESISTANT TO DISEASES BY
    THE DETOXIFICATION OF TOXINS   **115**
    Katsuyoshi Yoneyama and Hiroyuki Anzai

8. THE ROLE OF CELL WALL DEGRADING ENZYMES IN
    FUNGAL DISEASE RESISTANCE   **139**
    Karen Broglie, Richard Broglie, Nicole Benhamou,
    and Ilan Chet

9. PROTEINASE INHIBITORS IN THE POTATO
    RESPONSE TO WOUNDING   **157**
    Jose J. Sánchez-Serrano, Simone Amati, Christian Dammann,
    Marcus Ebneth, Karin Herbers, Thomas Hildmann,
    Ruth Lorberth, Salomé Prat, and Lothar Willmitzer

10. USING GENES ENCODING NOVEL PEPTIDES AND
    PROTEINS TO ENHANCE DISEASE RESISTANCE
    IN PLANTS   **175**
    Luis Destéfano-Beltrán, Pablito G. Nagpala,
    Selim M. Cetiner, Timothy Denny, and Jesse M. Jaynes

11. ROLE OF PHENOLICS IN PLANT DISEASE RESISTANCE   **191**
    J.P. Métraux and I. Raskin

**IV. APPLICATION OF BIOTECHNOLOGY FOR IMPROVING
BIOLOGICAL CONTROL**

12. GENETIC ENGINEERING OF MICROORGANISMS FOR
    IMPROVED BIOCONTROL ACTIVITY   **211**
    I. Chet, Z. Barak, and A. Oppenheim

13. THE GENETIC NATURE AND BIOCONTROL ABILITY
    OF PROGENY FROM PROTOPLAST FUSION IN
    TRICHODERMA   **237**
    G.E. Harman and C.K. Hayes

14. FUNGAL "KILLER" TOXINS AS POTENTIAL AGENTS
    FOR BIOCONTROL   **257**
    Y. Koltin, I. Ginzberg, and A. Finkler

15. BIOLOGICAL CONTROL OF PLANT PESTS AND
    PATHOGENS: ALTERNATIVE APPROACHES   **275**
    Jennifer A. Thomson

16. BIOLOGICAL ACTIVITIES OF BACTERIA USED IN
    PLANT PATHOGEN CONTROL   **291**
    Stephen T. Lam and Thomas D. Gaffney

17. DIAGNOSTIC TECHNIQUES FOR PLANT PATHOGENS   **321**
    S.A. Miller and T.R. Joaquim

18. APPLICATION OF DNA FINGERPRINTING FOR
    DETECTING GENETICAL VARIATION AMONG
    ISOLATES OF THE WHEAT PATHOGEN
    MYCOSPHAERELLA GRAMINICOLA   **341**
    A. Zilberstein, S. Pnini-Cohen, Z. Eyal, S. Shuster,
    J. Hillel, and U. Lavi

    INDEX   **355**

CONTRIBUTORS

SIMONE AMATI, Institut für Genbiologische Forschung Berlin GmbH, 1000 Berlin 33, Germany **[157]**

HIROYUKI ANZAI, Pharmaceutical Research Center, Meiji Seika Kaisha, Ltd., Kanagawa 222, Japan **[115]**

Z. BARAK, The Hebrew University of Jerusalem, Otto Warburg Center for Agricultural Biotechnology, Faculty of Agriculture, Rehovot 76100, Israel **[211]**

ROGER N. BEACHY, Division of Plant Biology, The Scripps Research Institute, La Jolla, CA 92037 **[89]**

NICOLE BENHAMOU, Department of Phytology, Laval University, Quebec, Canada **[139]**

ANDREW BENT, Department of Plant Pathology, University of California, Berkeley, CA 94720 **[65]**

KAREN BROGLIE, Experimental Station, E.I. duPont de Nemours & Co. Agricultural Products, Wilmington, DE 19880 **[139]**

RICHARD BROGLIE, Experimental Station, E.I. du Pont de Nemours & Co. Agricultural Products, Wilmington, DE 19880 **[139]**

SELIM M. CETINER, Cukurova University, Adana, Turkey **[175]**

ILAN CHET, The Hebrew University of Jerusalem, Otto Warburg Center for Agricultural Biotechnology, Faculty of Agriculture, Rehovot 76100, Israel **[xv,139,211]**

ALAN COLLMER, Department of Plant Pathology, Cornell University, Ithaca, NY 14853 **[39]**

CHRISTIAN DAMMANN, Institut für Genbiologische Forschung Berlin GmbH, 1000 Berlin 33, Germany **[157]**

The numbers in brackets are the opening page numbers of the contributors' articles.

TIMOTHY DENNY, Department of Plant Pathology, University of Georgia, Athens, GA [175]

LUIS DESTÉFANO-BELTRÁN, Laboratorium voor Genetica, Rijksuniversiteit Gent, B-9000 Gent, Belgium [175]

MARCUS EBNETH, Institut für Genbiologische Forschung Berlin GmbH, 1000 Berlin 33, Germany [157]

Z. EYAL, Department of Botany, The George S. Wise Faculty of Life Sciences, Tel Aviv University, Tel Aviv 69978, Israel [341]

A. FINKLER, Department of Molecular Microbiology and Biotechnology, Faculty of Life Sciences, Tel Aviv University, Tel Aviv 69978, Israel [257]

THOMAS D. GAFFNEY, Agricultural Biotechnology Research Unit, CIBA-Geigy Corporation, Research Triangle Park, NC 27709 [291]

A. GERA, Department of Virology, Agricultural Research Organization, The Volcani Center, Bet Dagan 50250, Israel [105]

I. GINZBERG, Department of Molecular Microbiology and Biotechnology, Faculty of Life Sciences, Tel Aviv University, Tel Aviv 69978, Israel [257]

G.E. HARMAN, Cornell University, Geneva, NY 14456 [237]

C.K. HAYES, Cornell University, Geneva, NY 14456 [237]

SHENG YANG HE, Department of Plant Pathology, Cornell University, Ithaca, NY 14853 [39]

BRUCE C. HEMMING, Microbe Inotech Laboratories, Inc., St. Louis, MO 63146 [15]

KARIN HERBERS, Institut für Genbiologische Forschung Berlin GmbH, 1000 Berlin 33, Germany [157]

THOMAS HILDMANN, Institut für Genbiologische Forschung Berlin GmbH, 1000 Berlin 33, Germany [157]

J. HILLEL, Department of Genetics, Faculty of Agriculture, The Hebrew University of Jerusalem, Rehovot 76100, Israel [341]

JOHN M. HOUGHTON, Stewart Technology Management, St. Louis, MO 63146 [15]

JESSE M. JAYNES, Department of Biochemistry, Louisiana State University and Agricultural and Mechanical College, Baton Rouge, LA 70803 [175]

T.R. JOAQUIM, Agri-Diagnostics Associates, Cinnaminson, NJ 01809 **[321]**

N.T. KEEN, Department of Plant Pathology and Graduate Group in Genetics, University of California, Riverside, CA 92521 **[65]**

Y. KOLTIN, Department of Molecular Microbiology and Biotechnology, Faculty of Life Sciences, Tel Aviv University, Tel Aviv 69978, Israel **[257]**

STEPHEN T. LAM, Agricultural Biotechnology Research Unit, CIBA-Geigy Corporation, Research Triangle Park, NC 27709 **[291]**

U. LAVI, Department of Horticulture Genetics and Breeding, Agricultural Research Organization, The Volcani Center, Bet Dagan 50250, Israel **[341]**

MAGDALEN LINDEBERG, Department of Plant Pathology, Cornell University, Ithaca, NY 14853 **[39]**

G. LOEBENSTEIN, Department of Virology, Agricultural Research Organization, The Volcani Center, Bet Dagan 50250, Israel **[105]**

J. LOGEMANN, Max-Planck-Institut für Züchtungsforschung, D-5000 Köln 30, Germany **[1]**

RUTH LORBERTH, Institut für Genbiologische Forschung Berlin GmbH, 1000 Berlin 33, Germany **[157]**

J.P. MÉTRAUX, Institut de Biologie Végétale et de Phytochimie, Université de Fribourg Suisse, CH-1700 Fribourg, Switzerland **[191]**

S.A. MILLER, Department of Plant Pathology, The Ohio State University, Wooster, OH 44691 **[321]**

PABLITO G. NAGPALA, Plant Molecular Biology, The Upjohn Co., Kalamazoo, MI **[175]**

A. OPPENHEIM, The Hebrew University, Hadassa Medical School, Jerusalem 90101, Israel **[211]**

S. PNINI-COHEN, Department of Botany, The George S. Wise Faculty of Life Sciences, Tel Aviv University, Tel Aviv 69978, Israel **[341]**

SALOMÉ PRAT, Institut für Genbiologische Forschung Berlin GmbH, 1000 Berlin 33, Germany **[157]**

I. RASKIN, AgBiotech Center, Rutgers University, Cook College, New Brunswick, NJ 08903 **[191]**

Jose J. Sánchez-Serrano, Institut für Genbiologische Forschung Berlin GmbH, 1000 Berlin 33, Germany [157]

J. Schell, Max-Planck-Institut für Züchtungsforschung, D-5000 Köln 30, Germany [1]

S. Shuster, Department of Botany, The George S. Wise Faculty of Life Sciences, Tel Aviv University, Tel Aviv 69978, Israel [341]

Brian Staskawicz, Department of Plant Pathology, University of California, Berkeley, CA 94720 [65]

Jennifer A. Thomson, Department of Microbiology, University of Cape Town, Private Bag Rondebosch 7700, South Africa [275]

Lothar Willmitzer, Institut für Genbiologische Forschung Berlin GmbH, 1000 Berlin 33, Germany [157]

Katsuyoshi Yoneyama, Laboratory of Plant Pathology, Faculty of Agriculture, Meiji University, Kanagawa 214, Japan [115]

A. Zilberstein, Department of Botany, The George S. Wise Faculty of Life Sciences, Tel Aviv University, Tel Aviv 69978, Israel [341]

PREFACE

Biotechnology, a relatively modern term, defined by the European Federation of Biotechnology as "the integrated use of biochemistry, microbiology and engineering sciences in order to achieve technological (industrial) application of the capabilities of microorganisms, cultured tissue cells and parts thereof," has been practiced for thousands of years. Wine production was described in the Old Testament in the Book of Isaiah.

Biotechnology, in conjunction with conventional breeding programs, could make significant contributions to sustainable agriculture, by producing improved crops that are more compatible with their environment. In recent years, biotechnology has developed rapidly as a practical means for improving economically important crops. Plant biotechnology research has made dramatic progress, and investments by the industrial countries has yielded product prototypes that are now being tested in the field.

There is growing concern, both in developed and developing countries, about the use of hazardous fungicides for controlling plant diseases. We now know that the accumulation of pesticides in the soil and water interferes with numerous biological activities. The fascinating developments in biotechnology on the one hand, and this concern on the other, have led to intensive research in agricultural biotechnology aimed at plant protection. This includes, among others, disease-free clones of fruits, vegetables, and ornamental crops; plants resistant to insects and microbial pathogens; herbicide-tolerant cultivars; and biopesticides.

Agricultural biotechnology offers potential benefits in plant production. Despite the development that has already been achieved, success stories, although significant, are few and far between. The realization of this potential requires more fundamental as well as applied research. Because of the interdisciplinary nature of biotechnology, it requires cooperation between scientists, farmers, and experts in a wide range of disciplines.

This book focuses on the rapidly emerging field of modern biotechnology as it relates to plant disease control. It consists of chapters contributed by leading experts directly involved in biotechnological research and development efforts in plant pathology. The book opens with the impact of biotechnology on both plant breeding and biological control and then covers the main subjects related to molecular genetic research in disease control, i.e., the introduction of disease resistance into plants by genetic engineering and the improvement of biological control by biotechnological means.

In a few years we will probably be able to exploit the potential of this biotechnological research for agricultural practice. I hope this book will be of interest to a wide audience and serve as a useful reference source to researchers and students interested in agricultural biotechnology.

I wish to thank Professor Ralph Mitchell for his stimulating ideas during the development of this book.

ILAN CHET
Rehovot, Israel

1

THE IMPACT OF BIOTECHNOLOGY ON PLANT BREEDING, OR HOW TO COMBINE INCREASES IN AGRICULTURAL PRODUCTIVITY WITH AN IMPROVED PROTECTION OF THE ENVIRONMENT

J. LOGEMANN
J. SCHELL

Max-Planck-Institut für Züchtungsforschung, D-5000 Köln 30, Germany

1. INTRODUCTION

It is often argued that some developed countries presently suffer from overproduction of agricultural products and, hence, that there is no need to intensify research aimed at further increasing agricultural productivity.

There are, however, three main reasons why it is imperative to stimulate research aimed at increasing plant productivity.

First of all it is well known that for many parts of the world, agriculture is a

Biotechnology in Plant Disease Control, pages 1–14
© *1993 Wiley-Liss, Inc.*

major and essential economic activity. No economic system that is not put to its maximum productivity will ever survive or be successful in a competitive market. The alternative to competitiveness is an endless series of more or less artificial subventions.

The second reason is very obvious and has been both used and misused in various discussions. Even moderate projections predict that the world population will reach a minimum of 8 billion people in the next 20–30 years (up from about 5 billion at this point of time). If we assume fast growing populations will use land in whatever way they can to produce food, and, if the land is used with very low productivity, this will lead to massive destruction of the environment on a world scale. This in turn will unavoidably result in a general deterioration of life. Both human solidarity and good common sense should therefore induce us to invest in research leading to improvement of agricultural productivity.

Third, although agriculture has made remarkable progress in the last decades, this success was achieved at a price. Input of mechanical energy sources, of chemicals, of improved methods of agronomy, and of vastly improved breeding methods are mostly responsible for this success. But the price for this high input agriculture, as we now realize, is a nonsustainable drain on the environment, leading ultimately to lasting damage. We cannot sustain this kind of agriculture, whatever its success, if the price for this success is an unacceptable destruction of the environment. This general goal is particularly important in relation to plant protection.

We therefore have a paradox. On the one hand, we have to increase the productivity of agriculture, on the other hand we cannot afford to continue to increase the drain on the environment, since soon all usable land will actually be in use. It is therefore imperative that we become more knowledgeable, acquire new technologies, and use all of our scientific and technological capacity to increase agricultural productivity without continuing to increase the negative impact of this agriculture on the environment.

For the moment we cannot see many different ways to achieve this goal. Probably the most realistic one will be to increase the impact of plant breeding in agriculture. That is to say, to modify the properties of plants in such a way that increased productivity can be achieved without a concomitant need for the kind of high input that is destructive to the environment. All available know-how will be needed if one is to achieve the proposed goal in time (i.e., in a few decades). Modern plant breeding relies on several different techniques ranging from crosses and selection by tissue cultures all the way to the integration in the plant hereditary program of well-defined genes (i.e., molecular plant breeding). This chapter shall focus primarily on the potential of the latter technique.

It is important to stress, however, that this technology cannot be considered or really understood by itself but has to be seen as a component of plant breeding. It is also not conceivable that any improvement in breeding will, by itself, solve the general problem if we do not remain aware of the impact of agriculture on the environment and carefully plan and organize agriculture accordingly. Whatever progress we make, there will be no way around the necessity to have bioconservation, that is to say to conserve the genetic and environmental potential and diversity that is available in our world.

Gene technology is an essential and crucial element in modern plant breeding. Our present capacity to introduce well-defined genes into plants originally developed from the study of certain types of naturally occurring plant tumors. The regulation of the growth properties of cells in these plant tumors is modified by the activity of oncogenes. The function of these oncogenes is to produce or modify the action of cellular growth factors. In the case of tumors induced by the plant pathogen *Agrobacterium tumefaciens,* it was found that the neoplasmic transformation is the direct consequence of the introduction by this soil bacteria of well-defined genes into the nucleus of plant cells. These transferred genes will lead to the deregulated production or modification of growth factors in the transformed plant cells. In fact, we are looking at a natural instance of genetic engineering. These bacteria indeed introduce defined genetic properties in plant cells with the result that the modified plant cells will proliferate and produce nutrients for the benefit of the bacteria. It was relatively straightforward to modify this system and to develop it as a general gene–vector system for plants by removing the oncogenes from the bacterial Ti plasmid (for *T*umor *i*nducing) and by replacing them by any gene of interest for research or for agricultural purposes. The soil bacteria (*A. tumefaciens*) will transfer to a large number of different plants any gene sequence that is introduced in the appropriate way into their gene transfer system. With this method, as with several others that have since been developed, it is now possible to introduce well-defined segments of DNA in a wide variety of crops (see Table 1.1). A major advantage of molecular breeding is that the source of the new genetic information to be introduced in plants is essentially unlimited. Thus genetic information from bacteria, other plants, animal cells, viruses, or yeasts (in short whatever source) can be added in a functional form to the genetic information already existing in crop plants. When done with appropriate care, this "new" genetic information will not modify the other properties of the crops, unless this was the aim of the exercise. Another advantage is that the addition of a "useful trait" by this method of breeding will not disrupt an "elite" phenotype. This can be a problem in conventional breeding because of genetic linkage between "beneficial" genes. In conventional breeding whole genomes are transferred and segregate, and it is often quite difficult to separate *beneficial* from *deleterious* genes when these are closely linked in one of the parents. By the introduction of single, well-defined genes, that usually do not disrupt the *elite* phenotype of given cultivar, this problem of conventional breeding can be avoided.

Therefore, this new technology represents an important additional possibility for plant breeding. One can leave the genetic properties of crop plants as they are and add to them well-defined new properties through the addition of single genes to the genetic program of plant cells.

TABLE 1.1. Transgenic Plants of Agricultural Importance

Maize	Rice
Celery	Asparagus
Sugarbeet	Oilseed rape
Turnip	Cucumber
Carrot	Soybean
Cotton	Sunflower
Walnut	Lettuce
Flax	Birds-foot trefoil
Tomato	Alfalfa
Tobacco	Petunia
French bean	Poplar
Potato	Clover
Cowpea	Pea
Pear	Cabbage

The use of gene vectors combined with the general methods known as recombinant DNA technology allows molecular breeders to remove pieces of DNA from one organism, study their function, and add them to the existing genetic properties of defined plants. An important element for the use of this system was an understanding of the general functional organization of plant genes. Mainly, one needed to answer the question What does it take for a foreign segment of DNA to become active after transfer and integration in the plant cellular nucleus? Indeed when, for example, bacterial DNA sequences were first introduced in plants in the late 1970s, they did not lead to the desired phenotypic effects. This was shown to be a result of these ''foreign'' genes not being ''expressed'' in their new cellular host, that is, no messenger ribonucleic acids (mRNAs) or proteins were made. Because the genetic code is universal and a given gene will code for the same protein whether that gene is part of its natural cellular host or part of new host cells, it became apparent that plant genes, as well as animal, viral, or yeast genes, carry specific ''regulatory'' sequences. These ''regulatory'' sequences govern gene expression (i.e., transcription to mRNA and subsequent translation in specific proteins). A foreign gene, i.e., a gene not of plant origin, that carries its own regulatory sequences will usually not be expressed in plant cells. It was, however, reasonably straightforward to show that one could circumvent the problem by constructing ''chimeric'' genes. These *chimeric* genes combine regulatory sequences derived from well-defined plant genes with the protein-coding sequences derived from *foreign* genes. A large number of experiments with various chimeric genes introduced in plants have shown that this simple strategy works and that one can construct plant genes with predictable and specific properties. For instance, consider the goal of constructing genes with which one could change the color of flowers or produce nematode controlling proteins in roots. The principle of the method involves taking a plant gene that is active in the part of the plant that one is interested in having the new property (e.g., flower or root), taking the regulatory sequences from this gene, and linking them to the protein coding sequence of a foreign gene that controls pigmentation (flower color) or nematode growth. Thus chimeric plant genes are constructed, which upon transfer to a particular plant, will convey onto the flower or the root of this plant the desired new property.

In summary, the following tools are now available. (1) We can isolate genes and introduce them in living plants such as they become a stable component of the genetic program of these plants. (2) Provided the foreign genes are appropriately restructured, they will usually function in a predictable fashion after their introduction in the plant cells.

One major advantage of molecular breeding occurs because when a particular gene has been isolated and reconstructed, it can be used in a wide variety of cultivars of different crops. Thus useful genes can first be tested in model plants that are easy to handle, such as tobacco, and later can be used in a large variety of crops.

To illustrate how rapidly this science is progressing, one should mention that transgenic plants were first obtained in the laboratory in 1983 and field tests with transgenic tobacco and tomatoes were first perfomed in 1987!

Certainly one of the most important goals of plant breeding must be to produce plants having properties that will help to stabilize yields and to render their cultivation sustainable from an ecological perspective. In consequence, significant research and development efforts are presently geared towards the production of plants with a high degree of tolerance or resistance to pests (insects, nematodes, etc.), diseases (viruses, fungi, or bacteria), or climatic stresses (drought, heat, cold, salinity, etc.). One can illustrate such attempts by discussing the following examples.

2. INSECT TOLERANCE

Some bacteria (e.g., *Bacillus thuringiensis* or B.t.) produce peptide toxins that bind to specific receptors in the intestine of insects and thus are specifically toxic to these insects. By taking the genes that code for these peptide toxins and expressing them in plants, several research groups have been able to produce transgenic tobaccos, tomatoes, potatoes, cotton, and so on, which express the bacterial toxin in their leaves, flowers, or other organs and are thereby effectively protected against attack by some insects. The insect larvae will feed from these plants, but as a result of the presence of the toxin their digestion is severely impaired, and they stop feeding, and ultimately die.

The results indicate that a significant measure of additional crop protection can be achieved by genetic means. It would, however, be unrealistic to think that biological protection can ultimately totally replace the use of chemical insecticides in intensive agriculture. However, the combination of breeding plants harboring different genes that reduce their sensitivity to pests, with responsible use of environmentally acceptable pesticides, can be expected to provide great advantages both in making agriculture more efficient and economical and in reducing its negative impact on the environment. Indeed, by combining several different methods to achieve pest control, one will drastically reduce the probability that any of these methods will soon become obsolete as a result of adaptation by the pest.

Thus, useful pesticides would not rapidly become inefficient and could be used at lower concentrations and the biological control achieved by products of a number of specific genes would not as readily be broken down. In this respect it might be of importance to mention that various other strategies aimed at the genetic containment of deleterious insect populations can now be investigated. For example, one could try to have crop plants produce insect-specific growth factors that would interfere with the reproduction of deleterious insects by preventing larval development into adult insects. The goal would be to combine for one and the same crop several different insect controlling properties. Thus, one could keep insect populations in check and by appropriate regulations and agronomical procedures one could prevent the total eradication of certain insect populations. The aim cannot be and does not have to be insect eradication but, instead, control over the size of insect populations.

3. VIRUS RESISTANCE

Very significant progress has recently been made in protecting crops against a number of viral diseases. It was found that if one breeds plants so that they will express a particular protein that is normally part of the envelope of the virus particle, this virus-derived protein will, also under field conditions, prevent infecting viruses from multiplying in these plants. This strategy appears to have general value and might well be used to control many different plant viral diseases. Other methods that have been shown to have potential are based on the interference with viral replication by the expression of specific anti-sense RNAs or of so-called satellite RNAs, or of so-called "Ribozymes." Ribozymes are RNA sequences capable of introducing breaks at specific sites in other RNA molecules. It is important to realize that no chemical agents are presently available that protect plants directly against viruses.

4. TOLERANCE TO FUNGAL DISEASES

The destruction of crop plants by fungal pathogen is a serious problem world-wide and annually leads to losses of about 15%. In order to prevent further losses, fungicids are used which are helpful in several (not all) cases but can also cause environmental problems. To prevent plants from being destroyed by fungal pathogens an alternative approach is to introduce genes into the plant genome which encode proteins with antifungal activity and therefore enable the plant to protect itself against fungal attack. Several genes with antifungal activity have been isolated from plants and from various soil organisms (bacteria and fungi). Such microorganisms often produce enzymes (e.g., chitinases) that attack the cell wall of plant pathogenic fungi. Chitinases, along with β-1,3-glucanases, have also been shown to be part of the natural mechanism used by a variety of plants to resist attack by pathogenic fungi. Obviously, those fungi that successfully attack a given crop (i.e., those known to be effective pathogens) have in turn somehow developed ways to resist or avoid the plant defense enzymes. By using chitinases (e.g., of bacterial origin), one can hope to break through this evolutionary mechanism of host–pathogen adaptation.

This approach might be successful. Indeed, strains of *Serratia marcescens* were isolated and shown to be effective in the biocontrol of a number of plant pathogenic fungi (e.g., *Sclerotium rolfsii*). Subsequently, it was shown that this biocontrol relied on the production and excretion of a chitinase. This chitinase (*chiA*) turned out to be very stable, resistant to heat, and to be able to inhibit fungal growth *in vitro*. A chimeric gene was constructed, introduced in a model plant, and shown to be active in the transgenic plants. These plants exhibited a markedly increased tolerance towards attack by the plant pathogen *Rhizocotonia solani*. The principle of introducing antifungal properties in defined crops to which the particular fungal pathogens are not adapted, was also at the basis of another approach. The seeds of several cereals are known to contain proteins that are toxic to some predators and some pathogens. Thus barley was shown to contain a *r*ibosome-*i*nhibiting *p*rotein (RIP: inhibits protein synthesis in target cells by specific RNA *N*-glycosidase modification of 28S rRNA), which while not toxic to plants, appeared to strongly inhibit the growth of a number of plant pathogenic fungi. Again, by transferring a chimeric gene capable of expressing barley RIP in the stem and roots of tobacco as a model plant, it was shown that the transgenic plants acquired the capacity to resist attack by *R. solani* (Fig. 1.1).

5. PROSPECTS FOR RESISTANCE TO BACTERIA AND NEMATODES

Because of the relative ease with which "foreign" proteins can be synthesized in plants from appropriate chimeric genes, it is realistic to predict that soon effective genes producing enzymes or other proteins and peptides that are toxic to a variety of plant pathogenic bacteria or nematodes, will be developed. It has, for

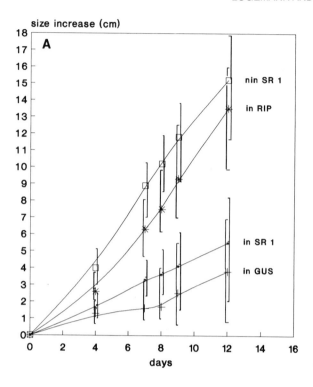

in GUS   in SR1   in RIP   nin SR 1

instance, been shown that the lymph glands of honeybees contain small proteins that have the capacity to insert themselves in the membranes of bacteria thus inactivating these bacteria. The bee gene coding for these antibacterial peptides could be isolated and used to express them in crops with the expectation that such transgenic crops will be tolerant to attack by some bacterial pathogens.

Another prospect for conveying resistance to bacterial pathogens is a chimeric gene expressing phage T4 lysozyme in plants in such a way that the lysozyme is targeted to move to the intercellular spaces of transgenic plants.

6. STRESS TOLERANCE

Environmental stresses severely affect the sustainability of agriculture. Plant molecular biologists are presently in the process of identifying and isolating genes that play a role in the physiological adaption of plants to environmental stresses, in an attempt to understand the mechanisms that are responsible for certain plants being able to tolerate extremes of temperature (high or low) or water and salt stresses better than others.

Several approaches are being used, among them one should mention:

1. Differential hybridization of complementary DNA (cDNA) libraries to plant mRNAs isolated from stressed and nonstressed tissues and from tolerant and sensitive cultivars.

2. Inactivation of stress tolerance by "gene tagging." Transposable DNA elements and T-DNA inserts can be used to inactivate genes. Subsequently, the transposable element or the T-DNA sequence are used as probes or "tags" to isolate the inactivated genes and with these the corresponding active genes. In this respect it is important to mention that transposable elements from corn have been shown to function in transgenic dicotyledonous plants. Transposable elements and T-DNA vectors have been developed specifically for the purpose of gene tagging. Thus it was possible not only to develop "tags" that can inactivate genes, but also tags that can activate silent genes and turn them into dominant, deregulated alleles.

3. Identification and isolation of complex loci involved in stress tolerance (or quantitative traits), by looking for their linkage to carefully mapped restriction

Fig. 1.1. A: RIP-transgenic plants are more protected against *R. solani* infection than control plants. The soil fungus *R. solani* causes disease on roots and lower stems of tobacco plants which leads to reduced growth and in extreme cases to plant death. Three independent tests on 15 tobacco plants of the same size (2 cm) for each of the following type were grown in soil inoculated with *R. solani* mycelia: control, R_1 *wun1*-GUS transgenic (in GUS); control nontransformed (in SRI); R1 *wun1*-RIP transgenic tobacco (in RIP). Average height for each plant type was measured at 4, 7, 8, 9, and 12 days as noted. The wun1-RIP plants grew more vigorously than the control plants. Statistical analysis demonstrated that growth differences were highly significant ($p = 0.001$ at 12 days). Individual *wun1*-RIP plants were found to grow as well as noninoculated, nontransformed plants (nin SRI). **B:** Size of representative plants removed from the soil and photographed 14 days after the beginning of the experiment. (Reproduced from Logemann et al., 1992, with permission of the publisher.)

fragment length polymorphism (RFLP mapping) sites. This method promises to be very valuable to most plant breeding programs. Much progress has recently been made in the identification of genes regulated by the plant hormone abscisic acid (ABA) and presumably involved in conveying tolerance to dessication to plants and to their seeds.

It is to be expected that the genetic control of complex physiological processes such as stress tolerance and yield will depend on the proper combination of a fairly large number of genes. Nevertheless, it is essential to study these genes because we might find (1) that some "key" genes regulate this complex process, (2) that relevant genes are linked in complex loci, or (3) that in some cases only a few genes are involved. It should be evident, however, that without the new cellular and molecular methods it would be unrealistic to hope to understand these complex processes.

7. NUCLEAR MALE STERILITY AND THE PRODUCTION OF HYBRID SEEDS

Self-pollination is a common feature of several economically important crops. This ability often counteracts the aim of plant breeders since they are interested in the production of hybrid seeds. One way to avoid self- or uncontrolled pollination is the use of male sterile plants. Traditionally cytoplasmic male sterile cultivars are used, in other cases male organs (anthers) have to be removed mechanically, which is a very time- and money-consuming procedure. Recently, new approaches have been developed to construct dominant chimeric genes causing male sterility. The idea is to construct genes that express a lethal function only in tissues involved in the formation of pollen. Promoter sequences derived from genes that are expressed only in anthers can be combined with DNA sequences coding for cell-lethal enzymes, such as RNAses or DNA restriction enzymes. "Restorer" genes can be constructed, which express proteins that will inactivate or compensate for the lethal effect of the dominant male sterility gene. This approach has recently been developed and tested for tobacco, rape seed, lettuce, Belgian endives, and even corn.

8. PROSPECTS

The examples provided above abundantly demonstrate the vast potential of molecular breeding for plant protection. Not only can one expect the growing crops to be protected in the field but one should also try to use similar approaches to biologically protect harvested crops against attack by various biological agents. Thus one can expect to have fruits, grains, vegetables, and so on, that will be more resistant to degradation by pests or diseases (thus dramatically reducing costs of storage and shipment), but also more resistant to intrinsic mechanisms of biological degradation. For example, it has recently been shown that the natural mechanism of ripening and spoilage of tomatoes can be controlled by expressing genes whose products interfere with the metabolism of plant growth factors (e.g., ethylene) responsible for fruit ripening or by controlling the expression of genes that stimulate ripening.

Once the defined genes are isolated and reconstructed their introduction into various crops, although sometimes difficult, can be achieved with relatively low-tech methods and does not require very heavy investments. This is of particular importance for crops of regional importance only (e.g., in the developing world). Even if it is true that in its initial stages genetic engineering was developed with high-input agriculture as an economic target, the methods and the genes that are being produced by this research will ultimately benefit agriculture worldwide, and will also help low-input regional agriculture.

9. CONCLUSION

Plant technology can significantly help to make intensive agriculture less damaging to the environment or to make low-input (organic) agriculture more productive. It should be possible to improve major crops (rice, wheat, corn, soybean, potato, etc.), as well as more regional crops.

If used wisely and responsible, there is no inherent danger in these new methods. It would be shortsighted and irresponsible not to make optimal use of these methods to relieve at least some of the tensions that already exist, and will dramatically increase in the near future, because of the intensive or unproductive use of land for agricultural production.

Last, but not least, these technologies are very likely to help increase the economic value of agriculture both in the developed and in the developing world.

REFERENCES

Crown Gall—Ti Plasmid Vectors

An G, Watson BD, Stachel S, Gordon MP, Nester EW (1985): New cloning vectors for the transformation of higher plants. EMBO J 4:277–284.

Bevan M (1984): Binary *Agrobacterium* vectors for plant transformation. Nucleic Acids Res 12: 8711–8721.

Fraley RT, Rogers SG, Horsch RB, Eichholtz DA, Flick JS, Adams S, Bittner M, Brand L, Fink C, Fry J, Galluppi J, Goldberg S, Hoffman N, Woo S (1985): The SEV system: A new disarmed Ti plasmid vector system for plant transformation. Bio/Technol 3:629–635.

Klee HJ, Rogers SG (1989): Plant transformation systems based on the use of *Agrobacterium tumefaciens*. In Schell J, Vasil IK (eds): Cell Culture and Somatic Cell Genetics of Plants, Vol. 6. New York: Academic Press, pp 2–25.

Koncz C, Schell J (1986): The promoter of T_L-DNA gene 5 controls the tissue-specific expression of chimeric genes carried by a novel type of *Agrobacterium* binary vector. Mol Gen Genet 204:383–396.

Stougaard J, Abildstein D, Marcker KA (1987): The *Agrobacterium rhizogenes* pRi TL segment as a gene vector system for the transformation of plants. Mol Gen Genet 207:251–255.

Zambryski P, Joos H, Genetello C, Leemans J, Van Montagu M, Schell J (1983): Ti plasmid vector for the introduction of DNA into plant cells without alteration of their normal regeneration capacity. EMBO J 2:2143–2150.

DNA-Mediated Gene Transfer

Hain R, Stabel P, Czernilofsky AP, Steinbiβ HH, Schell J (1985): Uptake, integration, expression and genetic transmission of a selectable chimeric gene by protoplasts. Mol Gen Genet 199:161–168.

Paszkowski J, Saul M, Potrykus I (1989): Plant gene vectors and genetic transformation: DNA mediated direct gene transfer to plants. In Schell J, Vasil IK (eds): Cell Culture and Somatic Cell Genetics of Plants, Vol. 6. New York: Academic Press, pp 51–68.

Paszkowski J, Shillito RD, Saul M, Mandak V, Hohn T, Hohn B, Potrykus I (1984): Direct gene transfer to plants. EMBO J 3:2717–2722.

Liposome

Deshayes A, Herrera-Estrella L, Caboche M (1985): Liposome-mediated transformation of tobacco mesophyll protoplasts by an *Escherichia coli* plasmid. EMBO J 4:2731–2737.

Microinjection

Crossway A, Oakes JV, Irvine JM, Ward B, Knauf VC, Shewmaker CK (1986): Integration of foreign DNA following microinjection into tobacco mesophyll protoplasts. Mol Gen Genet 203:179–185.

Neuhaus G, Spangenburg G, Mittelsten-Seneid O, Schweiger H-G (1987): Transgenic rape seed plants obtained by microinjection of DNA into microspore-derived embryoids. Theor Appl Genet 76:30–36.

Reich TJ, Eyer VN, Miki BL (1986): Efficient transformation of alfalfa protoplasts by the intranuclear microinjection of Ti plasmids. Bio/Technol 4:1001–1004.

Electroporation

Fromm M, Taylor LP, Walbot V (1985): Expression of genes transferred to monocot and dicot plant cells by electroporation. Proc Natl Acad Sci USA 82:5824–5828.

Pröls M, Schell J, Steinbiss H-H (1989): Critical evaluation of electromediated gene transfer and transient expression in plant cells. In Neumann E, Sowers EA, Jordan CA (eds): Electroporation and Electrofusion in Cell Biology. New York: Plenum Press, pp 367–375.

Particle Gun

Klein TM, Wolf ED, Wu R, Stanford JC (1987): High velocity microprojectiles for delivering nucleic acids into living cells. Nature (London) 327:70–73.

Gene Expression in Plants

Kuhlemeier Ch, Green PJ, Chua N-H (1987): Regulation of gene expression in higher plants. Annu Rev Plant Physiol 38:221–257.

Weising K, Schell J, Kahl G (1988): Foreign genes in plants: transfer, structure and expression. Annu Rev Genet 22:421–477.

BT—Insect Toxin

Delannay X, LaVallee BJ, Proksch RK, Fuchs RL, Sims SR, Greenplate JT, Marrone PG, Dodson RB, Augustine JJ, Layton JG, Fischhoff DA (1989): Field performance of transgenic tomato plants expressing the *Bacillus thuringiensis* var. *kurstaki* insect control protein. Bio/Technol 7:1265–1269.

Fischhoff DA, Bowdish KS, Perlak FJ, Marrone PG, McCormick SM, Niedermeyer JC, Dean DA, Kusano-Kretzmer K, Mayer EJ, Rochester DE, Rogers SG, Fraley RT (1987): Insect tolerant transgenic tomato plants. Bio/Technol 5:807–814.

Hilder VA, Gatehouse AMR, Sherman SE, Barker RF, Boulter D (1987): A novel mechanism of insect resistance engineered into tobacco. Nature (London) 330:160–163.

Höfte H, Whiteley HR (1989): Insecticidal crystal proteins of *Bacillus thuringiensis*. Microbiol Rev 53:242–255.

Vaeck M, Reynaerts A, Höfte H, Jansens S, DeBeuckeleer M, Dean C, Zabeau M, Van Montagu M, Leemans J (1987): Transgenic plants protected against insect attack. Nature (London) 328:33–37.

Viral Tolerance

Beachy RN, Loesch-Fries S, Tumer NE (1990): Coat protein-mediated resistance against virus infection. Annu Rev Phytopathol 28:451–474.

Gerlach W, Llewellyn D, Haseloff J (1987): Construction of a plant disease resistance gene from the satellite RNA of tobacco ringspot virus. Nature (London) 328:802–805.

Lawson C, Kaniewski W, Haley L, Rozman R, Newell C, Sanders P, Tumer NE (1990): Engineering resistance to mixed virus infection in a commercial potato cultivar: Resistance to potato virus X and potato virus Y in transgenic russet burbank. Bio/Technol 8:127–134.

Powell-Abel P, Nelson R, De B, Hoffmann N, Rogers S, Fraley R, Beachy R (1986): Delay of disease development in transgenic plants that express the tobacco mosaic virus coat protein. Science 232:738–743.

Tumer NE, O'Connell KM, Nelson RS, Sanders PR, Beachy RN, Fraley RT, Shah DM (1987): Expression of alfalfa mosaic virus coat protein gene confers cross-protection in transgenic tobacco and tomato plants. EMBO J 6:1181–1188.

Fungal Tolerance

Broglie K, Chet I, Holliday M, Cressman R, Biddle P, Knowlton S, Mauvais CJ, Broglie R (1991): Transgenic plants with enhanced resistance to the fungal pathogen Rhizoctonia solani. Science 254: 1194–1197.

Jach G, Logemann S, Wolf G, Oppenheim H, Chet I, Schell J, Logemann J (1992): Expression of a bacterial chitinase leads to improved resistance of transgenic tobacco plants against fungal infection. Biopractice 1:1–9.

Jones JDG, Dean C, Gidoni D, Gilbert D, Bond-Nutter D, Lee R, Bedbrook J, Dunsmuir P (1988): Expression of bacterial chitinase protein in tobacco leaves using two photosynthetic gene promoters. Mol Gen Genet 212:536–542.

Logemann J, Jach G, Tommerup H, Mundy J, Schell J (1992): Expression of a ribosome-inactivating protein (RIP) leads to fungal resistance in transgenic tobacco plants. Bio/Technology 10:305–308.

Resistance to Bacterial Diseases

Düring K, Porsch P, Fladung M, Lörz H (1991): Transgenic potato plants resistant to the phytopathogenic bacterium *Erwinia carotovora*. Submitted.

RFLP Mapping

Bonierbale MW, Plaisted RL, Tanksley SD (1988): RFLP maps based on a common set of clones reveal modes of chromosomal evolution in potato and tomato. Genetics 120:1095–1103.

Chao S, Sharp PJ, Worland AJ, Warham EJ, Koebner RMD, Gale MD (1989): TAG 78:495–504.

Gebhardt C, Ritter E, Debener T, Schachtschabel V, Walkemeier B, Uhrig H, Salamini F (1989): RFLP analysis and linkage mapping in *Solarium tuberosum*. TAG 78:65–75.

Helentjaris T (1987): A genetic linkage map for maize based on RFLP's. Trends Genet 3:217–221.

McCouch SR, Kochert G, Yu ZH, Wang ZY, Khush GS, Coffman WR, Tanksley SD (1988): Molecular mapping of rice chromosomes. TAG 76:815–829.

Zamir D, Tankley SD (1988): Tomato genome is comprised largely of fast-evolving, low copy-number sequences. Mol Gen Genet 213:254–261.

Genes Involved in Desiccation Tolerance

Baker J, Steele C, Dure L III (1988): Sequence and characterization of 6 Lea proteins and their genes from cotton. Plant Mol Biol 11:277–291.

Bartels D, Schneider K, Terstappen G, Piatkowski D, Salamini F (1990): Molecular cloning of ABA-modulated genes from the resurrection plant *Craterostigma plantagineum* which are induced during desiccation. Planta 181:27–34.

Close TJ, Kortt AA, Chandler PJ (1989): A cDNA based comparison of dehydration-induced proteins (dehydrins) in barley and corn. Plant Mol Biol 13:95–108.

Mundy J, Chua N-H (1988): Abscisic acid and waterstress induce the expression of a novel rice gene. EMBO J 7:2279–2286.

Male Sterility

Mariani C, De Beuckeleer M, Truettner J, Leemans J, Goldberg RB (1990): Induction of male sterility in plants by a chimaeric ribonuclease gene. Nature (London) 347:737–741.

Gene Tagging With Transposable Elements

Baker B, Coupland G, Federoff N, Starlinger P, Schell J (1987): Phenotypic assay for excision of maize controlling element *Ac* in tobacco. EMBO J 6:1547–1554.

Coen ES, Robbins TP, Almeida J, Hudson A, Carpenter R (1989): Consequences and mechanism of transposition in *Antirrhinum majus*. In Berg DE, Howe MM (eds): Mobile DNA. Washington DC: American Society of Microbiology, pp 413–436.

Gierl A, Saedler H, Peterson PA (1989): Maize transposable elements. Annu Rev Genet 23:71–85.

Masterson RV, Furtek D, Grevelding C, Schell J (1989): A maize *Ds* transposable element containing a dihydrofolate reductase gene transposes in *Nicotiana tabacum* and *Arabidopsis thaliana*. Mol Gen Genet 219:461–466.

Van Sluys MA, Tempé J, Federoff N (1987): Studies on the introduction and mobility of the maize activator element in *Arabidopsis thaliana* and *Daucus carota*. EMBO J 6:3881–3889.

Gene Tagging by T-DNA Insertion

Feldmann KA, Marks MD, Christianson ML, Quatrano RS (1989): A dwarf mutant of *Arabidopsis* generated by T-DNA insertion mutagenesis. Science 243:1351–1354.

Koncz C, Martini N, Mayerhofer R, Koncz-Kalman Z, Körber H, Redei GP, Schell J (1989): High-frequency T-DNA-mediated gene tagging in plants. Proc Natl Acad Sci USA 86:8467–8471.

Koncz C, Mayerhofer R, Koncz-Kalman Z, Nawrath C, Reiss B, Redei GP, Schell J (1990): Isolation of a gene encoding a novel chloroplast protein by T-DNA tagging in *Arabidopsis thaliana*. EMBO J 9:1337–1346.

Marks DM, Feldmann KA (1980): Trichome development in *Arabidopsis thaliana*. I. T-DNA tagging of the Glabrous 1 gene. Plant Cell 1:1043–1050.

INFLUENCE OF BIOTECHNOLOGY ON BIOCONTROL OF TAKE-ALL DISEASE OF WHEAT

BRUCE C. HEMMING

Microbe Inotech Laboratories, Inc., St. Louis, Missouri 63146

JOHN M. HOUGHTON

Stewart Technology Management, St. Louis, Missouri 63146

1. INTRODUCTION

Biotechnology products worth some $2.5 billion were sold worldwide during 1990. Current trends indicate that this amount is likely to quadruple by 1995. It is further expected that these numbers will more than double again by the year 2000. While it is acknowledged that human therapeutics and human diagnostics account for the greatest share of these amounts, biotechnology in the production of agricultural food and fiber most certainly will participate. Current estimates value biotechnology products at $500 million for animal agriculture and "several million dollars" for plant agriculture by the mid-1990s (Department of Commerce, 1990). One case in point is the opportunity to control take-all disease in cereal grain production.

Trends in biotechnology continue to demonstrate that this science is poised to make a major impact on society during the next decade and beyond. While the pharmaceutical area is perhaps receiving the greatest share of the publicity today,

Biotechnology in Plant Disease Control, pages 15–38
© 1993 Wiley-Liss, Inc.

biotechnology developments in the agricultural arena continue to progress. These developments may come not a moment too soon. Indications are that three major trends are destined to collide within the next several decades: world population growth, desire for improved diets worldwide, and the move to protect the environment. The major suppliers of food and fiber to world populations will demand ever more forcefully that agricultural biotechnology develop new avenues of production or at the very least enhance existing ones. Our current means of production will be inadequate on all counts: quantity, quality, and efficient use of natural resources.

If current trends continue:

The world population will increase to 7.2 billion people by the year 2010 (Hess, 1990). At this rate the world population will nearly double in size from its current level of 5 billion people over the next 40 years.

The above population increase will require a 40% increase in food by 2010 just to maintain the caloric intake required (Hess, 1990).

By the year 2030, this requirement will likely be for more than a doubling of food, fiber, and energy requirements, as standards of living rise.

Nutrition requirements of the food supply will go up.

The demand for industrial and pharmaceutical raw materials produced from crops will double.

The above increases will have to be met by an agricultural land base that at best will remain constant and more likely will decline under the pressure of an expanding population.

The increased demand for food and fiber will have to be met by fewer and fewer people directly engaged in agriculture as the industrial and service sectors of the world economy compete for workers.

Increased agricultural production will have to be accomplished in an environmentally sound manner. The imperative to clean up the environment and keep it clean is only going to get stronger.

And above all, the commercial application of biotechnology products will be dependent on the public's interest and the public's acceptance of the proposed uses of biotechnology. The ultimate success of this new technology will occur only if the public accepts and understands what biotechnology is and what it can realistically deliver. We must keep in mind that this understanding will have to be accomplished by a public that has little scientific knowledge and at best, an uneasy confidence in scientists.

Unquestionably, agriculture will respond to the increased demand for food and fiber if such explosive population growth occurs. Key food crops such as rice, wheat, corn, and soybean are already the subject of intense study as means are explored to enhance production while utilizing fewer resources. Biotechnology offers great hope to researchers for providing truly innovative solutions to some of the most serious production problems that have vexed growers for generations. Not the least of these problems has been the need to control soil-borne plant pathogens that cause extensive damage to some of our most critical crops, particularly cereal grains.

This chapter, which explores the impact of biotechnology and biocontrol on the control of soil-borne plant pathogens, indicates how this technology was developed for take-all disease. While biocontrol of take-all disease may hold great promise in increasing the harvestable yield in cereal grains, it will require the imagination of inventing a new technology and developing new products, while at the same time, overcoming the concern in the public arena for the intentional release of living microorganisms into the environment.

2. MAJOR TRENDS IN BIOTECHNOLOGY

Life sciences entered a new era with the advent of two fundamental events: (1) the discovery by Watson and Crick of deoxyribonucleic acid (DNA) some 38 years ago, and (2) the invention of gene splicing by Cohen and Boyer in 1973. While these inventions created whole new vistas from the perspective of the scientist, they also opened up an unanticipated public fervor and debate around the world on biotechnology and genetic engineering that was sparked, if not fueled, by Rachel Carson's Silent Spring, and that culminated in the first generation of environmentalists (Witt, 1990).

The application and deliberate release of genetically altered live microorganisms for pest control in crops has been an area especially vulnerable to public and regulatory concerns. Certainly, the impact of these concerns on the development of a potential biocontrol of take-all disease was felt throughout the major wheat growing areas of Europe, the United States, and other producing regions.

The specter of a deliberate release of live, genetically engineered organisms that reproduce themselves, that cannot be totally confined during outdoor field testing, and that afford no means for "undoing" the application has presented an unacceptable level of risk in the public's mind. At the current level of understanding, at least, the risks inherent in altering genetic properties of organisms have been a monumental barrier to the commercial development of wide scale live microbial treatments.

This concern over the utilization of biotechnology on a commercial level, one that is based on the public's *perception of risk,* will continue to exist until the misunderstandings are addressed in terms that can be understood by the public, and until the critical element of trust is reestablished between scientist–industry and the public (Lindheim, 1989). The challenge of today is to ensure that the technological tools created by science are used to enhance the public's standard of living and quality of life. In the specific instance of take-all disease, the challenge to science is to use new technologies to enhance biological control: a self-perpetuating, safe, inexpensive, environmentally sound disease control practice (Cate, 1990).

The search for a take-all biocontrol technology has richly benefitted from the synergism of scientists working in academia, government, and private industry. Biological control of plant diseases is an approach that is distinct from other control measures. Its technological base involves diverse scientific disciplines that uniquely address the natural enemies and antagonists of the targeted species (Cate, 1990). Presently, most methods utilized to influence and manipulate the plant microflora have been constrained because of limited knowledge of the factors and relationships in the rhizosphere and phyllosphere. Yet, such methods have offered positive indications of the possibilities of controlling economically important agricultural plant pathogens by bacterial antagonists (Hemming, 1990). Academic institutions, government, and industry must continue to work together if biocontrol approaches in agriculture are to bear fruit.

The need for cooperation among researchers is underscored by the climate of public mistrust that afflicts much of the world today, especially the United States. As a new science, biotechnology is meeting with an unusual degree of public suspicion, largely because of a basic lack of understanding of what the science is about.

To allay the suspicion and fear surrounding biotechnology will require an unprecedented educational effort on the part of scientists in all venues. But perhaps more importantly, it will necessitate a consensus, a setting of professional standards, if you will, to which all researchers in biotechnology adhere and to which the public can look for reassurance of scientific integrity and restraint. The best way to establish this common voice, it would seem, is to create cooperative research efforts among academia, industry, and government, which enable researchers to develop the science in an accepted, defensible manner that can be communicated effectively and convincingly to the public.

A precedent for this type of cooperation has already been established. A recent survey of biotechnology research and funding in the academic arena was conducted by the North Carolina Biotechnology Center (NCBC) (Bio-Tech, 1991) covering 4200 biotechnology faculty members supervising some 24,000 personnel in their laboratories across the United States. The survey reported that 99% of the scientists who responded to the survey would like to collaborate with or have their research funded by industry. Only 27% of those responding currently receive industrial funding. In a previous study on technology transfer, the NCBC reported that 75% of U.S. firms responding have research ties with U.S. universities. Ranked areas of interest were molecular biology, 53%; biochemistry, 35%; and cell biology, 28%. Basic research was favored over applied focus, 66%, with selected health care.

Both universities and industry have concerns about joining forces to collaborate on specific projects. Part of the reluctance stems from the divergent goals and recognition systems of the two groups. Academic concerns center on the fear that these so-called cooperative agreements would allow industry to subvert universities' freedom to conduct basic science research directed to fundamental discovery. Industrial scientists, on the other hand, while interested in carrying out some basic research, are more focused on the application of the science in an attempt to identify, develop, and commercialize products. Likewise, academic investigators are rewarded for their publication efforts, while industrial investigators are rewarded for their efforts in patents and products (Montague, 1991). These statements are generalizations. In the best relationships, it is realized that motives of both are quite similar differing by degree of polarization towards the two perspectives. Many company scientists may have the desire to publish and be recognized by peers as more than just "scientists from Company X," but may be represented by company lawyers or administrators who could be more easily persuaded to forego such recognition for an increase in other tangible or other types of intangible rewards. The university scientist may long for interaction with well-resourced company scientists. He/she may be willing to share some "limelight" in the literature for a reprieve from the funding dilemma faced by writing another research proposal. This may in turn be internally counterbalance by his/her own administration.

In actual practice, both large and small scale cooperative agreements have already been successful. Perhaps the largest and longest running industry–university collaboration, in the United States at least, exists between Monsanto Co. and Washington University in St. Louis, MO ($100 million over 12 years) (Culliton, 1990). Montague (1991) explored this relationship and suggests that it might serve as a model for others. First, he notes that, above all, there is a need to establish a high level of trust between the respective scientists and administrators. Second, the different cultures and needs of universities and industry must be recognized and honored. Third, both partners must gain from the collaboration, and fourth, the contract must be structured to adequately cover the details, but not at the expense of the intellectual and scientific exchange. The late H. A. Schneiderman, senior vice-president for research and development at Monsanto Co., was instrumental in the development of the cooperative research agreement with Washington University, and was adamant in his belief that the contract be as simple and straightforward as possible (Culliton, 1990).

The spirit of cooperation between universities and industry, if not its sustenance, is already present. With generous funding, a renewed commitment to scientific excellence and cooperation, and a concerted effort to bring the public into enlightened awareness and informed acceptance of biotechnology, this science can realize its potential to enhance the agricultural production of food and fiber

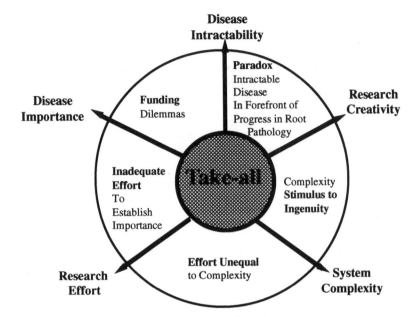

Fig. 2.1. Five major concerns in take-all studies and how they interact. Adapted from Home-Grown Cereals Authority (HGCA) Research Review No. 20, 148 pp. Compiled and Edited by Hornby D and Bateman GL, page 8, figure 1. By permission of Rothamsted Experimental Station. AFRC Institute of Arable Crops Research.

and enable the world's expanding populations to enjoy a better quality of life than now awaits them. The struggle to develop a biocontrol solution to the problem of take-all disease in cereal grains is as propitious a starting place as any. The market of opportunity must be large enough to risk $1.2 million per year in supporting five or six researchers for a few years to establish the possibility of a commercial product for the prospect of a larger return.

The recent report by Hornby and Bateman (1991) provides a pertinent correlation diagram (Fig. 2.1), which identifies ''five major concerns in the take-all studies and how they interact.'' Of interest is the combination of factors, such as funding, research effort, and research creativity. The university should be the seedbed for the riskier project ideas because the cost structure is significantly different than that in large agribusiness enterprises. Biotechnology has had an impact in bringing industrial and financial strength close to the university departments associated with the life sciences and also in helping to sustain an effort in a particular realm of research (i.e., biological control) through indirect support of symposia, seminars, and so on.

3. TAKE-ALL DISEASE: A CASE STUDY

In farming circles, take-all has always been notorious because of the continuing lack of chemical controls and commercial resistant wheat cultivars. Take-all is considered to be the most damaging root disease of wheat worldwide. "Take-all" suppressive soil containing microorganisms, which have been shown to reduce the level of disease caused by the agent fungus, provides a natural source of potential biocontrol agents (Cook and Rovira, 1976). Root colonizing fluorescent pseudomonads that produce antibiotics inhibitory to the soil-borne fungus, such as phenazine-1-carboxylic acid and related pigments, pyrrolnitrin, and diacetophenone derivatives (phloroglucinol derivatives) have been implicated in such a role (Gurusiddaiah et al., 1986; Homma et al., 1988; Reddi et al., 1969; Thomashow et al., 1988, 1990). An organism that colonizes wheat roots and produces an antibiotic at the site of disease may prove to be effective as a biofungicide, an alternative to chemical fungicides. Take-all presents a major challenge in crop production and plant pathology because of crop loss, farming practice (rotation) disruptions, and because no satisfactory chemical control or resistant variety option exists (Hornby and Bateman, 1991). Estimates provided by HGCA indicate that annual wheat yield losses attributable to take-all range from 3 to 20% in the United Kingdom, with greater than 50% of U.K. wheat fields at risk. For the U.S. Pacific Northwest, annual yield losses range from 10–50%, and in Western Australia, from 36–40% (Hornby and Bateman, 1991). Wheat is a major crop in each of the mainland states of Australia. For the 7-year period 1979–1980 to 1985–1986 average annual production was 15.6 megatonnes (Murray and Brown, 1987), making wheat the most important broadacre crop in Australia. When listed in order of potential losses, take-all ranks second among the six major diseases of wheat in Australia (Bureau of Agricultural Economics, 1988); however, if considered in order of present average annual losses (and given current control practices), it ranks on top as shown in Figure 2.2.

3.1. The Importance of the Marketplace Perspective

The information on potential and present losses has been used in Australia to estimate that if control of take-all could be achieved, it would add $81 million annually to wheat production in Australia alone (Brennan and Murray, 1988). While estimates of the annual worldwide market value for an effective take-all control product are elusive, generally accepted figures fall in the $150–250 M range. Such information is critical in providing practical people convincing data to be properly impressed with the importance of this type of research to the point of committing large sums of money for its support. Careful estimates obviously outrank vague guesses. Barriers to the general development of biopesticides are chiefly economic, not human hazard or environmental risks; however, this eco-

Australian Wheat Diseases
Listed in Order of Importance

By Potential Losses	By Present Average Annual Losses
1. Bunt	1. Take-all
2. Take-all	2. Septoria nodorum blotch
3. Stripe rust	3. Cereal Cyst Nematode
4. Septoria nodorum blotch	4. Black Point
5. Stem	5. Yellow spot
6. Cereal Cyst Nematode	

Fig. 2.2. Lists of Australian wheat diseases illustrating differences between potential and present losses, which may be interpreted to reflect the value of current control measures (i.e., Bunt is controlled, take-all is not). (Data derived from Brennan and Murray, 1988.)

nomic picture makes the area susceptible to the perception of risk with its attendant costs. In specific instances, the private sector is not willing to invest in research to discover and develop environmentally compatible biopesticides because their high pathogen specificity usually limits market potential. This commonly is cited as a problem with specific disease targets; however, this does not severely handicap take-all disease as a target, since the stand alone market potential is significant. The caveat is that with the control of the take-all fungus, *Pythium* or *Rhizoctonia* based root disease problems may be more significant in this niche. This leads to the possible conclusion that a mixture of chemical fungicide and biological control agent could lead to a one-two punch to win the round and control these root diseases. The use of a chemical treatment to drop the initial fungal inoculum, followed by a more persistent biological inhabitant of the rootzone for further later season control, is another concept considered by researchers in this field. Efficacy and economics will determine the approach.

In Australia, 95% of the current control methods for take-all are either cultural practices or crop rotation with the remaining 5% by pesticides. The value of current control methods of take-all in wheat in this area of the world has been estimated based on current expenditures to be about $4.7/acre (Brennan and Murray, 1988). The overall potential cost of the disease is estimated at $8.2/acre. Within the four states of Victoria, South Australia, Western Australia, and Southern New South Wales, generally a moderate to very severe incidence of take-all is likely each year (Murray and Brown, 1987) with the acres grown about 22.4 million. Within these figures, it would probably be likely that agribusinesses would set a potential price on a controal product near $4.20/acre. The question naturally arises Can a biocontrol agent meet such criteria?

To answer these and other similarly important questions related to biocontrol research directed towards controlling take-all disease of wheat requires information from a number of disciplines. The interdisciplinary approach uses skills and knowledge from plant pathology, entomology, nematology, weed science and the basic support sciences of economics, genetics, chemistry, molecular biology, microbiology, physiology, ecology, systematics, and systems science. These disciplines can be applied in addressing problems and opportunities using a project team approach. The complexity of the take-all problem can be the stimulus to ingenuity by integrating multifunctional groups addressing the same product–market target. Figure 2.3 presents one of the many possible interactive group arrange-

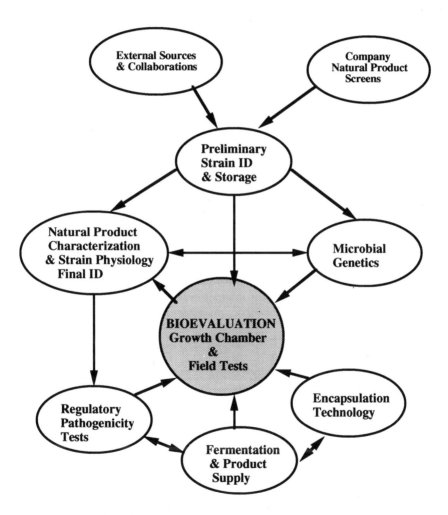

Fig. 2.3. Interrelationships between functional tasks of a biofungicide research team where arrows indicate material and information flow.

ments on this disease problem. The central issue being whether an efficacious agent is under development and whether a potential product can be shown to be functional in applied field tests. The impact of biotechnology in this research arena comes primarily from microbial genetics to wheat represents a forefront of plant science research awaiting further developments. Biotechnology also opened industrial sources of funding for researchers helping to alleviate some of the funding dilemmas for take-all research. The combination of private industry, government, and university laboratories has been tapped to bring the resources and effort more in line with the complexity of the problem to be solved.

3.2. Technology Development—"An Area of Cooperation"

The inoculation of genetically modified bacteria into natural ecosystems to control insects and diseases, enhance development of agronomically important plant varieties, for waste treatment, pollution control, and mineral leaching could provide significant benefits. The ability of certain pseudomonad soil bacteria to efficiently colonize the root surfaces of a variety of crop species has made them prime candidates for use in plant growth promotion and pest control, and specifically for the possibility of controlling take-all in wheat. Many of these naturally occurring soil bacteria are benign root-associated organisms, being quite efficient in taking nutritive advantage of the normal exudates given off by the roots of healthy plants. Moreover, there are some studies indicating that certain strains of pseudomonads may have a definite beneficial effect on plants as well (Burr et al., 1978; Kloepper et al., 1980).

It is of prime interest to eventually employ root-colonizing pseudomonads as efficient delivery vehicles for plant-beneficial agents, such as natural pesticides, fungicides, or plant hormones, the use of which would not be practical or even possible without the precise, targeted, continuous input afforded by these bacteria. Critical evaluation of the performance capabilities of such bacteria is essential, even to determine which are the most efficient host organisms to use for given environmental conditions or crop species. For this analysis to be realistic, it is absolutely crucial to determine pesticidal effectiveness of the introduced bacteria under actual field conditions.

To genetically "mark" strains for environmental inoculation, isolates of test bacteria have been typically selected that are resistant to the antibiotics rifampicin (Rif) and nalidixic acid (Nal). This phenotype can provide an adequate means of detection and tracking for some bacteria by reculturing strains from soil or leaf wash samples on media containing these antibiotics. Unfortunately this practice suffers from the inability of many such "marked" bacteria, having both Rif-resistance (Rif^r) and Nal-resistance (Nal^r) phenotypes to maintain their competitive ability and environmental persistence. In our experience, strains that are Rif^r only, generally do not appear to be significantly impaired in their root-colonizing efficiency, although some lack of competitiveness has been observed. However, the background level of resistance among soil microbes to either Rif or Nal alone can appreciably limit detection capabilities, particularly when the number of bacteria to be monitored is low ($<10^3$/g of soil) relative to the total bacterial population (often $>10^9$/g of soil). For instance, it is not atypical to encounter a naturally occurring Rif^r population of 10^3–10^5 cfu/g of soil. Native bacteria naturally resistant to other antibiotics, such as kanamycin, neomycin, and streptomycin, can often be found at these high background levels in soil and water samples.

The *lac*ZY-microbial tracking system developed by Monsanto Co. largely bypasses these problems and is particularly effective for the pseudomonads (Drahos, 1986). The marking system relies on the introduction and expression of the *Escherichia coli lac* operon genes *lac*Z and *lac*Y. The unexpected inability of pseudomonads tested to date (>11, 530) to efficiently utilize lactose as a sole carbon source permits the metabolic selection of marked strains on defined laboratory media before and after inoculation into the rhizosphere. Furthermore, methods were used that allowed the *lac*ZY genes alone to be introduced directly into the host bacterial chromosome (Barry, 1986). This provides great stability of these genes within the engineered strain under natural, nonselective field conditions. In addition, these insertion methods effect an essentially permanent gene transfer into the chromosome, providing a high degree of contrainment for the *lac*ZY element within the engineered strain. Finally, chromosomal introduction of these genes, which have been sequenced and their products well characterized, may eventually aid the implementation of direct bacterial detection methods, such as nucleic acid and antibody hybridization probes, once these techniques become more practical in application.

A biofungicide project team was formed at Monsanto Co., which initiated in April, 1990 and again in the fall of the same year (Monsanto, 1990), collaborative small-scale field tests (<1.5 acres per site) of host (Ps. Q2-87 and QC-65) and *lac*ZY (*E. coli* K-12) engineered (Ps. Q2-87AL6 and Ps. QC-65AL7) *Pseudomonas aureofaciens* organisms (also called *P. fluorescens* biotype E). The nonengineered host strains, Q2-87 and QC-65, are very similar to Ps. 2-79RN, which had been evaluated by a number of academic researchers (Thomashow et al., 1988) in field studies for several years and which served as the host in the successful *lac*ZY marker system tests of 1988–1989 (Monsanto, 1988, 1989). Results of the earlier studies with Ps. 2-79RN had shown this naturally occurring soil microbe to be suppressive against take-all, a disease primarily of wheat, which is caused by the fungus *Gaeumannomyces graminis* var. *tritici* (Ggt).

As early as 1976, a role for fluorescent pseudomonads in suppression of wheat take-all disease was suggested (Cook and Rovira, 1976). Smiley (Smiley, 1978) showed that the population of antibiotic-producing pseudomonads in soil was higher when wheat only was grown than when wheat was rotated with other crops. A *P. aureofaciens* 2-79 (Ps. 2-79) strain was originally isolated (Weller and Cook, 1983) from the roots of wheat growing in soil from a field where the wheat root disease take-all caused by Ggt had spontaneously declined (Shipton, 1975). This Ps. 2-79 strain was selected for resistance to the antibiotics Rif and Nal. When cells of this bacterium are applied to wheat seed, the bacteria effectively colonize the wheat roots and suppress the Ggt pathogen on the roots (Weller, 1983; Weller and Cook, 1983). Additional studies (Gurusiddaiah et al., 1986) provided evidence suggesting that the Ggt suppressive activity of *P. aureofaciens* strain 2-79RN may be a dimer of phenazine-carboxylic acid. More thorough studies (Brisbane et al., 1987 showed that this antibiotic is a monomer of phenazine-1-carboxylate. This antibiotic showed excellent activity against several species of fungi, including the wheat pathogens Ggt, *Rhizoctonia solani,* and *Pythium aristosporum.* There were no toxic symptoms when mice received the phenazine carboxylic acid antibiotic by oral doses up to 464 mg/kg (Gurusiddaiah et al., 1986).

The small-scale field tests were carried out in collaboration with (1) Purdue University scientists at Purdue's Agronomy Farm in Vincennes, Knox County, IN; (2) USDA–ARS researchers working with Washington State University at the Department of Plant Pathology's Disease Research Plots, which are located in Pullman, WA, and (3) Montana State University faculty/staff at the university's A. H. Post Research Agronomy Farm located approximately 5 miles west of Bozeman, Gallatin County, MT. The main objective of these field tests was to further confirm that these *lac*ZY marked fluorescent pseudomonads demonstrate potential as biological control agents under actual field conditions in three separate geographic areas and soil types (Indiana, Washington, and Montana) and during different seasonal periods (spring and fall). Control of the soil-borne pathogen, Ggt, the causal agent of take-all disease, was determined by the level of root infection by the Ggt fungus and by examination of disease incidence and wheat yield. In addition to these tests, additional field tests were conducted by the team using native nonengineered strains at the locations of Monsanto Co. research farms in Monmouth, IL; Loxley, AL, and Troy, MO. The total number of field sites in 1990 numbered 12, which included sites in 5 states and 1 foreign country (Australia).

A small-scale field test of a similar *lac*ZY marked *P. aureofaciens* strain (Ps. 3732RNL11) was previously conducted in 1987–1989 collaboratively by Monsanto and Clemson University in South Carolina (Monsanto, 1987; Monsanto and Clemson University, 1989). The findings obtained through harvest of three test crops demonstrated the efficacy of the *lac*ZY tracking system in the field and provided important information concerning dissemination and survival of soil inoculated pseudomonads. This test was followed by concurrently run tests in 1988–1989 at both Clemson University and with the USDA–ARS Personnel at Washington State University in Pullman, WA. These field tests confirmed as reported to the U.S. Environmental Protection Agency (EPA–OPP Designation No.: 524-NMP-006) that the chromosomal *lac*ZY marker provided an accurate and practical means to monitor the survival and location of pseudomonad organisms under actual field conditions in two widely geographic areas (South Carolina and Washington), and there was minimal spread of the inoculated bacterial strains to adjacent non-inoculated wheat plants, as was also noted in the first Clemson University test of similar *lac*ZY-marked strains. Excellent wheat root colonization was observed at both test sites even with variances in environmental and agronomic practices, such as growing season, soil type, test period, and wheat cultivar. Unfortunately, there was no significant suppression of the Ggt (take-all) disease by any of the inoculated strains used, Ps. 2-79, Ps. 2-79RN, or Ps. 2-79RNL3. Consequently, a primary purpose of the more recent tests was to evaluate new marked strains having potential for significant suppression of Ggt (take-all) disease.

The *lac*ZY marker is a sensitive, selectable tracking system that was initially tested, as indicated, in fluorescent pseudomonads. The product of the *lac*Z gene, β-galactosidase, cleaves lactose sugar into the sugars, glucose and galactose. Fluorescent pseudomonads do not make β-galactosidase and subsequently cannot efficiently use lactose as a sole energy source. Similarly, some nonfluorescent pseudomonads also do not make β-galactosidase. Moving the *lac*Z gene into these soil bacteria allows them to grow on media where lactose is the only energy source and to cleave a chromogenic dye called X-Gal. When X-Gal is cleaved by the *Pseudomonas* bacteria, brilliant blue-colored rather than yellow-white colonies appear. The *lac*Y gene, which codes for the enzyme lactose permease, allows pseudomonad organisms to take up lactose sugar efficiently. The combination of these two genes (the *lac*ZY marker system) enables the detection of as few as 1–10 *Pseudomonas* bacteria in a gram of natural field soil or root washing, when recovered on defined lactose growth medium. It is the inclusion of these genes in fluorescent pseudomonads that provides a uniquely retrievable handle (fluorescent, blue-colored *P. aureofaciens* colonies growing on lactose media), which makes the system very useful for field studies. The inclusion of the *lac*ZY marker in the naturally occurring microbial fungicide (Ps. 2-79RN) strain to form the *lac*ZY marked Ps. 2-79RNL3 strain permitted a rapid and very effective evaluation of wheat root colonization by Ps. 2-79RNL3 and the level of wheat root fungus infection [i.e., take-all (Ggt) disease] as described for our earlier field tests. Marking with the *lac*ZY marker of the nonfluorescent Rif resistant pseudomonad strain (Ps. 2140RL3) imported from Australia provides this unique handle for more easily retrieving information on this organism.

3.3 Research Development—"Blending of the Disciplines"

In association with its biofungicide project team, Monsanto developed a new microencapsulation technology for delivery of living microorganisms together with the wheat seed at the time of planting. Of interest is the fermentation and formulation costs and the importance of reducing application rates. These topics and the introduction of the new encapsulation technology for microbial pesticides were presented at the UCLA Symposium on Biological Control (Baker and Henis, 1990). The 1990 field trials represented the first field tests of the combination of *lac*ZY marked strains with the new encapsulation technology, the combination of subjects was reviewed by the United States Environmental Protection Agency (US-EPA) prior to the initiation of the tests. The encapsulation beads carrying the microorganisms were prepared by mixing the microorganisms with an aqueous nonionic polymer [poly(vinyl alcohol), MW range 10,000–125,000 and from 85–100% hydrolyzed] solution in which the polymer is present at a concentration of at least 3% w/v. Beads are formed by adding this mixture dropwise into a water-immiscible nonsolvent for the polymer maintained at a temperature sufficient to rapidly freeze the bead, but not so slow as to cause freeze-fracture of the polymer bead. The polymer beads containing the microorganisms are then dried to remove substantially all unbound water. Alumina (up to 30%) is added as a density agent to increase bead density to about 1.8/ml. The encapsulated microorganisms (10^8–10^9 microorganisms per gram of beads) are in a dormant state and, generally less than 1.5 logs of cells is lost during the encapsulation. Beads made with poly(vinyl alcohol) polymer are substantially insoluble in water below 40°C. The encapsulated microbe formulation is a free-flowing material that can be delivered either dry or wet using standard agricultural application equipment. With regards to potential for environmental impact of the beads, it is noteworthy that soil *Pseudomonas* organisms have been isolated that effectively degrade poly(vinyl alcohol) (Hashimoto and Fujita, 1985; Sakai et al., 1985; Shimao et al., 1985; Shimao et al., 1986). The primary components of the formulation for encapsulation are FDA GRAS (generally regarded as safe) materials and were field tested previously using native pseudomonad strains.

The microbe is likely to be the single most expensive part of any microbial pesticide; therefore, it is very important to protect this asset. An adequate formulation and stable, flexible delivery system becomes essential. A desirable microbial formulation for biological control would permit high cell loading using materials compatible with soils and plants, biodegradable, and easy for the applicator to handle and deliver. There are a number of formulations and formulation technologies in development and a few have actually reached the market place. The old standbys like peat and clay mixes tend to fall under the category of inexpensive but not very protective of living cells (a process that loses three to five logs of viable cells is expensive.) Newer alginate capsule techniques look promising, but retention of adequate viability in a stored product is still problematic.

The tests conducted employed methylcellulose coatings of bacteria onto the winter wheat seeds, as was used in early tests, in addition to the new technology. Cell losses following encapsulation run on the order of just one log unit. As mentioned, beads made for rhizosphere delivery have a densifying agent (alumina) added to help them flow well in conventional planting equipment. The beads are quite sturdy and handle well during planting. The generation of data on this combination will greatly facilitate the inclusion of the safety factors inherent in these technologies in delivery of living biological control agents in agricultural settings. With successful generation of data and further studies into the shelf life of such formulations, the vagarious nature of the weather and other time dependent factors may be circumvented or greatly ameliorated to avoid the loss of fermentations prepared for agricultural field introductions, not to mention the minimization of batch-to-batch variations in preparations for the regimen of field tests and, ultimately, of commercial products.

In these tests, the level of risk to humans and the environment associated with the proposed small-scale field tests was negligible. This conclusion was based on the innocuous nature of the host strains, their similarlity to previously field tested Ps. 2-79RNL3 and Ps. 3732RNL11, the nonpathogenic character of the permanently inserted *lac*ZY marker element, and the inherent safety of materials of the formulations including microencapsulation beads. In view of the negligible risk potential for these tests and the availability to the US-EPA of a significant data base from laboratory and field studies with host *P. aureofaciens* 2-79RN and *lac*ZY marked *P. aureofaciens* 3732RNL11 organisms, the US-EPA could reasonably and defensibly conclude that approval following review of these field tests was appropriate. The strength of industrial approach to formulation, attention to regulatory matters, and project coordination coupled with public and governmental entities in conducting the tests exhibits on a sufficient scale the advantages of the interdisciplinary project team approach for working towards controlling this complex disease. A summary of the tests using recombinant strains conducted by scientists then associated with Monsanto Co. is provided in Table 2.1.

3.4 International Cooperation

A nonfluorescent pseudomonad was originally isolated from natural field soil in Australia and shown to be a potential take-all disease control agent. The nonengineered host strain, Ps. 2140, had been evaluated by Australian Government researchers [The Australian Commonwealth Scientific and Industrial Research Organization (CSIRO), Division of Soils] in field studies for at least 2 years. The strain was confirmed to be a *Pseudomonas corrugata* strain and marked in Monsanto laboratories with the *lac*ZY genes, as part of a collaborative project between the two parties. The specified *P. corrugata* 2140 strain was shown to be a minor plant pathogen of tomato. This strain is somewhat similar [within the same ribosomal ribonucleic acid (rRNA) homology group, *Pseudomonas* group I] to the previously introduced field tested strains *P. auerofaciens* 2-79RN, 3732RN, Q2-87A strains. Strains of the *P. auerofaciens* species are not plant pests or plant pathogens. This fact was confirmed as part of a previous in-depth review by USDA–APHIS (Animal and Plant Health Inspection Service) and EPA-OTS of a Monsanto Co. proposal for field testing of *lac*ZY marked *P. aureofaciens* (also

TABLE 2.1. Genetically Engineered Microbial Field Tests Conducted by Multidisciplinary Team With Monsanto Co.

Strain	Description	Reviewing Agencies	Test Purpose	Test Status		Final report Completion date	Monsanto Contacts
				Submission date	Initiation Date and Location		
Ps. 3732RN	BT gene *Tn5* genomic insertion in *P. fluorescens*	EPA-Office of Pesticide Programs	Efficacy of insect control on corn roots	April, 1985	Still Pending (Previously Proposed site: St. Charles, MO)		L.W. Watrud
Ps. 3732RNL3	*lacZY Tn7* based genomic insertion in *P. fluorescens* 3732RN	EPA-Office of Toxic Substances— TSCA regulations: Pre-Manufacture Notification No. P-87-1291 and Concurrence of Nonpathogenicity of Ps. 3732RN by USDA-APHIS	Bacterial colonization and monitoring test in a wheat-soybean-wheat rotation (18-month test) in collaboration with Clemson University, South Carolina	June 18, 1987	Nov. 2, 1987 Edisto Research and Education Center near Blackville, SC	July 17, 1989	E.J. Brandt D.J. Drahos B.C. Hemmig G.F. Barry
PS. 2-79RNL3	*lacZY Tn7* based genomic insertion in *P. aureofaciens* 2-79RN (*P. fluorescens* biotype E) a native antifungal strain	EPA-Office of Pesticide Programs— FIFRA regulations: Notification of small-scale field tests (Designation No. 524-NMP-006)	Biocontrol efficacy test and marker test in an antifungal strain against Take-all Disease of wheat in collaboration with Washington State University/ USDA-ARS and Clemson University	July 20, 1988	1988 at Spillman Farm in WA 1988 at Edisto Site near Blackville, SC	Dec. 7, 1989	D.J. Drahos E.J. Brandt B.C. Hemming
Ps. Q2-87AL6 and Ps. QC-67AL7	*lacZY Tn7* based genomic insertion in *P. fluorescens* and *aureofaciens* strains	EPA-Office of Pesticide Programs— FIFRA regulations: Notification of small-scale field tests (Designation No. 524-NMP-011)	Biocontrol efficacy test using marked strains in both Spring and Winter wheat at 3 widely different soil and geographical sites in conjunction	March 6, 1990	May, 1990 at Vincennes, IN May, 1990 at Pullman, WA Sites at IN, WA and MT were planted fall, 1990	Final report not required by agency	B.C. Hemming E.J. Brandt

Organism	Modification	Agency	Description		Date	Status	Contact
P. fluorescens biotypes I, II, III and P. putida	lacZY Tn7 based genomic insertion in any of these strains excepting P. marginalis	EPA-Office of Pesticide Programs—Registration Division—Generic Notification	This request for broad approval or generic notification would apply to Monsanto and its collaborators for future testing of these type strains in non-contained labs, greenhouses and small-scale field sites exempting the need for obtaining an Experimental Use Permit.	with Washington State University/USDA–ARS, Purdue University, and Montana State University	June 20, 1989 with completed responses on Feb. 28, 1990	Published in Federal Register May 22, 1990 with public comment period ended June 21, 1990 Under review and involved with scope oversight issues	E.J. Brandt D.J. Drahos B.C. Hemming
2140RL3	lacZY Tn7 based genomic insertion in a non-fluorescent P. corrugata strain of Australia	Australian Genetic Manipulation Advisory Committee (GMAC) for site test in the state of South Australia	Biocontrol efficacy test in marked strain in winter wheat in collaboration with CSIRO Division of Soils and Monsanto Australia in South Australia.		April, 1990	Sept., 1990 Approval granted June 18, 1990 GMAC Review Harvested, Spring 1991	Maarten Ryder (CSIRO) B.C. Hemming

called *P. fluorescens* biotype E) strain, Ps. 3732RNL11 (EPA–OTS PMN No. P-87-1292). The *P. corrugata* taxon, however, is known to contain strains exhibiting pathogenicity on tomato, particularly greenhouse grown tomato plants where virulent strains of this minor pathogen cause a stem and pith necrosis. Plant pathogenicity studies with Ps. 2140 were initially performed by CSIRO scientists in Australia and this strain did not exhibit pathogenicity on tomato plants. However, subsequent testing at Monsanto Co., St. Louis and also by CSIRO in Australia and by Dr. Felix Lukezic at Pennsylvania State University using the susceptible disease check, *Lycopersicon esculentum* (Tomato) cv. Mountain Pride, did demonstrate the plant pathogenicity of this strain under specific conditions.

Although this genus and species (*P. corrugata*) is known to exist in U.S. soils, the nonfluorescent pseudomonad host strain, Ps. 2140, was imported from Australia to Missouri under an APHIS permit from the laboratories of Dr. Albert Rovira, Dr. Maarten Ryder, and the late Dr. Andrew Simon of the CSIRO Division of Soils, Adelaide, S.A., Australia, following their selection and field tests for the biocontrol of take-all disease in Australian wheat fields. Interstate shipping permits were obtained for strain shipments between St. Louis, MO, the three cooperative universities in Montana, Washington, and Indiana, as well as, for Alabama, the site of a small scale (<0.5 acre), which had commenced a year earlier under authorization of the APHIS to conduct a small-scale test at Monsanto's Loxley Farm in Baldwin County, AL of two strains of *P. corrugata* (one of which was Ps. 2140). This test was completed with no indications of plant disease or survival of the strain after the completion of the harvest of the wheat. Susceptible tomato plants had been transplanted to the site to act as possible disease indicator plants.

The Institutional Biosafety Committee (IBC) of Monsanto Co. and each of the three cooperating universities (Purdue, Washington State, and Montana State) reviewed and provided approval of the earlier collaborative field studies prior to their initiation. The correspondence from the four IBCs concerning the result of their reviews were provided to the oversight agencies. Monsanto's IBC reviewed and approved the registration document prepared by CSIRO Division of Soils (Ryder, 1990) using data supplied both from CSIRO and Monsanto scientists entitled "Application for Field Release of a Genetically Engineered Strain of *Pseudomonas* for the Purpose of Testing a Microbial Tracking System," which provided a proposal to test the *lac*ZY-marked *P. corrugata* 2140RL3 strain in South Australia on property controlled by the University of Adelaide. Monsanto's IBC approval was contingent upon approval by the Australian Genetic Manipulation Advisory Committee (GMAC) to conduct the test. The Australian GMAC approval was provided on June 18, 1990.

In support of the proposed small-scale field tests (<1.5 acres each) in Indiana, Washington, and Montana of the *lac*ZY marked microorganism tests, Monsanto Co. and the collaborating universities had previously prepared a public information program with the goal of establishing a favorable public opinion climate, with a focus in the area around the proposed test sites. Similar *lac*ZY-marker tests have been previously approved and are under test by these collaborators. Native fluorescent pseudomonad tests had been previously conducted at all sites. A community relations program was carried out, including key contacts to local, county, state and appropriate federal officials, and civic leaders from the area around each of the test sites. The program was initiated collaboratively upon receipt of the field test approval by each institution's biosafety committee.

In addition, Monsanto Co. worked with the universities in terms of news media preparedness and materials for release at the appropriate time to all general and specialized media anticipated to have an interest in the collaborative small-scale field tests and the microbial genetic engineering on which the tests were based. These programs were successful in bringing to light the universities' strengths to their regional constituencies through the respective university public relations office. Similar preparations were made by Monsanto Australia in relation to the field test conducted by CSIRO in South Australia.

3.5 Strain Identification and Characterization

Maarten H. Ryder (CSIRO) under authorization of a USDA/APHIS importation permit secured by Monsanto scientists, sent the Australian strain to St. Louis for characterization. Dr. Ryder reported that the strain was originally isolated from the rhizosphere of wheat plants grown in pots in a continuous wheat soil (slightly acid red-brown earth), which was collected from Wagga Wagga, New South Wales, Australia. Artificial take-all inoculum had been added to the soil before the wheat was sown. The bacterium was isolated on a semiselective medium for fluorescent pseudomonads but was shown to be nonfluorescent on King's medium B or *Pseudomonas* F agar. As a part of the collaborative agreement between Monsanto and CSIRO, Dr. Ryder, under the tutelage of Monsanto scientists marked the strain with the genetic metabolic marker genes in Monsanto laboratories. In return, he provided detailed information relating to the strain's behavior as a control agent. The Monsanto Co. then exported the marked strain to Australia for field testing. Additional isolates for which comparative information was collected by the Monsanto Co. included the fluorescent pseudomonad strains designated as *P. aureofaciens* 2-79RN (originally, also obtained from the rhizosphere of wheat plants grown in the State of Washington by USDA scientists, as a spontaneously derived rifampicin and nalidixic acid resistant fluorescent pseudomonad strain) and *P. auerofaciens* 3732RN (originally, obtained in 1981 from soil planted to corn at a Monsanto-owned research farm, located in St. Charles County Missouri. Thirteen native U.S. *Pseudomonas* strains were also included in pathogenicity studies conducted at Monsanto Co. These strains were isolated in 1982 from root washings of apparently healthy soybean plants (*Glycine max* cv. Williams) grown near Sun Prairie, WS, St. Charles, MO, or Hoopeston, IL and subsequently identified by gas chromatography–fatty acid methyl ester (GC–FAME) analysis as belonging to the taxon, *P. corrugata*. Nonfluorescent pseudomonads selected for comparison purposes were obtained from the American Type Culture Collection (ATCC) and included ATCC No. 29736 *P. corrugata*, ATCC NO. 19302 *Pseudomonas gladioli* pv. *alliicola*, ATCC No. 10248 *Pseudomonas gladioli* pv. *gladioli*, ATCC No. 17460 *Pseudomonas cepacia*, and ATCC No. 25416 *P. cepacia* (Preceptrol culture).

All strains discussed have been examined using the Hewlett–Packard 5898A Microbial Identification System (MIS). The MIS analyzes and identifies microorganisms isolated in pure culture on artificial media through use of a standardized sample preparation procedure and GC to yield qualitatively and quantitatively reproducible fatty acid composition profiles. The fatty acid composition is then compared to a library of reference organisms stored in the system's computer to determine the identify of the strain. The manufacturer's recommended culture conditions were closely observed, since deviations may possibly alter the fatty acid composition, resulting in poor similarity matches to those of a representative number of reference strains. This instrumental method has proven very useful for monitoring and identification of bacteria from the environment (Hemming, 1988).

Confidence in the reliability of this instrumentation and software for identification of pseudomonads was reinforced by its performance in the identification of well-characterized reference strains of a collection that includes a number of strains originally obtained as reference stocks from the ATCC. Of 47 such pseudomonad reference strains, 41 strains (87%) were correctly identified to species or biotype. Of the six unidentified or misidentified strains, three represented species unlisted in the library data file of the system; namely, two *Pseudomonas delafieldii* strains (*acidovorans* group) and one *Pseudomonas methanolica* strain. In summary, 17 of 19 species, which included four biotypes of *P. aureofaciens*, were correctly identified. In addition, on the order of 9 of the 13 strains previously isolated in 1982 from the roots of healthy soybean plants in the United States, which were identified by the GC–FAME method, have subsequently been confirmed as *P. corrugata* in phytopathogenicity tests on tomato conducted at Monsanto Co. The 13 *P. corrugata* isolates hybridize specifically, in colony lifts, along with other characterized bacterial isolates belonging to the rRNA homology group 1 pseudomonads (including the fluorescent pseudomonads) to a DNA fragment consisting of 360 bp of the *P. aeruginosa* 23S rRNA gene (Festl et al., 1986). In these studies, this probe hybridized only with the DNA from this homology group screened from 597 characterized, soil isolates. This is evidence of *P. corrugata* in normal agricultural soils in the United States.

3.6. Future Potential

More reliable and rapid detection, differentiation, and enumeration methods are required for greater quantitative analysis of *in situ* natural product synthesis, disease scaling, and in dealing with both the target organisms (take-all fungal strains) and the biocontrol of take-all disease will be manifested in the use of genetic manipulations of control agents. Unfortunately, the projects at Monsanto Co. have fallen victim to the U.S. economic recession commencing in 1990 and a realignment in strategy by a company whose interests in the biotechnological applications in the pharmaceutical world take on an even greater market priority. The future directions of biocontrol of take-all in terms of delivering useful products will undoubtedly require further developments. The developments in the use of high-resolution image analysis on desktop computers will permit more rapid quantitative analysis of disease or will facilitate analysis for quality control in manufacturing the product (i.e., the size distribution of microencapsulated bacteria).

Furthermore, a sensitive and extremely rapid method of differentiation and monitoring recombinant organisms is under development (Hughes et al., 1989). The method is based on a conventional enzymatic procedure known commonly as aminopeptidase profiling. Modification of the conventional assay through chemical as well as instrumental means has produced an assay that has reduced bacterial cell concentrations 20,000-fold, reduced the incubation time from 20 h to 3 min, and has decreased the total turn-around time for identification from 2.5 days to approximately 6 h. This methodology has been used in differentiating between closely related species of *Pseudomonas* and genetically modified *P. aureofaciens*. In addition, adaptation of the laser based aminopeptidase assay for examination of the products of β-galactosidase has proven sensitive enough to differentiate between the wild-type *P. aureofaciens* (Ps. 3732RN), this strain containing a single *lac*ZY gene inserted into the bacterial chromosome (Ps. 3732RNL11), and the wild-type strain transformed to harbor multiple copies of a plasmid bearing the *lac*ZY insert (Ps. 3732RNpMON5003). This methodology is adaptable to tracking phytopathogenic fluorescent pseudomonads in infection, survival and transmission studies, or similar organisms used in bioremediation and biodegradation studies.

In addition to the technological developments necessary for future success, a scientifically sound and politically stable regulatory and registration environment will be required. Public awareness and education in the nature of biological control is necessary to implement improvements not only in productivity, but perhaps more importantly for the environmental compatibility of our agricultural technologies. Finally, the similarities between the biocontrol project discussed and research in bioremediation are obvious. Each project relies on bacterial agents that must be characterized, identified, produced and delivered, or introduced into the environment. The option to use native indigenous organisms or engineered organisms often exists. Unfortunately, the status of regulatory development in bioremediation lags behind the progress described in this chapter. This fact is evident because not a single systematic application of any microbe for hazardous waste treatment has been field tested in the environment. This legacy of biocontrol of take-all may ultimately impact bioremediation. Basic research methodologies may simultaneously affect both areas. It is clear, however, that industry is not going to develop such agents, unless they know they can use them.

ACKNOWLEDGMENTS

We wish to acknowledge the researchers at universities and industrial companies worldwide who have pursued studies in biocontrol of take-all disease for the improvement of crop production in agriculture. Without their untiring efforts to create the ''discipline interfaces'' that are necessary to new technology and new product development, this chapter would not have been possible.

REFERENCES

Baker CA, Henis JMS (1990): Commercial production and formulation of microbial biocontrol agents. In Baker RR and Dunn PE (eds): New Directions in Biological Control: Alternatives for Suppressing Agricultural Pests and Diseases. New York: Alan R. Liss, pp 333–344.

Barry GF (1986): Permanent insertion of foreign genes into the chromosomes of soil bacteria. Bio/Technol 4:446–449.

Bio-Tech Transfer (1991): Academic biotechnologists want more alliances with industry, survey finds. Biotechnol Week. Oct 7, 29.

Brennan JP, Murray GM (1988): Australian Wheat Diseases. Agric Sci (Australian Institute of Agricultural Science), Dec., 26–35.

Brisbane PG, Janik LJ, Tate ME, Warren RFO (1987): Revised structure for the phenazine antibiotic from *Pseudomonas fluorescent* 2-79 (NRRL B-15132). Antimicrobial Agents Chemother 31:1967–1971.

Bureau of Agricultural Economics (1988): Q Rev Rural Econ, March 10 (1), 5–7.

Burr TJ, Schroth MN, Suslow T (1978): Increased potato yield by treatment of seedpieces with specific strains of *Pseudomonas fluorescens* and *P. putida*. Phytopathology 68:1377–1383.

Cate R (1990): Biological control of pests and diseases: integrating a diverse heritage. In Baker RR and Dunn PE (eds): New Directions in Biological Control: Alternatives for Suppressing Agricultural Pests and Diseases. New York: Alan R. Liss, pp 333–344.

Cook RJ, Rovira AD (1976): The role of baceria in the biological control of *Gaeummanomyces grammis* var. *triticii* by suppressive soils. Soil Biol Biochem **8**, 269–273.

Culliton BJ (1990): Monsanto renews ties to Washington University. News & Comment. Science March 2, 247:1027.

Drahos DJ, Hemming BC, McPherson S (1986): Tracking recombinant organisms in the environment: β-galactosidase as a selectable non-antibiotic marker for fluorescent pseudomonads. Bio/Technol 4:439–444.

Department of Commerce (1990): U.S. Industrial Outlook. Washington, DC: US Government Printing Office, p 20–24.

Festl H, Ludwig W, Schleifer K (1986). Appl Environ Microb 52, 1190–1194.

Gurusiddaiah S, Weller DM, Sarkar A, Cook RJ (1986): Characterization of an antibiotic produced by a strain of *Pseudomonas fluorescens* inhibitory to *Gaeummanomyces graminis* var. *tritici* and *Pythium* spp. Antimicrob Agents Chemother 29:488–499.

Hashimoto S, Fujita M (1985): Isolation of a bacterium requiring three amino acids for polyvinyl alcohol degradation. J Ferment Technol 63:471–474.

Hemming BC (1988): Monitoring and identification of bacteria from agricultural environments by gas chromatography fatty acid methyl ester (GCFAME) analysis. In Lange CT (ed): Bioinstrumentation, St. Louis Mathematics and Science Education Center, University of Missouri—St. Louis, pp 57–70.

Hemming BC (1990): Bacteria as antagonists in biological control of plant pathogens. In Baker RR, Dunn PE (eds): New Directions in Biological Control: Alternatives for Suppressing Agricultural Pests and Diseases. New York: Alan R. Liss, pp 223–242.

Hess CE (1990): Issues facing the U.S. farm community: Farm structure, environmental stewardship, and international competition. Prepared remarks given at U.S. Agriculture: Forces reshaping the future. Kline & Company symposium, Chicago, IL.

Homma Y, Sato A, Hirayama F, Konno K, Shirahama H, Suzui T (1988): Production of antibiotics by *Pseudomonas cepacia* as an agent for biological control of soilborne plant pathogens. Soil Biol Biochem 21:723–728.

Hornby D, Bateman GL (1991): Take-all disease of Cereals. Home-Grown Cereals Authority (HGCA). Rothamsted APRC Institute of Arable Crops Research: Research Review No. 20, 148 pages.

Hughes KD, Lytle FE, Roseman TS, Huber DM, Hemming BC. (1989): Differentiation of genetically engineered bacteria with laser based aminopeptidase profiling,'' 1989 Pittsburgh Conference & Exposition, Atlanta, GA, March 6–10, 1989.

Kloepper JW, Schroth MN, Miller TD (1980): Effects of rhizosphere colonization by plant growth-promoting rhizobacteria on potato plant development and yield. Phytopathology 70:1078–1082.

Lindheim JB (1989): Address to the Society of the Chemical Industry: Brussels, Belgium.

Monsanto Company (1987): Biotechnology Premanufacturing Notification proposal for field testing of host *Pseudomonas fluorescens* (Ps. 3732RN) and genetically engineered *Pseudomonas fluorescens* (Ps. 3732RNL11). A collaborative study by Monsanto Company and Clemson University (June 18, 1987), Vols. 1 and 2, EPA-OTS PMN Document No. P-87-1292.

Monsanto Company (1988): Small-Scale Field Testing of a *lac*ZY (*E. coli* K-12) Engineered Microbial Fungicide: Evaluation of Wheat Root Colonization By Ps. 2-79RNL3 and Level of Root Infection By Take-All (Ggt) Disease, Vols 1–3, EPA-OTS submitted to the Fungicide and Herbicide Branch Registration Division EPA-Office of Pesticide Programs, on July 20, 1988.

Monsanto Company (1990): Small-Scale Field Testing of *lac*ZY (*E. coli* K-12) Engineered Microbial Fungicides, Ps. Q2-87AL6 and Ps. QC-65AL7: Evaluation for Efficacy as Biological Control Agents Against Take-All (Ggt) Disease of Wheat,'' March 6, 1990 and April 13, 1990, Vol 1 and supplement, EPA-OTS Designation No. 524-NMP-011.

Monsanto Company (1989): Final field study report: Small Scale Field Testing of a *lac*ZY (E. coli K-12) Engineered Microbial Fungicide EPA-OPP Designation No.: 524-NMP-006.

Monsanto Company and Clemson University (1989): Final Report: Field Testing of Host *Pseudomonas aureofaciens* (Ps. 3732RN) and Genetically Engineered *Pseudomonas aureofaciens* (Ps. 3732RNL11). A Collaborative Study By Monsanto Company and Clemson University. EPA Premanufacture Notification No. P-87-1292. July 17, 1989, Vol I of VI Part 7: Final Report (through week 83).

Montague MJ (1991): A university–industry partnership approach to technology transfer. July/August Product Process Innovation 1(4): 19–23.

Murray GM, Brown JF (1987): The incidence and relative importance of wheat diseases in Australia. Aust Plant Path 16(2): 34–37.

Reddi PK, YP Khudyakov, Borovkov AA (1969): *Pseudomonas fluorescens* strain 26-0, producing phytotoxic substances. Mikrobiologiya. 38: 909–913.

Ryder MH (1990): Application for field release of a live genetically engineered strain of *Pseudomonas* for the purpose of testing a microbial tracking system. Glen Osmond, South Australia: CSIRO Division of Soils, 42 p.

Sakai K, Hamada N, Watanabe Y (1985): A new enzyme, β-diketone hydrolase: a component of a poly(vinyl alcohol)-degrading enzyme preparation. Agric Biol Chem 49: 1901–1902.

Shimao M, Fujita I, Kato N, Sakazawa C (1985): Enhancement of pyrroloquinoline quinone production and polyvinyl alcohol degradation in mixed continuous cultures of *Pseudomonas putida* VM15A and *Pseudomonas* sp. Stain VM15C with mixed carbon sources. Appl Environ Microbiol 49: 1389–1391.

Shimao M, Ninomiya K, Kuno O, Kato N, Sakazawa C (1986): Existence of a novel enzyme, pyrroloquinoline quinone-dependent polyvinyl alcohol dehydrogenase, in a bacterial symbiont, *Pseudomonas* sp. strain VM15C. Appl Environ Microbiol 51:268–275.

Shipton PJ (1975): Take-all decline during cereal monoculture. In Bruehl GW (ed). Biology and control of soil-borne plant pathogens. St. Paul: American Phytopathological Society. pp 137–144.

Smiley RW (1978): Colonization of wheat roots by *Gaeummanomyces graminis* inhibited by specific soil microorganisms and ammonium-nitrogen. Soil Biol Biochem 10: 175–179.

Thomashow LS, Weller DM (1988): Role of a phenazine antibiotic from *Pseudomonas fluorescens* in biocontrol of *Gaeumannomyces graminis* var. *triticii*. J. Bacteriol 170:3499–3508.

Thomashow LS, Weller DM, Bonsall RF, Pierson III LS (1990): Production of the antibiotic phenazine-1-carboxylic acid by fluorescent *Pseudomonas* species in the rhizosphere of wheat. Appl Environ Microbiol 56: 908–912.

Weller DM (1983): Colonization of wheat roots by a fluorescent pseudomonad suppressive to Take-All. Phytopathology 73: 1548–1553.

Weller DM, Cook RJ (1983): Suppression of Take-All of wheat by seed treatments with fluorescent pseudomonads. Phytopathology 73: 463–469.

Witt Steven C (1990) Center for Science Information. Biotechnology, Microbes and the Environment. San Francisco, CA: Center for Science Information, 219 p.

≡3

PROTEIN SECRETION BY PLANT PATHOGENIC BACTERIA

SHENG YANG HE
MAGDALEN LINDEBERG
ALAN COLLMER

Department of Plant Pathology, Cornell University, Ithaca, New York 14853

1. INTRODUCTION

Many bacteria have an impact on agriculture because of the proteins they secrete. These extracellular proteins can be beneficial in processes like biomass utilization, or they can be devastating when acting as virulence factors in plant diseases. Biochemical genetic studies begun in the mid-1980s revealed that some plant pathogenic, gram-negative bacteria secrete a surprising variety of plant cell wall degrading enzymes and that these enzymes (particularly the pectic enzymes) contribute significantly to virulence (Daniels et al., 1988). These studies also revealed the essential role of protein secretion pathways in pathogenesis. As will be discussed later, mutants of *Erwinia chrysanthemi, Erwinia carotovora,* and *Xanthomonas campestris* pv. *campestris* with deficiencies in individual plant cell wall degrading enzymes are only partially reduced in virulence. In contrast, mutants with deficiencies in protein secretion are generally nonpathogenic. Similarly, random mutants that have completely lost virulence are more commonly found to be deficient in the secretion rather than the synthesis of plant cell wall degrading enzymes. Thus, the secretion pathways are more vital to virulence and more vulnerable to mutation than any of the proteins that travel them.

Biotechnology in Plant Disease Control, pages 39–64
© *1993 Wiley-Liss, Inc.*

Despite its importance in plant (and animal) pathogenesis, protein secretion by gram-negative bacteria is still a poorly understood process. The gram-negative bacterial cell is bounded by an inner (cytoplasmic) membrane, a periplasmic space that contains the peptidoglycan cell wall layer, and an outer membrane. Periplasmic and outer-membrane proteins are *exported* across the inner membrane via the *sec* gene-dependent, general export pathway. Extracellular proteins are *secreted* to the cell exterior by pathways that also cross the outer membrane. Whereas all gram-negative bacteria must export proteins to the periplasm and outer membrane for cell viability, only some bacteria appear able to secrete proteins to the cell exterior. Laboratory strains of *Escherichia coli,* for example, lack a protein secretion apparatus, and this limits their utility in producing proteins for biotechnological purposes. In contrast, some plant pathogenic bacteria efficiently secrete an extraordinary array of proteins.

To understand the secretion process well enough to either enhance the production of proteins that are beneficial in agriculture and biotechnology, or to interfere with the production of proteins that contribute to pathogenesis, we should know the signals that target proteins for secretion, the components and location of the secretion apparatus, and the mechanism by which that apparatus recognizes and translocates proteins across the cell envelope. This chapter will provide an update on progress toward those objectives. We will first explore the structure of the cell envelope of gram-negative bacteria and the evidence that there are at least two general pathways for protein secretion across the envelope. Each of these pathways has been most extensively explored with a human pathogen. The first pathway is used by hemolytic strains of *E. coli* to secrete hemolysin. The second is used by *Klebsiella* spp. to secrete pullulanase. We will summarize the work with these two models before turning to work on the secretion of extracellular virulence proteins by four model plant pathogens: *X. c. campestris, Pseudomonas solanacearum, E. carotovora,* and *E. chrysanthemi.* Finally, we will discuss the significance of protein secretion pathways in bacterial pathogenesis and the potential benefits of a better understanding of macromolecular traffic across the bacterial envelope.

2. THE ENVELOPE OF GRAM-NEGATIVE BACTERIA AND PROTEIN TRANSLOCATION PATHWAYS

2.1. Envelope Structure

The structure of the cell envelope of *E. coli* and *Salmonella typhimurium* is well described in Neidhardt et al. (1987). Only features relevant to protein secretion will be discussed in this chapter. The spatial relationship between the inner and outer membranes demands consideration because secreted proteins must cross these barriers to reach the cell exterior. A similar problem is encountered by eukaryotes in the import of nuclear-encoded proteins into the mitochondrial matrix. In that case, protein translocation appears to occur at contact sites between the outer and inner membranes (Attardi and Schatz, 1988). Although similar "adhesion zones," or "Bayer junctions," have been proposed to occur in bacterial envelopes (Bayer, 1968), their physiological existence is now questioned. Improved electron microscopy techniques employing freeze substitution rather than chemical fixation reveals that the two membranes, each 7–11 nm wide, are separated by a periplasmic space that is uniformly 11–15 nm wide, and, most importantly, lacks adhesion zones (Kellenberger, 1990; Graham et al., 1991). That the two membranes are functionally separate in the secretion process is further supported by evidence, discussed below, that extracellular proteins of *Klebsiella oxytoca, P. solanacearum,* and *E. chrysanthemi* cross the inner and outer membranes in two distinct steps.

The outer membrane has four distinguishing features that are potentially relevant to protein secretion mechanisms. First, the outer membrane is a lipid bilayer that is asymmetric because the outer lipid monolayer contains lipopolysaccharide, whereas the inner monolayer contains phospholipid. Second, lipoproteins and other proteins in the outer membrane keep the membrane tightly associated with the peptidoglycan layer, which in turn forms a defined and continuous outer layer in the periplasm. Third, the outer membrane contains porins that render it permeable to low molecular weight hydrophilic compounds. The *E. coli* porins are not gated, and there is no proton motive force across the outer membrane (Sen et al., 1988). However, Wong and Buckley (1989) reported that a proton motive force across the outer membrane of *Aeromonas salmonicida* is involved in protein secretion. This observation remains to be tested with other gram-negative bacteria that secrete proteins. Fourth, because of the anionic membrane-derived oligosaccharides in the periplasm, there is both a Donnan and an osmotic potential across the outer membrane. The outer membrane is an effective exclusion barrier against a variety of lytic proteins that could damage the bacterial cell, and it effectively retains periplasmic proteins—even when extracellular proteins are being actively secreted.

2.2 The Multiple Pathways Into and Across the Cell Envelope

The proteins synthesized by a gram-negative bacterium may potentially be localized to any of five cellular compartments: the cytoplasm, inner membrane, periplasm, outer membrane, and cell exterior. The Sec-mediated general export pathway is used by most proteins that are inserted in the inner membrane, by all proteins targeted to the periplasm or the outer membrane, and as the first step in the secretion of many extracellular proteins. The Sec-mediated pathway is described succinctly by Bassford et al. (1991) and Model and Russel (1990) and in more detail by Randall and Hardy (1989), Saier et al. (1989), Bieker and Silhavy (1990), Schatz and Beckwith (1990), Kumamoto (1991), and Wickner et al. (1991). Briefly, proteins are targeted for export by an N-terminal signal peptide that functions to retard folding of the newly synthesized protein and that may interact (along with domains in the mature protein) with cytoplasmic chaperone proteins like SecB and with membrane-associated SecA. Subsequent translocation of the protein across the inner membrane is dependent on SecA ATPase activity, the proton motive force, and at least four more Sec proteins in the inner membrane. Following translocation, the signal peptide is cleaved by a signal peptidase, and the protein is sorted to a location in either of the membranes or in the periplasm by sequences in the mature protein. The bacterial general export pathway has been studied with rigor because powerful genetic selections for Sec phenotypes are available and because activity of the Sec proteins can be studied *in vitro* with reconstituted proteoliposomes. Neither of these tools is available for studying secretion or translocation across the outer membrane. It is also sobering to note that, despite the powerful tools applied to study of the general export pathway, the functions of the signal peptide are still not completely resolved, it is still not known whether exported proteins pass directly through the membrane lipid or through a proteinaceous pore, and even the primacy of the Sec pathway in general export has recently been challenged (Randall and Hardy, 1989; Bassford et al., 1991; Rapoport, 1991).

In contrast to the universal pathway used by bacteria to export proteins across the inner membrane, multiple pathways are used to secrete proteins to the cell exterior (Hirst and Welch, 1988; Pugsley et al., 1990a). Colicins, for example, are released by a mechanism that involves partial cell lysis. This mechanism will not be considered further, because the proteins are leaked rather than secreted from the cell. However, it is important to note that *E. carotovora* may release pectin lyase (an enzyme that cleaves highly methyl esterified pectic polymers) by a similar mechanism (Chatterjee et al., 1991). Pectin lyase is produced by many strains of *E. carotovora* and *E. chrysanthemi* following treatment with DNA-damaging agents (Tsuyumu and Chatterjee, 1984), and pectin lyase induction is RecA dependent and accompanied by cell lysis (Zink et al., 1985). Furthermore, the sequence of the *E. carotovora* subsp. *carotovora* 71 *pnlA* gene reveals no N-terminal signal peptide, and the protein is produced by *Erwinia* without removal of any N-terminal sequences (Chatterjee et al., 1991).

Two other pathways are used by a variety of gram-negative bacteria for true secretion of proteins, that is, without cell lysis or release of periplasmic or cytoplasmic marker proteins. The secretion of hemolysin and related cytolysins by gram-negative bacteria involves a pathway that is independent of the general export pathway. Among plant pathogens, *E. chrysanthemi* has been shown to secrete proteases by this mechanism. In contrast, the secretion of pullulanase by *K. oxytoca* involves a different pathway that is Sec dependent. Most virulence proteins produced by plant pathogenic bacteria appear to be secreted by Sec-dependent pathways. In Sections 2.3 and 2.4, progress on characterizing the pathways of hemolysin and pullulanase secretion will be summarized, respectively.

2.3. Secretion of Hemolysin by the Sec-Independent Pathway

Hemolysin is a virulence determinant in extraintestinal *E. coli* infections and is secreted via a Sec-independent pathway. The structural (*hlyA*), secretion (*hlyB*, *hlyD*), and activation (*hlyC*) genes are clustered as one operon in the chromosome or on a plasmid in hemolytic strains (Wagner et al., 1983; Holland et al., 1990). Hemolysin (HlyA) does not have an N-terminal signal peptide (Felmlee et al., 1985a); instead a novel targeting signal located within the last 53 C-terminal amino acids is essential for the direct translocation of HlyA across both cell membranes without any periplasmic intermediate (Koronakis et al., 1989). The C-terminal targeting signal is not conserved at the primary sequence level, but has multiple secondary features: an amphiphilic α-helix immediately proximal to a cluster of charged residues and a weakly hydrophobic sequence rich in hydroxylated residues at the C-terminus. A membrane-bound translocator specific for hemolysin secretion, composed of HlyB, HlyD (Felmlee et al., 1985b; Gray et al., 1989), and TolC (Wandersman and Delepelaire, 1990), has been identified and characterized. The HlyB translocator is a putative adenosine triphosphate (ATP)-binding protein located in the inner membrane and has a mammalian homolog, P-glycoprotein (Higgins et al., 1986). The HlyD translocator is also an inner-membrane protein but has an extensive periplasmic domain (Holland et al., 1990). The TolC translocator is an *E. coli* outer membrane protein that was shown to be required for HlyA secretion (Wandersman et al., 1990). The precise assembly of the translocator in the *E. coli* envelope and the role of individual components (HlyB, HlyD, or TolC) in the secretion of HlyA remain unclear. Recent studies indicate that this secretion pathway is widely distributed among gram-negative bacteria and is used for the secretion of a subset of extracellular proteins, especially hemolysins, leukotoxins, cytolysins, and some proteases. The *Erwinia* proteases are the only extracellular proteins that are known to be secreted from plant pathogens by this pathway, and the important work on the *Erwinia* proteases will be discussed in Section 4.

2.4 Secretion of Pullulanase by the Sec-Dependent Pathway

Pullulanase is a starch-debranching enzyme that is secreted by *K. oxytoca* and other species of *Klebsiella*. Most aspects of pullulanase secretion are representative of the Sec-dependent pathway (Pugsley et al., 1990a), although there are some unusual features resulting from the lipoprotein nature of pullulanase (Pugsley et al., 1986). Pugsley et al. extensively characterized the pullulanase secretion pathway in both *K. oxytoca* and recombinant *E. coli* cells (see Pugsley et al.,

1990a for a review). Secretion of pullulanase in *E. coli* was accomplished by cloning a 22-kb DNA fragment containing both the structural gene (*pulA*) and 14 secretion genes (*pulS, pulC-O*), which flank both sides of *pulA* (d'Enfert et al., 1987; Puglsey et al., 1990a). In the absence of the secretion genes, prepullulanase is processed by signal peptidase II (specific for lipoproteins), acylated at the cys residue (Pugsley et al., 1986), and then localized to both the inner and outer membranes in *E. coli* (Pugsley et al., 1990b). In *K. oxytoca* or in *E. coli* cells carrying the functional secretion gene cluster, however, the mature pullulanase is exposed on the cell surface during exponential growth and slowly released into the medium, in an aggregated form, in the stationary phase (Michaelis et al, 1985; d'Enfert et al., 1987). The cell surface association and formation of extracellular aggregates appear due to the presence of the acyl group at the pullulanase N-terminus, because a naturally unacylated form of pullulanase is rapidly secreted into the medium without aggregate formation (Kornacker et al., 1989).

The Sec dependency of pullulanase secretion is supported by the observation that mutations in several *sec* genes abolish the extracellular secretion of pullulanase in *E. coli* even in the presence of the functional secretion gene cluster (Pugsley et al, 1991a). The secretion process thus appears to involve two steps: translocation of prepullulanase across the inner membrane and translocation of the acylated, mature protein across the outer membrane. The two steps can be uncoupled by first allowing export of pullulanase across the cytoplasmic membrane and then inducing expression of the *pulC-O* operon (Pugsley et al., 1991b).

In is not known how pullulanase is recognized by the PulC-O secretion machinery. The N-terminal one-half of pullulanase appears important for secretion, because protein fusions between the N-terminal portion of pullulanase and β-lactamase (a periplasmic protein in *E. coli*) are secreted by *E. coli* cells carrying the *pulC-O* operon (Kornacker et al., 1990). The specific roles of most of the 14 secretion pathway proteins and the interaction between these proteins and pullulanase during secretion are also not understood. Many extracellular proteins (e.g., plant cell wall degrading polysaccharidases of plant pathogenic bacteria, aerolysin of *Aeromonas hydrophila,* and cholera toxin of *Vibrio cholerae*) are secreted by a similar mechanism, and, as discussed below, it is becoming increasingly evident that the Sec-dependent pathway is highly conserved and widely distributed among gram-negative bacteria.

3. THE EXTRACELLULAR VIRULENCE PROTEINS OF PLANT PATHOGENIC BACTERIA

Most bacterial pathogens of plants are gram negative. The extracellular proteins and secretion pathways of four bacteria in this group are now being actively explored. An important feature of the protein secretion pathways of the plant pathogens is that they transport a menagerie of extracellular proteins, rather than just one, as was described above for hemolysin and pullulanase. In this section, we will inventory the arsenal of proteins secreted by each of these pathogens and briefly summarize the evidence that the proteins have a role in pathogenesis. Most of the proteins are plant cell wall degrading enzymes, primarily pectic enzymes. Reviews discussing plant cell wall structure and the general involvement of degradative enzymes in pathogenesis are available elsewhere (McNeil et al., 1984; Collmer and Keen, 1986; Kotoujansky, 1987; Daniels et al., 1988). Structural aspects of these proteins related to potential targeting signals and data on secretion-deficient mutants will be discussed in Sections 5.1 and 5.5.

Xanthomonas campestris pv. *campestris* causes spreading "black rot" lesions on leaves of cruciferous plants. The bacterium secretes endoglucanase, amylase, two proteases, and three isozymes of pectate lyase (Dow et al., 1989a). Genes encoding protease, endoglucanase, and one pectate lyase isozyme have been cloned and used to construct marker-exchange mutants (Tang et al., 1987; Gough et al., 1988; Dow et al., 1989b). Although virulence assessments differ with inoculation technique (Shaw and Kado, 1988), the endoglucanase- and protease-deficient mutants have been shown to be reduced in virulence (Gough et al., 1988; Dow et al., 1990). A mutant deficient in pectate lyase isozyme I showed no reduction in virulence (Dow et al., 1989b), but the production of three pectate lyase isozymes confounds easy analysis of the role of this enzyme in virulence.

Pseudomonas solanacearum is a soil bacterium that causes wilt diseases in a variety of important crop plants. The bacterium secrete endoglucanase, polygalacturonase, and one or more *exo*-polygalacturonases; genes encoding the first two enzymes have been cloned (Schell, 1987; Roberts et al., 1988; Schell et al., 1988; Denny et al., 1990; Allen et al., 1991). Strains deficient in endoglucanase, the major polygalacturonase, or both enzymes were constructed by marker-exchange mutagenesis. A reduction in virulence is observed with each mutant. That is, wilt symptoms in susceptible tomato plants are produced more slowly by the mutants than by the wild type (Roberts et al., 1988; Schell et al., 1988). But the effect of the polygalacturonase mutation can be observed only when low levels of inoculum are used, and is less than the effect of endoglucanase mutations (Denny et al., 1990).

Eriwinia carotovora and *E. chrysanthemi,* the "soft-rot erwinias," cause diseases involving extensive maceration of parenchymatous tissues in a wide array of plants. These bacteria secrete a variety of degradative enzymes, including proteases, cellulases, and most importantly for virulence, several different pectic enzymes (Kotoujansky, 1987). Both species produce multiple isozymes of pectate lyase, and many strains also produce pectin lyase (whose unusual production was discussed above). The two species appear to differ in the types of hydrolytic pectic enzymes they make. *Erwinia carotovora* produces polygalacturonase, an endo-cleaving enzyme, whereas *E. chrysanthemi* produces *exo*-poly-α-D-galacturonosidase, which removes dimers from the nonreducing end of pectic polymers (Ried and Collmer, 1986). The involvement of the *Erwinia* pectic enzymes in maceration has been extensively studied with purified enzymes, recombinant *E. coli* strains expressing cloned *Erwinia* genes, and *Erwinia* strains containing directed mutations in pectic enzyme structural genes (Bateman and Basham, 1976; Collmer and Keen, 1986; Kotoujansky, 1987; Daniels et al., 1988). This previously reviewed work and more recent studies afford several relevant conclusions: (1) endo-cleaving pectic enzymes, regardless of reaction mechanism or substrate preference, are generally able to macerate plant tissues, (2) maceration efficiency is associated with high protein pI, with the *E. chrysanthemi* acidic pectate lyase isozyme PelA having no apparent maceration activity, and (3) pectic enzymes like PelA or pectin methylesterase, which contribute little to maceration, are nevertheless important for systemic invasion of whole plants (Boccara et al., 1988; Boccara and Chatain, 1989).

4. SEC-INDEPENDENT SECRETION OF PROTEASES BY *ERWINIA* SPP.

Early genetic studies with *E. chrysanthemi* revealed that mutations affecting the secretion of pectic enzymes and cellulase did not affect protease secretion (Andro et al., 1984; Thurn and Chatterjee, 1985). This indicated that protease and the other degradative enzymes were secreted by different pathways. A 16-kb chromosomal DNA fragment was cloned from *E. chrysanthemi* EC16 and shown to mediate both the production and secretion of a protease in *E. coli* (Barras et al., 1986). Subsequent work showed that this DNA fragment contained clustered genes for a protease inhibitor and three proteases (A, B, and C) and 4-kb region required for secretion (Dahler et al., 1990).

A similar gene cluster was cloned from *E. chrysanthemi* B374, and an 8.5-kb subclone was shown to produce and secrete two proteases (B and C) in *E. coli* (Wandersman et al., 1987). A DNA sequence analysis of the 8.5-kb fragment revealed at least six genes: *prtB* and *prtC* (proteases B and C), *inh* (protease inhibitor), and *prtD, prtE,* and *prtF* (protease secretion) (Letoffe et al., 1990). Proteases B and C lack typical signal peptides at the N-termini, and the three secretion genes share significant sequence similarity with the *E. coli* hemolysin secretion genes *hlyB, hlyD,* and *tolC* (Letoffe et al., 1990). Furthermore, protease B was shown to be recognized and secreted by the hemolysin secretion machinery, albeit at low efficiency (Delepelaire, et al., 1990). Thus the secretion machineries for *E. chrysanthemi* proteases and *E. coli* hemolysin are homologous and functionally interchangeable. Much less is known about protease secretion in *E. carotovora*. A similar secretion mechanism may operate, since it has been shown that mutations in the Sec-dependent pathway in this bacterium do not affect protease secretion (Hinton and Salmond, 1987; Murata et al., 1990).

5. SEC-DEPENDENT SECRETION OF PLANT CELL WALL DEGRADING ENZYMES BY PLANT PATHOGENIC BACTERIA

5.1. Mutagenesis and Cloning of Secretion Pathway Genes

Molecular biological techniques have greatly facilitated the study of the Sec-dependent secretion pathway in plant pathogenic bacteria, particularly by enabling the mutagenesis, cloning, and comparison of genes encoding pathway components. Initial investigations of the genetics of secretion by *X. c. campestris*, *E. carotovora*, and *E. chrysanthemi* involved identification of mutants deficient in their ability to secrete multiple extracellular enzymes. These mutants continued to synthesize degradative enzymes but retained them in the periplasm and will be referred to as Out⁻ mutants (Andro et al., 1984) in the remainder of this chapter. This observation indicate the presence of secretion genes that were independent of enzyme structural genes and required for the translocation of multiple proteins across the outer membrane but not the inner membrane. Subsequently, genes involved in the secretion pathway have been identified, cloned, and sequenced from all three bacteria.

Daniels et al. (1984a) reported the isolation of nonpathogenic mutants of *X. c. campestris*. Cosmid pIJ3000 was identified on the basis of its ability to restore pathogenicity to one of these mutants (Daniels et al., 1984b) and was subsequently shown to contain a 10-kb region, composed of 4.2 and 5.8-kb subregions, that is necessary for the secretion of protease, endoglucanase, amylase, and pectate lyase (Dow et al., 1987). Sequence analysis of the 5.8-kb subregion revealed a putative promoter sequence and five open reading frames, designated *xpsE, xpsF, xpsG, xpsH,* and *xpsI* (Dums et al., 1991).

Erwinia carotovora Out⁻ mutants deficient in their ability to secrete several extracellular proteins have been isolated by several approaches. Chatterjee et al., (1985) identified nitrosoguanidine-induced mutants of *E. carotovora* subsp. *carotovora* ECC193 that were unable to secrete pectate lyase, polygalacturonase, and cellulase. Hinton and Salmond (1987) used Tn*phoA* as a reporter for genes encoding exported proteins and isolated Out⁻ mutants of *E. carotovora* subsp. *carotovora* (Manoil and Beckwith, 1985). Cosmid clones complementing some of these Out⁻ mutants were subsequently isolated (Gibson et al., 1988). Pirhonen et al. (1991) identified Out⁻ *E. carotovora* subsp. *carotovora* strain SCC3193 mutants among transposon mutants strongly reduced in virulence on axenic tobacco seedlings. Murata et al. (1990) screened directly for transposon-induced Out⁻ mutants of *E. carotovora* subsp. *carotovora* 71 unable to secrete polygalacturonase, cellulase, and pectate lyases. A cluster of *out* genes on a 15.7-kb DNA fragment cloned from *E. carotovora* subsp. *carotovora* 71 restored the secretion phenotype to a variety of strain 71 and *E. chrysanthemi* Out⁻ mutants.

Out$^-$ mutants of *E. chrysanthemi* have also been isolated by transposon muta-genesis (Ando et al., 1984; Thurn and Chatterjee, 1985; Ji et al., 1987). As with the *E. carotovora* Out$^-$ mutants, cellulase and pectic enzymes accumulate in the periplasm of these mutants, while protease secretion is unaffected. By con-structing Mu-*lac* fusions, Ji et al. (1987) were able to show that the expression of at least one *out* gene was not affected by the presence of polygalacturonate or galacturonate in the medium or by growth phase. The altered periplasmic protein profiles of two *E. chrysanthemi* Out$^-$ mutants suggest that two different Out proteins are located in the periplasm. Thurn and Chatterjee (1985) observed the loss of a 35-kD periplasmic protein in three Out$^-$ strain EC16 mutants. In con-trast, Ji et al. (1987) observed the loss of a 65-kD protein from the periplasmic protein profiles of two Out$^-$ derivatives of strain 3937.

Ji et al. (1989) cloned an *out* gene, designated as *outJ*, from *E. chrysanthemi* 3937. The Tn*phoA* insertions that were introduced into the cloned DNA and then marker exchanged into strain 3937 were used to define a 2.5-kb region that encoded exported protein(s) and was required for secretion. By expressing the cloned DNA with the T7 promoter system in *E. coli*, they were able to observe the 83-kD precursor and 81-kD mature protein product of *outJ*. The relationship between the product of the cloned *outJ* gene and the 65-kD protein missing from *outJ* mutants is unclear at this point.

He et al. (1991a) isolated a cluster of *out* genes from *E. chrysanthemi* by mobi-lizing an *E. chrysanthemi* cosmid library into several transposon-induced secre-tion mutants. Five cosmids were found to restore secretion competency to all mutants tested. Transposon mutagenesis of one of these cosmids, pCPP2006, indi-cated that genes essential for the secretion phenotype are clustered in a 12-kb region. Subsequent nucleotide sequencing of a 2.4-kb subclone from this region revealed two complete and two incomplete open reading frames, designated *outH*, *outJ*, and *outK* (He et al., 1991a).

In contrast to the *K. oxytoca pulS* and *pulC-O* operon genes, the Sec-dependent secretion pathway genes in plant pathogenic bacteria are not linked to the struc-tural genes of the secreted proteins. Nevertheless, the secretion genes of *X. c. campestris, E. carotovora,* and *E. chrysanthemi* EC16 appear to by clustered as in *K. oxytoca* (Dow et al., 1987; Murata et al., 1990; He et al., 1991a). Simi-larly, most Out$^-$ mutations in *E. chrysanthemi* 3937 map to the *outJ* region of the chromosome, although some mutations with a partial Out$^-$ phenotype map elsewhere (Hugouvieux-Cotte-Pattat et al., 1989; Ji et al., 1989).

The remarkable ability of plant pathogenic bacteria to secrete a variety of pro-teins suggests that the envelopes of these bacteria might be extensively adapted for this function. However, He et al., (1991a) found that a cluster of cloned *E. chrysanthemi* EC16 *out* genes is sufficient to enable *E. coli* to selectively secrete all four pectate lyase isozymes, *exo*-poly-α-D galacturonosidase, and pectin methylesterase. As yet, no Sec-dependent secretion pathway genes cloned from gram-negative bacteria other than *K. oxytoca* or *E. chrysanthemi* EC16 have been shown to possess this ability. The failure of secretion gene clusters from other bacteria to function in *E. coli* may result from poor expression of these genes in *E. coli* or from essential genes being located elsewhere in the genome.

5.2. Conservation of Sec-Dependent Secretion Pathway Components

The observation that *E. carotovora* subsp. *carotovora out* genes could complement *E. chrysanthemi out* mutations suggested that these closely related bacteria share similar Sec-dependent secretions pathways (Murata et al., 1990). The DNA sequence data have now confirmed the genetic relatedness of these secretion gene clusters and revealed their membership in a super family of transport associated genes. The similarities in deduced amino acid sequence of a variety of secretion related proteins are presented in Table 3.1.

Nucleotide sequence data from the *E. chrysanthemi* EC16 *out* genes, the first to be reported from a plant pathogenic bacterium, reveal a remarkable similarity between four *out* genes and secretion genes *pulH*, *pulI*, *pulJ*, and *pulK* from the *K. oxytoca* pullulanase secretion operon (He et al., 1991a). Because of this simi-

TABLE 3.1. Reported Similarities of Proteins in the *K. oxytoca* Sec-Dependent Secretion Pathway With Proteins Involved in Macromolecular Traffic Across the Envelope of Other Eubacteria

Species	Protein		Percent Identity to *K. oxytoca* Pul Secretion Proteins[a]												
K.o.	Pul	S	C	D	E	F	G	H	I	J	K	L	M	N	O
E.ch.	Out		C	D	E	F	G	H	I	J	K	L	M		O
			37	67	73	67	77	34	51	50	44	38	37		62
X.c.c.	Xps				E	F	G	H	I	J*					
					44	33	28	24	23	32					
P.a.	Xcp											Y	Z		A
												20	28		48
P.a.	Pil				B	C									D
					30	26									47
					T										
					23										
B.s.	Com				G1	G2									C
					28	14									31
A.t.	Vir				B11										
					13										

[a]The number below each protein indicates the percentage identity in deduced amino acid sequence with the corresponding pullulanase secretion pathway protein shown on the first line. An asterisk indicates that the complete sequence has not yet been determined. Species abbreviations: *K.o.* is *K. oxytoca*; *E.ch.* is *E. chrysanthemi*; *x.c.c.* is *X. c. campestris*; *P.a.* is *P. aeruginosa*; *B.s.* is *B. subtilis*; *A.t.* is *A. tumefaciens*. References and additional similarities are given in the text.

larity, each *out* gene was given the same suffix as the corresponding pullulanase secretion gene (thus, the *outJ* cloned from strain 3937 is probably not a homolog of the gene designated as *outJ* from strain EC16). Subsequent DNA sequencing has revealed that the *E. chrysanthemi* EC16 *out* cluster is colinear with the entire *pulC-O* operon, except there is no *pulN* homolog (Lindeberg and Collmer, 1992). The *xps* genes of *X. c. campestris* and *out* genes *E. carotovora* subsp. *carotovora* also share extensive similarity with genes in the *K. oxytoca pul* operon. The *xps* genes on a 5.8-kb fragment cloned from *X. c. campestris* are highly similar to *pulE, pulF, pulG, pulH,* and *pulI* (Dums et al., 1991), and *out* genes on a DNA fragment from *E. carotovora* subsp. *carotovora* SCI193 are homologous and colinear with *pulC-O* (Salmond GPC, personal communication). Protein secretion-associated genes similar in sequence to the *pul* operon have also been reported recently in *P. aeruginosa* (Filloux et al., 1990; Bally et al., 1991).

Whitchurch et al. (1990) noted additional similarities in transport-related genes from several bacteria. As summarized in Table 3.1, similarities with deduced Pul secretion protein sequences can be found with three pilin assembly proteins in *P. aeruginosa* (Nunn et al., 1990), with a twitching motility protein (PilT) in *P. aeruginosa* (Nunn and Lory, 1991), with three Com proteins involved in DNA uptake by *Bacillus subtilis* (Albano et al., 1989; Mohan et al., 1989), and with the VirB11 protein of *Agrobacterium tumefaciens*, a membrane protein involved in T-DNA transfer (Ward et al., 1988). Furthermore, d'Enfert et al. (1989) previously noted a sequence similarity between PulD and the pIV protein of filamentous phages, an outer membrane protein that plays a role in assembly and extrusion of phage particles (Brissette and Russel, 1990). Michiels et al., (1991) recently sequenced the *yscC* gene of *Yersinia enterocolitica*, whose product is necessary for the secretion of Yop proteins (virulence factors in diseases caused by *Yersinia* spp.). They noted the presence of two domains that are highly conserved between PulD, pIV, and YscC.

5.3 Potential Functions of Sec-Dependent Secretion Pathway Components

The observed similarities between the proteins used by different bacteria for macromolecular traffic across the bacterial envelope suggest that clues regarding function from one system will be useful in understanding the others. As discussed in Section 2.4, the *K. oxytoca* secretion genes are currently the best characterized. Of particular importance, Pugsley et al. determined that PulD and PulS are outer-membrane proteins, PulE is cytoplasmic, and the other secretion proteins appear to be associated with the inner membrane, most with extensive periplasmic domains (d'Enfert et al., 1989; d'Enfert and Pugsley, 1989; Pugsley and Reyss, 1990; Reyss and Pugsley, 1990). Similarity in sequence and hydropathy profiles suggest that the homologous secretion proteins in other bacteria are similarly located (Filloux et al., 1990; Nunn et al., 1990; Bally et al., 1991; Dums et al., 1991; He et al., 1991a; Whitchurch et al., 1990).

No biochemical function has been demonstrated for any of the Pul secretion proteins, but provocative clues have come from work with *P. aeruginosa* and *A. tumefaciens*. Recent studies on *pilD*, the *P. aeruginosa* pilin assembly gene sharing high homology with *pulO*, have shown it to encode a leader peptidase embedded in the inner membrane (Nunn and Lory, 1991). Distinct from leader peptidases I and II, the *pilD* protein cleaves a six to seven amino acid leader sequence from the N-terminus of type IV fimbrial subunits. Inhibition of this leader peptidase does not affect localization of the fimbrial subunits but does disrupt their assembly into functioning pili (Nunn and Lory, 1991). Peptidase activity by either the *pulO* gene product or its homologs has not been tested. However, such activity is highly plausible given the presence of type IV fimbriae subunitlike leader sequences at the N-termini of PulG, PulI, and PulJ (Nunn and Lorry, 1991; Whitchurch et al., 1990). The important role of this protein is supported by the observation that *xcpA*, a gene required for protein secretion by *P. aeruginosa,* appears to be the same gene as *pilD* (Strom et al., 1991).

Clues regarding the function of PulE and its homologs came with the discovery of type A and B nucleotide binding sites imbedded in these proteins (Christie et al., 1989; Dums et al., 1991; Whitchurch et al., 1990). Because proteins with a wide variety of activities bind mononucleotides, further characterization is necessary. The only PulE homolog yet to be investigated biochemically is the *A. tumefaciens* VirB11 protein, which has been shown to bind ATP and to possess ATPase and autophosphorylation activity (Christie et al., 1989). The energy requirements for the Sec-dependent pathways are unknown. The observations with VirB11 raise the possibility that PulE may have an ATPase activity coupled with an energy-requiring step in secretion. Whatever role PulE and its homologs play in the secretion machinery, there are indications that its function is at the core of the secretion process: This gene is one of the most highly conserved of the secretion genes.

Finally, it is tempting to speculate that PulD and its homologs are directly involved in the translocation of proteins across the outer membrane. PulD is present in the outer membrane and is conserved in other secretion systems. The conservation of two domains in PulD, the filamentous phage pIV protein and the *Y. enterocolitica* YscC protein is particularly intriguing, because these proteins are involved in rather different secretion processes: Sec-dependent secretion of pullulanase, phage assembly and extrusion, and secretion of Yop proteins by a novel pathway that is unlike that used by either pullulanase or hemolysin (Michiels et al., 1991).

5.4 Biochemical Studies of the Sec-Dependent Secretion Pathway

Two initial observations suggested that Sec-dependent export across the inner membrane is the first step in the secretion of plant cell wall degrading enzymes by plant pathogenic bacteria. First, all of these proteins have N-terminal signal peptides resembling those of exported *E. coli* proteins. Expression of genes encoding these enzymes in *E. coli* results in removal of the signal peptides from the preproteins and accumulation of the mature proteins in the periplasm (Keen and Tamaki, 1986; Lei et al., 1987; Guiseppi et al., 1988; Lei et al., 1988; Plastow,

1988; Tamaki et al., 1988; Hinton et al., 1989a; Reverchon et al., 1989; Gough et al., 1990; He and Collmer, 1990; Hinton et al., 1990; Huang and Schell, 1990a and b). Second, secretion pathway mutants accumulate mature plant cell wall degrading enzymes in the periplasm, suggesting that translocation across the outer membrane is the step that is blocked (Andro et al., 1984; Thurn and Chatterjee, 1985; Dow et al., 1987; Ji et al., 1987). Several additional insights into the secretion process have been made with plant pathogenic bacteria.

5.4.1. Xanthomonas campestris pv. campestris

Dow et al. (1987) reported that the pectate lyase isozyme profile of X. c. campestris is identical for the extracellular fraction of wild-type cells and the periplasmic fraction of Out$^-$ cells. Pectate lyase in Out$^-$ mutants is not membrane associated, which provides further evidence that secretion occurs in two steps and involves a periplasmic intermediate (Dow et al., 1989a). They also noted the lack of any differences in the membrane proteins of wild-type and Out$^-$ cells.

5.4.2. Pseudomonas solanacearum

Pseudomonas solanacearum secretes polygalacturonase and endoglucanase by significantly different processes. Whereas secretion of PglA appears similar to the secretion of pectic enzymes in *E. carotovora, E. chrysanthemi,* and *X. c. campestris* (Huang and Schell, 1990a), secretion of endoglucanase presents some remarkable features. The precursor of endoglucanase (ppEgl, 48 kD) was shown to possess a 45 amino acid signal peptide at the N-terminus, of which the first 19 amino acids are typical of a lipoprotein signal peptide and are processed by signal peptidase II in *E. coli* (Huang et al., 1989; Huang et al., 1990b). The processed, acylated endoglucanase intermediate (pEGL, 46 kD) is located inside *E. coli* cells and is mostly membrane bound (Huang et al., 1990b). Since extracellular endoglucanase has a molecular weight of 43 kD, the acylated intermediate is presumably processed again by removal of additional N-terminal sequences during its translocation across the outer membrane of *P. solanacearum*. Secretion mutants defective in the secretion of the two enzymes have not been isolated, so it is not known whether their secretion pathways share common steps. The processing steps that accompany endoglucanase secretion have not been observed in any other gram-negative bacteria. It will be interesting to see whether *out*-like mutants can be isolated in this bacterium and whether the *out* genes are responsible for the processing and/or covalent modifications of endoglucanase.

5.4.3. Erwinia carotovora

Chatterjee et al. (1979) studied the kinetics of pectate lyase secretion in *E. carotovora* by measuring enzyme activity in culture fluids and disrupted cells during bacterial growth. They concluded that only 30–60% of pectate lyase activity is extracellular in *E. carotovora* subsp. *carotovora* ICPB EC153 and that secretion is delayed relative to the rapid increase in enzyme synthesis following addition of the inducer, polygalacturonate. However, analysis of the kinetics of pectate lyase secretion in *E. carotovora* is complicated by the subsequent finding that the bacterium produces periplasmic isozymes of pectate lyase. Indeed, several genes encoding periplasmic isozymes have now been cloned from strains of *E. carotovora* and have been assigned to a distinct family of pectate lyases according to their similar nucleotide sequences (Hinton et al., 1989a; Trollinger, 1989).

5.4.4 *Erwinia chrysanthemi*

The Sec-dependent secretion pathway has been more extensively studied in *E. chrysanthemi* than in any other plant pathogenic bacterium. The ability of *E. chrysanthemi* to efficiently and rapidly secrete plant cell wall degrading enzymes was initially reported by Chatterjee et al. (1979). Recently, we further character-ized the pectate lyase secretion process in *E. chrysanthemi* EC16, using the most alkaline isozyme, PelE, as a model extracellular protein (He et al., 1991b). Pulse-chase experiments indicate that PelE is secreted quite rapidly. During logarith-mic growth, virtually all of the newly synthesized PelE is secreted to the cell exterior within 2 min. Thus, the secretion of PelE from *E. chrysanthemi* is much more rapid than the secretion of polygalacturonase from *P. solanacearum* (Huang and Schell, 1990a), of aerolysin from *Aeromonas hydrophila* (Wong et al., 1989), and of heat-labile enterotoxin from *V. cholerae* (Hirst et al., 1987). We also obtained several lines of evidence for the involvement of the Sec pathway in the extracel-lular secretion of PelE in Out$^+$ *E. coli* cells. Most importantly, mature PelE is present transiently in a cell-bound fraction during active secretion, the processing of pre-PelE is SecA dependent, not Out dependent, and mature PelE or PelE-PhoA hybrids accumulate in the periplasm rather than the cytoplasm of cells overexpressing *pelE* or containing *pelE::phoA* fusions (He et al., 1991b). Ji et al. (1987) reported that secretion results in no apparent covalent modifications of pectic enzymes: the five pectate lyase isozymes produced by *E. chrysanthemi* 3937 have the same isoelectric focusing profile whether obtained from the extra-cellular fraction of wild-type cells or the periplasm of an Out$^-$ mutant. But the possibility of transient, covalent modifications cannot be ruled out. The model for PelE secretion shown in Figure 3.1 is consistent with these observations and with the model proposed by Pugsley et al. (1990a) for pullulanase secretion from *K. oxytoca*. It should also be noted that Mildenhall et al. (1988) reported that high concentrations (1.4% w/v) of LiCl and NaCl cause pectate lyase to accumu-late in the periplasm of *E. chrysanthemi*. The enzyme that accumulates in the periplasm under these conditions is released nonspecifically along with acid phos-phatase, a periplasmic marker protein, upon resuspension of washed cells in water. Mildenhall et al. (1981) had previously determined the salt concentrations that inhibit growth of *E. chrysanthemi* and the effects of various salts and nonionic solutes on pectate lyase secretion (Mildenhall and Prior, 1983). These observa-tions suggest a relationship between the water content of host tissues and effec-tive secretion of pectate lyase, but the basis for this effect on secretion is unknown.

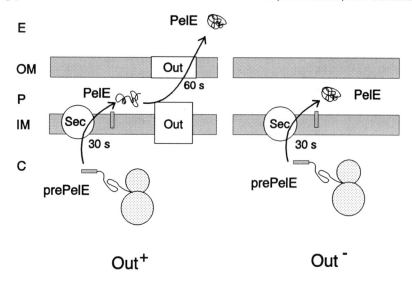

Fig. 3.1. Steps in the secretion of pectate lyase isozyme PelE by the Out pathway of *E. chrysanthemi* EC16. The PrePelE is translocated from the cytoplasm (C) across the inner membrane (IM) to the periplasm (P) by the Sec pathway. The signal peptide is removed in the periplasm, and mature PelE is translocated across the outer membrane (OM) to the exterior (E) of the cell by the Out system. Pulse-labeling experiments indicate that mature PelE appears in a cell-bound fraction (presumably the periplasm) within 30 s after synthesis and then in the medium 60 s later. In Out_ cells, PelE accumulates in the periplasm. The PelE proteins in the medium of Out⁺ cells and the periplasm of Out⁻ cells appear identical, but the form of the enzyme in the Out pathway remains unknown. The location of the Out proteins is inferred from the location of homologous *K. oxytoca* pullulanase secretion proteins, as presented by Pugsley et al. (1990a). Many of the Pul secretion proteins are in the inner membrane with extensive periplasmic domains, but PulE is cytoplasmic, and at least PulD and PulS are associated with the outer membrane. Evidence that two of the *E. chrysanthemi* Out proteins may be periplasmic is presented in the text.

5.5. The Search for Targeting Signals in Extracellular Proteins

Targeting signals for the general export and Sec-independent pathways are located at the N-terminus and C-terminus, respectively, of the exported or secreted protein (as discussed previously). Sections 5.5.1–5.5.3 will describe efforts to identify the elusive signals that target proteins for secretion by the Sec-dependent pathway.

5.5.1. Studies With Hybrid Proteins

The N-terminal signal peptide targets proteins for export across the inner membrane in the first step of the Sec-dependent secretion pathway. Secretion signals that sort proteins into the second step of the pathway could also reside in the signal peptide, or more likely, they could reside in the mature protein sequences. The signal peptides of extracellular proteins are generally similar to those of the

E. coli exported proteins, having one or two charged amino acids at the very N-terminus followed by a hydrophobic core region and a signal peptide processing site (Randall et al., 1989). But other properties not revealed by primary sequences or predicted secondary structure could have a role in targeting. One interesting feature of some of the pectic enzyme signal peptides is the presence of multiple lysine/arginine residues at the N-terminus (Hinton et al., 1989a; He and Collmer, 1990). Pectate lyase isozyme PelE of *E. chrysanthemi,* for example, has two arginine and two lysine residues in that region (Keen and Tamaki, 1986). To test whether these multiple charged residues or other features in the PelE signal peptide contribute to the second-step targeting, we removed the first two of the four charged amino acids by fusing the remaining PelE signal sequence to the intact signal sequence of β-lactamase, a periplasmic protein (He et al., 1991b). The hybrid signal peptide was correctly processed, and the mature PelE was secreted by *E. chrysanthemi* (He et al., 1991b). Thus, it appears that the secretion signal resides in the mature portion of the protein.

The location of secretion signals in the mature *P. solanacearum* polygalacturonase PglA and the *E. chrysanthemi* pectate lyase isozyme PelE has been explored by making gene fusions with *E. coli* periplasmic proteins (Huang and Schell, 1990a; He et al., 1991b). The PglA-PhoA and PelE-PhoA fusion proteins were not secreted to the medium, although one PelE-PhoA fusion was found to be associated with the *E. chrysanthemi* outer membrane in an *out* dependent manner (He et al., 1991b). About 70% of the PelE polypeptide and an even greater portion of PglA were present in these fusions. Secretion of these hybrids may have failed because the secretion signal is in the C-terminus of PelE and PglA or because PhoA blocks translocation across the outer membrane, possibly as a result of dimerization.

It is now apparent that the secretion signal may not be a simple primary sequence but a higher order structure formed through interactions of relatively distant amino acid residues. There are examples of the existence of higher order structures forming before an extracellular protein is translocated across the outer membrane. Hirst and Holmgren (1987) reported that *V. cholerae* enterotoxin subunits are assembled into a holotoxin in the periplasm before being secreted. Recently, Pugsley et al. (1991b) detected an intracellular secretion intermediate of pullulanase with substantial tertiary structure, including at least one disulfide bond. That the secretion intermediates have higher order structures strongly suggests that the secretion signal may be conformational. This is further supported by the comparison of protein sequences of the pectic enzymes of *Erwinia* spp., which is discussed in Section 5.5.2.

5.5.2. Comparison of the Amino Acid Sequences of Secreted Proteins

Hinton et al. (1989a) aligned the sequences of most of the *Erwinia* pectate lyase isozymes and assigned them to three distinct families, PelADE, PelBC, or periplasmic Pel, according to their relative sequence similarities. Proteins in the

PelADE family share considerable sequence similarity with those of the PelBC family, but proteins in neither of these families have sequence similarity with those in the periplasmic Pel family (Hinton et al., 1989a). This observation extended the previous reports by Keen and co-workers that two highly conserved regions reside in six sequenced genes encoding extracellular Pel isozymes of *E. chrysnathemi* and *E. carotovora* (Keen and Tamaki, 1986; Tamaki et al., 1988). However, the sequence similarities do not extend to pectin methylesterase, *exo*-poly-α-D-galacturonosidase, and cellulase, which are all secreted by the Out pathway in *E. chrysanthemi* (Hinton et al., 1989a; He and Collmer, 1990). Furthermore, pullulanase shares no sequence similarities with any of the *Erwinia* pectic enzymes and cellulase (Hinton et al., 1989a). Thus, the sequence similarities identified among the Pel isozymes are probably important in their common catalytic function but not in their secretion.

5.5.3. Specificity of the Sec-Dependent Secretion Pathway

The lack of any conserved sequences among proteins secreted by Sec-dependent pathways contrasts with the high similarity of the secretion pathway genes among different bacteria. This conservation suggests that a bacterium possessing an active Sec-dependent secretion pathway may be able to secrete the extracellular proteins of other bacteria. However, Py et al. (1991) obtained evidence against this with the observation of ''species-specific'' secretion of cellulase in *E. carotovora* and *E. chrysanthemi*. The *E. chrysanthemi* CelZ could not be secreted by *E. carotovora* SCRI193. Similarly, the *E. carotovora* CelV was not secreted by *E. chrysanthemi* 3937. In a study on the specificity of pectate lyase secretion by *E. chrysanthemi*, *E. carotovora*, and *K. oxytoca*, we found that the *E. chrysanthemi* EC16 PelE was not secreted by *E. carotovora* subsp. *carotovora* ATCC15713 or by *K. oxytoca* ATCC15050 (He et al., 1991a). Furthermore, the PelI of *E. carotovora* subsp. *carotovora* was not secreted by *E. coli* cells harboring the *E. chrysanthemi out* gene cluster, although these cells were able to secrete PelE and other pectic enzymes of *E. chrysanthemi*. These observations suggest that among the secretion pathway proteins there is at least one functionally variable component that controls entry of proteins into the pathway. The *E. carotovora* and *E. chrysanthemi* Out pathways for secretion of plant cell wall degrading enzymes provide a promising system for exploring this hypothesis.

6. PROTEIN SECRETION AND PLANT PATHOGENESIS

The expected importance of the Sec-dependent secretion pathway in the virulence of *X. c. campestris*, *E. carotovora*, and *E. chrysanthemi* has been supported by the results of two approaches to isolating relevant mutants. Mutants identified on the basis of their inability to secrete proteins in culture have been found to be severely reduced in virulence (Andro et al., 1984; Chatterjee et al., 1985; Murata et al., 1990). Conversely, collections of mutants that are severely reduced in virulence have been found to contain a substantial proportion that are deficient in protein secretion (Daniels et al., 1984b; Handa et al., 1986; Hinton et al., 1989b; Pirhonen et al., 1991). As discussed previously, it is now clear that the products of several *out* genes are required for the secretion of any proteins by the Sec-dependent pathway and that this pathway is traveled by a variety of virulence proteins. Thus, it is not surprising that attenuated mutants with protein secretion deficiencies are commonly encountered when working with these bacteria.

The strong effect of *out* mutations on the virulence of *E. chrysanthemi* EC16 has provided evidence that the bacterium produces extracellular virulence proteins in addition to those previously characterized. Mutants with directed deletions or insertions in the genes encoding all of the pectic enzymes secreted by this bacterium in culture were found to retain a residual capacity to macerate plant tissues (Ried and Collmer, 1988; He and Collmer, 1990). The discrepancy between the effect on virulence of *out* mutations and of pectic enzyme gene mutations suggests that *E. chrysanthemi* EC16 can produce additional pectic enzymes under certain conditions. Subsequently, we have found that strain EC16 secretes a second set of extracellular pectate lyase isozymes in planta or when grown on medium containing plant extracts (Collmer et al., 1991).

Protein secretion pathways may also provide clues in the search for disease factors in bacteria where the molecular basis for pathogenicity is unknown. An important example is the *Pseudomonas syringae* pathovars. Although toxins contribute significantly to the virulence of many *P. syringae* strains, toxin production is generally not required for these bacteria to cause watersoaked lesions or multiply in planta (Gross, 1991). On the other hand, the elusive product(s) of the *hrp* (*h*ypersensitive *r*esponse and *p*athogenicity) genes are essential for these abilities (Willis et al., 1991). Do extracellular proteins contribute to the pathogenicity of these bacteria, and is protein secretion important in the deployment of any biologically active product of the *hrp* genes?

To address these questions, we have used Tn*phoA* as a reporter for exported proteins to identify two *P. s. syringae* 61 *hrp* genes, *hrpH* and *hrpI*, that encode envelope proteins (Huang et al., 1991). DNA sequence analysis of *hrpH* revealed an open reading frame encoding an 82-kD preprotein with a typical N-terminal signal peptide (Huang et al., 1992). Most importantly, the protein contains a domain that is conserved in several proteins that are located in the outer membrane and are involved in protein translocation (Michiels et al., 1991). These include the *K. oxytoca* PulD, *Y. enterocolitica* YscC, and the gene IV protein of filamentous phages. Mutations in this *hrp* gene abolish the ability of *P. s. syringae* 61 to elicit the hypersensitive response in tobacco, a nonhost, or to multiply or produce disease in bean, a host (Huang et al., 1991). These observations suggest that protein secretion may be vital to the pathogenicity of *P. syringae* and that, in general, a better understanding of macromolecular traffic across the bacterial envelope will accelerate elucidation of the molecular basis of plant–bacterial interactions.

There are two other general ways in which protein secretion research could benefit agriculture and biotechnology. First, secretion pathway components may provide targets for a novel class of antibacterial agents that would act by attenuating virulence rather than by interfering with general metabolic functions in bacteria. Second, a rigorous understanding of the secretion process could enable the engineering of bacteria that not only produce high levels of useful, heterologous proteins, but also secrete them. This method would overcome a major limitation in the capacity of *E. coli* to produce biotechnologically useful proteins. It could also enable ecologically specialized bacteria to deliver useful, heterologous proteins to targeted environments, thereby providing a new tool for biological control of plant diseases and a variety of other uses.

7. RECENT DEVELOPMENTS

The suspected importance of protein secretion in bacterial elicitation of the hypersensitive response has been supported by the discovery that *Erwinia amylovora* produces a cell-envelope associated protein, harpin, which is required for the bacterium to cause fire blight disease symptoms in pear (a host) or the hypersensitive response in tobacco (a nonhost) (Wei et al., 1992). The purified 44-kD protein can cause tobacco leaf tissue to collapse in a manner typical of the hypersensitive response induced by living bacteria. The gene encoding harpin, *hrpN*, is located adjacent to a cluster of *hrp* genes that are conserved in a broad array of plant pathogenic bacteria (Laby and Beer, 1992). Mutations in many of these genes, including *hrpI*, inhibit the ability of whole bacterial cells, but not sonicated cells, to elicit the hypersensitive response (Wei Z-M et al., unpublished), which suggests that the mutations effect the secretion rather than the synthesis of harpin.

DNA sequence analysis of the *hrpI* genes from *E. amylovora* (Wei Z-M et al., in preparation) and *P. s. syringae* 61 (Huang H-C et al., in preparation) has revealed a remarkable similarity with the inner membrane-spanning LcrD protein of *Yersinia* spp. (Plano et al., 1991; Viitanen et al., 1990) and related proteins from other gram-negative bacteria with a suspected role in protein translocation (Galán et al., 1992). The similarity of HrpI and HrpH (discussed in the previous section) with proteins required for the virulence of *Yersinia* spp. is intriguing. Recent findings reviewed by Lory (1992) indicate that a novel pathway, distinct from that used by either pullulanase or hemolysin, is involved in the secretion of the Yop virulence proteins. Of particular interest is the observation that both the *Yersinia* Yop proteins and the *E. amylovora* harpin lack N-terminal signal peptides. These observations support the testable hypothesis that many of the conserved *hrp* genes in plant pathogenic bacteria encode a *Yersinia*-like protein translocation apparatus required for the secretion of proteins essential to disease.

ACKNOWLEDGMENT

This work was supported by NSF grant DCB-9106431.

REFERENCES

Albano M, Brietling R, Dubnau DA (1989): Nucleotide sequence and genetic organization of the *Bacillus subtilis comG* operon. J Bacteriol 171:5386–5404.

Allen C, Huang Y, Sequeira L (1991): Cloning of genes affecting polygalacturonase production in *Pseudomonas solanacearum*. Mol Plant Microbe Interact 4:147–154.

Andro T, Chambost JP, Kotoujansky A, Cattaneo J, Bertheau Y, Barras F, Van Gijsegem F, Coleno A (1984): Mutants of *Erwinia chrysanthemi* defective in secretion of pectinase and cellulase. J Bacteriol 160:1199–1203.

Attardi G, Schatz G (1988): Biogenesis of mitochondria. Annu Rev Cell Biol 4:289–333.

Bally M, Ball G, Badere A, Lazdunski A (1991): Protein secretion in *Pseudomonas aeruginosa*: the *xcpA* gene encodes an integral inner membrane protein homologous to *Klebsiella pneumoniae* secretion function protein PulO. J Bacteriol 173:479–486.

Barras F, Thurn KK, Chatterjee AK (1986): Export of *Erwinia chrysanthemi* (EC16) protease by *Escherichia coli*. FEMS Microbiol Lett 34:343–348.

Bassford P, Beckwith J, Ito K, Kumamoto C, Mizushima S, Oliver D, Randall L, Silhavy T, Tai PC, Wickner B (1991): The primary pathway of protein export in *E. coli*. Cell 65:367–368.

Bateman DF, Basham HG (1976): Degradation of plant cell walls and membranes by microbial enzymes. In Heitefuss R, Williams PH (eds): Encyclopedia of Plant Physiology New Series, Physiological Plant Pathology, Vol. 4. New York: Springer-Verlag, pp 316–355.

Bayer ME (1968): Areas of adhesion between wall and membrane of *Escherlchia coli*. J Gen Microbiol 53:395–404.

Bieker KL, Silhavy TJ (1990): The genetics of protein secretion in *E. coli*. Trends Genet 6:329–334.

Boccara M, Chatain V (1989): Regulation and role in pathogenicity of *Erwinia chrysanthemi* 3937 pectin methylesterase. J Bacteriol 171:4085–4087.

Boccara M, Diolez A, Rouve M, Kotoujansky A (1988): The role of individual pectate lyases of *Erwinia chrysanthemi* strain 3937 in pathogenicity on Saintpaulia plants. Physiol Mol Plant Pathol 33:95–104.

Brissette JL, Russell M (1990): Secretion and membrane integration of a filamentous phage-encoded morphogenetic protein. J Mol Biol 211:565–580.

Chatterjee AK, Buchanan GE, Behrens MK, Starr MP (1979): Synthesis and excretion of polygalacturonic acid trans-eliminase in *Erwinia, Yersinia and Klebsiella* species. Can J Microbiol 25:94–102.

Chatterjee A, McEvoy JL, Chambost JP, Blasco F, Chatterjee AK (1991): Nucleotide sequence and molecular characterization of *pnlA*, the structural gene for damage-inducible pectin lyase of *Erwinia carotovora* subsp. *carotovora* 71. J Bacteriol 173:1765–1769.

Chatterjee AK, Ross LM, McEvoy JL, Thurn KK (1985): pULB113, an RP4::mini-Mu plasmid, mediates chromosomal mobilization and R-prime formation in *Erwinia amylovora, E. chrysanthemi,* and subspecies of *E. carotovora*. Appl Environ Microbiol 50:1–9.

Christie PJ, Ward JE Jr, Gordon MP, Nester EW (1989): A gene required for transfer of T-DNA to plants encodes an ATPase with autophosphorylating activity. Proc Natl Acad Sci USA 86:9677–9681.

Collmer A, Bauer DW, He SY, Lindeberg M, Kelemu S, Rodriquez-Palenzuela P, Burr TJ, Chatterjee AK (1991): Pectic enzyme production and bacterial pathogenicity. In Hennecke H, Verma DPS (eds): Advances in Molecular Genetics of Plant-Microbe Interactions, Vol. 1. Dordrecht: Kluwer Academic Publishers, pp 65–72.

Collmer A, Keen NT (1986): The role of pectic enzymes in plant pathogenesis. Annu Rev Phytopathol 24:383–409.

d'Enfert C, Pugsley AP (1987): A gene fusion approach to the study of pullulanase export and secretion in *Escherichia coli*. Mol Microbiol 1:159–168.

d'Enfert C, Pugsley AP (1989): *Klebsiella pneumoniae pulS* gene encodes an outer membrane lipoprotein required for pullulanase secretion. J Bacteriol 171:3673–3679.

d'Enfert C, Reysse I, Wandersman C, Pugsley AP (1989): Protein secretion by Gram-negative bacteria: Characterization of two membrane proteins required for pullulanase secretion by *Escherichia coli* K-12. J Biol Chem 264:17462–17468.

d'Enfert C, Ryter A, Pugsley AP (1987): Cloning and expression in *Escherichia coli* of the *Klebsiella pneumoniae* genes for production, surface localization and secretion of the lipoprotein pullulanase. EMBO J 6:3531–3538.

Dahler GS, Barras F, Keen NT (1990): Cloning of genes encoding extracellular metalloproteases from *Erwinia chrysanthemi* EC16. J Bacteriol 172:5803–5815.

Daniels MJ, Barber CE, Turner PC, Cleary WG, Sawczyc MK (1984a): Isolation of pathogenicity mutants of *Xanthomonas campestris*. J Gen Microbiol 130:2447–65.

Daniels MJ, Barber CE, Turner PC, Sawczyc MK, Byrde RJW, Fielding AH (1984b): Cloning of genes involved in pathogenicity of *Xanthomonas campestris* pv. *campestris* using the broad host range cosmid pLAFRI. EMBO J 3:3323–28.

Daniels MJ, Dow JM, Osbourn AE (1988): Molecular genetics of pathogenicity in phytopathogenic bacteria. Annu Rev Phytopathol 26:285–312.

Delepelaire P, Wandersman C (1990): Protein secretion in gram-negative bacteria. The extracellular metalloprotease B from *Erwinia chrysanthemi* contains a C-terminal secretion signal analogous to that of *Escherichia coli* α-hemolysin. J Biol Chem 265:17118–17125.

Denny TP, Carney BF, Schell MA (1990): Inactivation of multiple virulence genes reduces the ability of *Pseudomonas solanacearum* to cause wilt symptoms. Mol Plant Microbe Interact 3:293–300.

Dow JM, Clarke BR, Milligan DE, Tang J-L, Daniels MJ (1990): Extracellular proteases from *Xanthomonas campestris* pv. *campestris,* the black rot pathogen. Appl Environ Microbiol 56:2994–2998.

Dow JM, Daniels MJ, Dums F, Turner PC, Gough C (1989a): Genetic and biochemical analysis of protein export from *Xanthomonas campestris*. J Cell Sci Suppl 11:59–72.

Dow JM, Milligan DE, Jamieson L, Barber CE, Daniels MJ (1986b): Molecular cloning of a polygalacturonate lyase gene from *Xanthomonas campestris* pv. *campestris* and role of the gene product in pathogenicity. Physiol Mol Plant Pathol 35:113–120.

Dow JM, Scofield G, Trafford K, Turner PC, Daniels MJ (1987): A gene cluster in *Xanthomonas campestris* pv. *campestris* required for pathogenicity controls the excretion of polygalacturonate lyase and other enzymes. Physiol Mol Plant Pathol 31:261–271.

Dums F, Dow JM, Daniels MJ (1991): Structural characterization of protein secretion genes of the bacterial phytopathogen *Xanthomonas campestris* pathovar *campestris*: relatedness to secretion systems of other gram-negative bacteria. Mol Gen Genet 229:357–364.

Felmlee T, Pellett S, Lee EY, Welch RA (1985a): *Escherichia coli* hemolysin is released extracellularly without cleavage of a signal peptide. J Bacteriol 163:88–93.

Felmlee T, Pellett S, Welch RA (1985b): Nucleotide sequence of an *Escherichia coli* chromosomal hemolysin. J Bacteriol 163:94–105.

Filloux A, Bally M, Ball G, Akrim M, Tommassen J, Lazdunski A (1990): Protein secretion in gram-negative bacteria: transport across the outer membrane involves common mechanisms in different bacteria. EMBO J 9:4323–4329.

Galán JE, Ginocchio C, Costeas P (1992): Molecular and functional characterization of the *Salmonella* invasion gene *invA*: homology of InvA to members of a new protein family. J Bacteriol 174: 4338–4349.

Gibson M, Whitcome D, Reeves P, Hinton J, Ellard F, Salmond GPC (1988): Identification of genes and proteins required for pathogenicity determinant secretion by *Erwinia*. In Palacios R, Verma DPS (eds): Molecular Genetics of Plant–Microbe Interactions. St. Paul: APS Press, pp 255–256.

Gough CL, Dow JM, Barber CE, Daniels MJ (1988): Cloning of two endoglucanase genes of *Xanthomonas campestris* pv. *campestris*: Analysis of the role of the major endoglucanase in pathogenesis. Mol Plant–Microbe Interact 1:275–281.

Gough CL, Dow JM, Keen J, Henrissat B, Daniels MJ (1990): Nucleotide sequence of the *engXCA* gene encoding the major endoglucanase of *Xanthomonas campestris* pv. *campestris*. Gene 89:53–59.

Graham LL, Harris R, Villiger W, Beveridge TJ (1991): Freeze-substitution of gram-negative eubacteria: General cell morphology and envelope profiles. J Bacteriol 173:1623–1633.

Gray L, Baker K, Kenny B, Mackman N, Naigh R, Holland IB (1989): A novel C-terminal signal sequence targets *Escherichia coli* haemolysin directly to the medium. J Cell Sci Suppl 11:45–57.

Gross DC (1991): Molecular and genetic analysis of toxin production by pathovars of *Pseudomonas syringae*. Annu Rev Phytopathol 29:247–278.

Guiseppi A, Cami B, Aymeric JL, Ball G, Creuzet N (1988): Homology between endoglucanase Z of *Erwinia chrysanthemi* and endoglucanase of *Bacillus subtilis* and alkalophilic *Bacillus*. Mol Microbiol 2:159–164.

Handa AK, Bressan RA, Korty AG, Jayaswal RK, Charles DJ (1986): Isolation and characterization of pectolytic nonpathogenic mutants of *Erwinia carotovora* subsp. *carotovora* (ECC). In Civerolo EL, Collmer A, Davis RE, Gillaspie AG (eds): Plant Pathogenic Bacteria: Proceedings of the Sixth International Conference on Plant Pathogenic Bacteria, Maryland, June 2–7, 1985. Dordrecht: Martinus Nijhoff Publishers, pp 212–217.

He SY, Collmer A (1990): Molecular cloning, nucleotide sequence and marker-exchange mutagenesis of the *exo*-poly-α-D-galacturonosidase-encoding *pehX* gene of *Erwinia chrysanthemi* EC16. J Bacteriol 172:4988–4995.

He SY, Lindeberg M, Chatterjee AK, Collmer A (1991a): Cloned *Erwinia chrysanthemi out* genes enable *Escherichia coli* to selectively secrete a diverse family of heterologous proteins to its milieu. Proc Natl Acad Sci USA 88:1079–1083.

He SY, Schoedel C, Chatterjee AK, Collmer A (1991b): Extracellular secretion of pectate lyase by the *Erwinia chrysanthemi* Out pathway is dependent upon Sec-mediated export across the inner membrane. J Bacteriol 173:4310–4317.

Higgins CF, Hiles ID, Salmond GPC, Gill DR, Downie JA, Evans IJ, Holland IB, Gray L (1986): A family of related ATP-binding subunits coupled to many distinct biological processes in bacteria. Nature (London) 323:448–450.

Hinton JCD, Gill DR, Lalo D, Plastow GS, Salmon GPC. (1990): Sequence of the *peh* gene of *Erwinia carotova*: homology between *Erwinia* and plant enzymes. Mol Microbiol 4:1029–1036.

Hinton JCD, Salmond GPC. (1987): "Use of Tn*phoA* to enrich for extracellular enzyme mutants of *Erwinia carotovora* subspecies *carotovora*. Mol Microbiol 1:381–386.

Hinton JCD, Sidebotham JM, Gill DR, Salmond GPC. (1989a): Extracellular and periplasmic isoenzymes of pectate lyase from *Erwinia chrysanthemi* subspecies *carotovora* belong to different gene families. Mol Microbiol 3:1785–1795.

Hinton JCD, Sidebotham JM, Hyman LJ, Perombelon MCM, Salmond GPC (1989b): Isolation and characterization of transposon-induced mutants of *Erwinia* subsp. *atroseptica* exhibiting reduced virulence. Mol Gen Genet 217:141–148.

Hirst TR, Holmgren J (1987): Conformation of protein secreted across bacterial outer membranes: A study of enterotoxin translocation from *Vibrio cholerae*. Proc Natl Acad Sci USA 84:7418–7422.

Hirst TR, Welch RA (1988): Mechanisms for secretion of extracellular proteins by Gram-negative bacteria. Trends Biochem Sci 13:265–269.

Holland B, Blight MA, Kenny B (1990): The mechanism of secretion of hemolysin and other polypeptides from Gram-negative bacteria. J Bioenergetics Biomembranes 22:473–491.

Huang H-C, He SY, Bauer DW, Collmer A (1992): The *Pseudomonas syringae* pv. *syringae* 61 *hrpH* product: an envelope protein required for elicitation of the hypersensitive response in plants. J Bacteriol 174: (in press).

Huang H-C, Hutcheson SW, Collmer A (1991): Characterization of the *hrp* cluster from *Pseudomonas syringae* pv. *syringae* 61 and Tn*phoA* tagging of genes encoding exported or membrane-spanning Hrp proteins. Mol Plant–Microbe Interact 4:469–476.

Huang J, Schell MA (1990a): DNA sequence analysis of *pglA* and mechanism of export of its polygalacturonase product from *Pseudomonas solanacearum*. J Bacteriol 172:3879–3887.

Huang J, Schell MA (1990b): Evidence that extracellular export of the endoglucanase encoded by *egl* of *Pseudomonas solanacearum* occurs by a two-step process involving a lipoprotein intermediate. J Biol Chem 265:11628–11632.

Huang J, Sukordhaman M, Schell MA (1989): Excretion of the *egl* gene product of *Pseudomonas solanacearum*. J Bacteriol 171:3767–3774.

Hugouvieux-Cotte-Pattat N, Reverchon S, Robert-Baudouy J (1989): Expanded linkage map of *Erwinia chrysanthemi* strain 3937. Mol Microbiol 3:573–581.

Ji J, Hugouvieux-Cotte-Pattat N, Robert-Baudouy J (1987): Use of Mu-*lac* insertions to study the secretion of pectate lyases by *Erwinia chrysanthemi*. J Gen Microbiol 133:793–802.

Ji J, Hugouvieux-Cotte-Pattat N, Robert-Baudouy J (1989): Molecular cloning of the *outJ* gene involved in pectate lyase secretion by *Erwinia chrysanthemi*. Mol Microbiol 3:285–293.

Keen NT, Tamaki S (1986): Structure of two pectate lyase genes from *Erwinia chrysanthemi* EC16 and their high level expression in *Escherichia coli*. J Bacteriol 168:595–606.

Kellenberger E (1990): The 'Bayer bridges' confronted with results from improved electron microscopy methods. Mol Microbiol 4:697–705.

Kornacker MG, Boyd A, Pugsley AP, Plastow GS (1989): *Klebsiella pneumoniae* strain K21: evidence for the rapid secretion of an unacylated form of pullulanase. Mol Microbiol 3:497–503.

Kornacker MG, Pugsley AP (1990): The normally periplasmic enzyme β-lactamase is specifically and efficiently translocated through the *Escherichia coli* outer membrane when it is fused to the cell-surface enzyme pullulanase. Mol Microbiol 4:1101–1109.

Koronakis V, Koronakis E, Hughes C (1989): Isolation and analysis of the C-terminal signal directing export of *Escherichia coli* hemolysin protein across both bacterial membranes. EMBO J 8:595–605.

Kotoujansky A (1987): Molecular genetics of pathogenesis by soft-rot erwinias. Annu Rev Phytopathol 25:405–430.

Kumamoto CA (1991): Molecular chaperones and protein translocation across the Escherichia coli inner membrane. Mol Microbiol 5:19–22.

Laby RJ, Beer SV (1992): Hybridization and functional complementation of the *hrp* gene cluster from *Erwinia amylovora* strain Ea321 and DNA of other bacteria. Mol Plant-Microbe Interact 5: (in press).

Lei SP, Hin HC, Wang SS, Wilcox G (1988): Characterization of the *Erwinia carotovora pelA* gene and its product pectate lyase A. Gene 62:159–164.

Lei SP, Lin HC, Wang SS, Callaway J, Wilcox G (1987): Characterization of the *Erwinia carotovora pelB* gene and its product pectate lyase. J Bacteriol 169:4379–4383.

Letoffe S, Delepelaire P, Wandersman C (1990): Protease secretion by *Erwinia chrysanthemi*: the specific secretion functions are analogous to those of *Escherichia coli* α-haemolysin. EMBO J 9:1375–1382.

Lindeberg M, Collmer A (1992): Analysis of eight out genes in a cluster required for pectic enzyme secretion by *Erwinia chrysanthemi*. Protein sequence comparison with secretion genes from other gram-negative bacteria. J Bacteriol 174: (in press).

Lory S (1992): Determinants of extracellular protein secretion in gram-negative bacteria. J Bacteriol 174:3423–3428.

Manoil C, Beckwith J (1985): Tn*phoA*: A transposon probe for protein export signals. Proc Natl Acad Sci USA 82:8129–8133.

McNeil M, Darvill AG, Fry SC, Albersheim P (1984): Structure and function of the primary cell walls of plants. Annu Rev Biochem 53:625–663.

Michaelis S, Chapon C, D'Enfert C, Pugsley AP, Schwartz M (1985): Characterization and expression of the structural gene for pullulanase, a maltose-inducible secreted protein of *Klebsiella pneumoniae*. J Bacteriol 164:633–638.

Michiels T, Vanooteghem J-C, de Rouvroit CL, China B, Gustin A, Boudry P, Cornelis GR (1991): Analysis of *virC*, an operon involved in the secretion of Yop proteins by *Yersinia enterocolitica*. J Bacteriol 173:4994–5009.

Mildenhall JP, Lindner WA, Prior BA, Tutt K (1988): Elevation and release of cell-associated pectate lyase in *Erwinia chrysanthemi* by lithium and sodium chloride. Phytopathology 78:213–217.

Mildenhall JP, Prior BA (1983): Water relations of *Erwinia chrysanthemi*: intracellular and extracellular pectate lyase production. J Gen Microbiol 129:3019–3025.

Mildenhall JP, Prior BA, Trollope LA (1981): Water relations of *Erwinia chrysanthemi*: growth and extracellular pectic acid lyase production. J Gen Microbiol 127:27–34.

Model P, Russel M (1990): Prokaryotic secretion. Cell 61:739–741.

Mohan S, Aghion J, Guillen N, Dubnau D (1989): Molecular cloning and characterization of *comC*, a late competence gene of *Bacillus subtilis*. J Bacteriol 171:6043–6051.

Murata H, Fons M, Chatterjee A, Collmer A, Chatterjee AK (1990): Characterization of transposon insertion Out⁻ mutants of *Erwinia carotovora* subsp. *carotovora* defective in enzyme export and of a DNA segment that complements out mutations in *E. carotovora* subsp. *carotovora, E. carotovora* subsp. *atroseptica*, and *Erwinia chrysanthemi*. J Bacteriol 172:2970–2978.

Neidhardt FC, Ingraham JL, Low KB, Magasanik B, Schaechter M, Umbarber HE. (1987): *Escherichia coli* and *Salmonella typhimurium*. Washington, DC: American Society for Microbiology,

Nunn D, Bergman S, Lory S (1990): Products of three accessory genes, *pilB, pilC,* and *pilD,* are required for biogenesis of *Pseudomonas aeruginosa* pili. J Bacteriol 172:2911–2919.

Nunn D, Lory S (1991): Product of the *Pseudomonas aeruginosa* gene *pilD* is a prepilin leader peptidase. Proc Natl Acad Sci USA 88:3281–3285.

Pirhonen M, Saarilahti H, Karlsson M-J, Palva ET (1991): Identification of pathogenicity determinants of *Erwinia carotovora* subsp. *carotovora* by transposon mutagenesis. Mol Plant–Microbe Interact 4:276–283.

Plano GV, Barve SS, Straley SC (1991): LcrD, a membrane-bound regulator of the *Yersinia pestis* low-calcium response. J Bacteriol 173:7293–7303.

Plastow GS (1988): Molecular cloning and nucleotide sequence of the pectin methyl esterase gene of *Erwinia chrysanthemi* B374. Mol Microbiol 2:247–254.

Pugsley AP, Chapon C, Schwartz M (1986): Extracellular pullulanase of *Klebsiella pneumoniae* is a lipoprotein. J Bacteriol 166:1083–1088.

Pugsley AP, d'Enfert C, Reyss I, Kornacker MG (1990a): Genetics of extracellular protein secretion by gram-negative bacteria. Annu Rev Gen 24:67–90.

Pugsley AP, Kornacker MG, Poquet I (1991a): The general protein-export pathway is directly required for extracellular pullanase secretion in *Escherichia coli* K12. Mol Microbiol 5:343–352.

Pugsley AP, Kornacker MG, Ryter A (1990b): Analysis of the subcellular location of pullulanase produced by *Escherichia coli* carrying the *pulA* gene from *Klebsiella pneumoniae* strain UNF5023. Mol Microbiol 4:59–72.

Pugsley AP, Poquet I, Kornacker MG (1991b): Two distinct steps in pullulanase secretion by *Escherichia coli* K12. Mol Microbiol 5:865–873.

Pugsley AP, Reyss I (1990): Five genes at the 3′ end of the *Klebsiella pneumoniae pulC* operon are required for pullulanase secretion. Mol Microbiol 4:365–379.

Py B, Salmond GPC, Chippaux M, Barras F (1991) Secretion of cellulases in *Erwinia chrysanthemi* and *E. carotovora* is species-specific. FEMS Microbiol Lett 79:315–322.

Randall LL, Hardy SJ S. (1989): Unity in function in the absence of consensus in sequence: role of leader peptides in export. Science 243:1156–1159.

Rapoport TA (1991): A bacterium catches up. Nature (London) 349:107–108.

Reverchon S, Huang Y, Bourson C, Robert-Baudouy J (1989): Nucleotide sequences of the *Erwinia chrysanthemi ogl* and *pelE* genes, negatively regulated by the *kdgR* gene product. Gene 85:125–134.

Reyss I, Pugsley AP (1990): Five additional genes in the *pulC-O* operon of the gram-negative bacterium *Klebsiella oxytoca* UNF5023 which are required for pullulanase secretion. Mol Gen Genet 222:176–184.

Ried JL, Collmer A (1986): Comparison of pectic enzymes produced by *Erwinia chrysanthemi, Erwinia carotovora* subsp. *carotovora,* and *Erwinia carotovora* subsp. *atroseptica*. Appl Environ Microbiol 52:305–310.

Ried JL, Collmer A (1988): Construction and characterization of an *Erwinia chrysanthemi* mutant with directed deletions in all of the pectate lyase structural genes. Mol Plant–Microbe Interact 1:32–38.

Roberts DP, Denny TP, Schell MA (1988): Cloning of the *egl* gene of *Pseudomonas solanacearum* and analysis of its role in phytopathogenicity. J Bacteriol 170:1445–1451.

Saier MH Jr., Werner PK, Müller M (1989): Insertion of proteins into bacterial membranes: mechanism, characteristics, and comparisons with the eucaryotic process. Microbiol Rev 53:333–366.

Schatz PJ, Beckwith J (1990): Genetic analysis of protein export in *Escherichia coli*. Annu Rev Gen 24:215–248.

Schell MA (1987): Purification and characterization of an endoglucanase from *Pseudomonas solanacearum*. Appl Environ Microbiol 53:2237–2241.

Schell MA, Roberts DP, Denny TP (1988): Cloning of the *pglA* gene of *Pseudomonas solanacearum* and its involvement in phytopathogenicity. J Bacteriol 170:4501–4508.

Sen K, Hellman J, Nikaido H (1988): Porin channels in intact cells of *Escherichia coli* are not affected by Donnan potentials across the outer membrane. J Biol Chem 263:1182–1187.

Shaw JJ, Kado CI (1988): Whole plant wound inoculation for consistent reproduction of black rot of crucifers. Phytopathology 78:981–986.

Strom MS, Nunn D, Lory S (1991): Multiple roles of the pilus biogenesis protein PilD: Involvement of PilD in excretion of enzymes from *Pseudomonas aeruginosa*. J Bacteriol 173:1175–1180.

Tamaki SJ, Gold S, Robeson M, Manulis S, Keen NT (1988): Structure and organization of the *pel* genes from *Erwinia chrysanthemi* EC16. J Bacteriol 170:3468–3478.

Tang JL, Gough CL, Barber CE, Dow JM, Daniels MJ (1987): Molecular cloning of protease gene(s) from *Xanthomonas campestris* pv. *campestris*: expression in *Escherichia coli* and role in pathogenicity. Mol Gen Genet 210:443–448.

Thurn KK, Chatterjee AK (1985): Single site chromosomal Tn5 insertions affect the export of pectolytic and cellulolytic enzymes in *Erwinia chrysanthemi*. Appl Environ Microbiol 50:894–898.

Trollinger D, Berry S, Belser W, Keen NT (1989): Cloning and characterization of a pectate lyase gene from *Erwinia carotovora* EC153. Mol Plant–Microbe Interact 2:17–25.

Tsuyumu S, Chatterjee AK (1984): Pectin lyase production in *Erwinia chrysanthemi* and other soft-rot *Erwinia* species. Physiol Plant Pathol 24:291–302.

Viitanen A-M, Toivanen P, Skurnik M (1990): The *lcrE* gene is part of an operon in the *lcr* region of *Yersinia enterocolitica* 0:3. J Bacteriol 172:3152–3162.

Wagner W, Vogel M, Goebel W (1983): Transport of hemolysin across the outer membrane of *Escherichia coli* requires two functions. J Bacteriol 154:200–210.

Wandersman C, Delepelaire P (1990): TolC, an *Escherichia coli* outer membrane protein required for hemolysin secretion. Proc Natl Acad Sci USA 87:4776–4780.

Wandersman C, Delepelaire P, Letoffe S, Schwartz M (1987): Characterization of *Erwinia chrysanthemi* extracellular proteases: cloning and expression of the protease genes in *Escherichia coli*. J Bacteriol 169:5046–5053.

Ward JE, Akiyoshi DE, Regier D, Datta A, Gordon MP, Nester EW (1988): Characterization of the *virB* operon from an *Agrobacterium tumefaciens* Ti plasmid. J Biol Chem 263:5804–5814.

Wei Z-M, Laby RJ, Zumoff CH, Bauer DW, He SY, Collmer A, Beer SV (1992): Harpin, elicitor of the hypersensitive response produced by the plant pathogen *Erwinia amylovora*. Science 257: 85–88.

Whitchurch CB, Hobbs M, Livingston SP, Krishnapillai V, Mattick JS (1990): Characterisation of a *Pseudomonas aeruginosa* twitching motility gene and evidence for a specialised protein export system widespread in eubacteria. Gene 101:33–34.

Wickner W, Driessen AJM, Hartl F-U (1991): The enzymology of protein translocation across the *Escherichia coli* plasma membrane. Annu Rev Biochem 60:101–124.

Willis DK, Rich JJ, Hrabak EM (1991): *hrp* genes of phytopathogenic bacteria. Mol Plant–Microbe Interact 4:132–138.

Wong KR, Buckley JT (1989): Proton motive force involved in protein transport across the outer membrane of *Aeromonas salmonicida*. Science 246:654–655.

Zink RT, Engwall JK, McEvoy JL, Chatterjee AK (1985): *recA* is required in the induction of pectin lyase and carotovoricin in *Erwinia carotovora* subsp. *carotovora*. J Bacteriol 164:390–396.

PLANT DISEASE RESISTANCE GENES: INTERACTIONS WITH PATHOGENS AND THEIR IMPROVED UTILIZATION TO CONTROL PLANT DISEASES

N.T. KEEN

Department of Plant Pathology and Graduate Group in Genetics, University of California, Riverside, California 92521

ANDREW BENT
BRIAN STASKAWICZ

Department of Plant Pathology, University of California, Berkeley, California 94720

1. INTRODUCTION

The incorporation of resistance genes into important crop plants is the major disease control method utilized in worldwide production agriculture. This strategy epitomizes effective biological disease control— resistance is heritable and, therefore, inexpensive and permanently available once introduced. It also does not require the introduction of foreign agents into the biosphere as with the use of chemical pesticides or bio-control microorganisms. Furthermore, genetically determined resistance is generally effective when plants are grown in a wide range of climates, soil types, and cropping regimes, is relatively easy to introduce into

Biotechnology in Plant Disease Control, pages 65–88
© 1993 Wiley-Liss, Inc.

new cultivars, and generally affords high levels of resistance to disease. Despite these appealing attributes, problems have historically been encountered in the use of genetic disease resistance. The most serious problems have been the unavailability in certain plant species of resistance genes against particular pathogens, and the emergence in some cases of new virulent pathogen races able to "overcome" a previously effective resistance gene.

The major thesis of this chapter is that the problems above can be dealt with by deployment of large numbers of disease resistance genes. This method will require the molecular cloning of numerous resistance genes from many different plants that are targeted against several pests and pathogens. Then, it will be possible to transform various combinations of these genes into susceptible cultivars, including species entirely unrelated to the plants from which the resistance genes were originally isolated. This approach would permit the use of previously unavailable resistance specificities in particular crop plants and should lower the frequency with which newly virulent pathogen races become predominant.

Unfortunately, disease resistance genes have not yet been molecularly cloned from any plant. While waiting for this discovery to happen, plant biologists have amused themselves by transforming plants with several other genes that antagonize pathogens and pests. Indeed, some of these approaches show considerable experimental promise for controlling diseases, as will be described elsewhere in this book. Efforts to clone disease resistance genes are also well underway in a number of laboratories, and the first successes are expected in the near future. Insight into the molecular basis of disease resistance is also being gained through studies on pathogen genetics. Results from these studies indicate that development of disease and pest resistant plants should be more efficient in the future. With increasing popular opinion against the use of pesticides of any kind in agriculture, there is great interest in the development of genetically resistant plants. This chapter will examine some of the progress made in this field and assess possible approaches.

2. CHARACTERISTICS OF DISEASE RESISTANCE AND DISEASE RESISTANCE GENES

Plants possess many mechanisms for natural resistance to potential pests and pathogens. These mechanisms include structural and morphological features, such as tissue and organ color and shape, the nature of cuticular and other coverings, and the degree of differentiation of cell walls. Plants are also frequently armed with toxic or potentially toxic preformed chemicals that may repel invaders (Hahlbrock and Scheel, 1987). Certain insect pests and pathogens have evolved mechanisms that permit the avoidance of such preformed plant barriers and may therefore be considered as virulence factors (see e.g., Dickman et al., 1989). If preformed resistance mechanisms fail, pathogen attack typically causes a number of induced defense responses. The hypersensitive reaction (HR), a complex response associated with resistance to numerous pathogen species (Kiraly, 1980) is foremost among these responses. If the HR is not induced rapidly enough or if the defensive systems deployed are detoxified by a pathogen (VanEtten et al., 1989), then the plant will be susceptible to attack and disease occurs. The business of the entomologists and plant pathologists is to minimize or prevent this outcome in large scale, production agriculture. Engineered deployment of the HR offers a promising, broadly effective mechanism through which to achieve this goal.

The HR occurs in most, if not all, higher plant species against all major groups of pathogens. It is an inducible, active defense response that is initiated only after initial pest or pathogen attack, when structural and morphological defenses have failed (Kiraly, 1980). The HR superficially involves the necrosis of several plant cells at the infection site. While there is debate as to whether these dead plant cells *per se* supply factors that account for resistance to an invading pathogen, an antagonistic environment forms at the HR site that is sufficient to curtail further growth and spread of the pathogen. Additional components of the HR include the production of toxic chemicals, called phytoalexins, the establishment of inducible mechanical barriers (such as lignin or hydroxyproline-rich glycoproteins) and the production of antagonistic proteins (such as protease inhibitors, glucanases, and chitinases) (Bowles, 1990; Collinge and Slusarenko, 1987; Dixon and Lamb, 1990; Hahlbrock and Scheel, 1987). Much evidence now documents the occurrence during hypersensitive reactions of specific activation of several plant genes, which accounts for the formation of these barriers. Unfortunately, for most of these responses there is no strong evidence indicating a causal role in resistance. The construction of gain-of-function and loss-of-function plant lines for these responses should permit more incisive determination of their relative roles.

3. HOW PATHOGENS INTERACT WITH PLANTS CARRYING DEFINED DISEASE RESISTANCE GENES

Since the HR is not expressed by normal plant tissues, it must be induced following pathogen infection. This reaction occcurs in at least two different ways. The first permits the plant to recognize pathogen biotypes producing *specific elicitors*. As will be discussed below, these elicitors result directly or indirectly from the action of pathogen avirulence genes and are believed to be perceived by plant receptors encoded by disease resistance genes. Numerous plant–pathogen interactions follow this genetically defined pattern, which leads to agriculturally useful resistance. The second mechanism for induction of the HR is by plant recognition of *general elicitors* that are typically liberated by all biotypes of a pathogen taxon and recognized by all members of a plant species. The HR-like responses can be produced by application of general elicitors, but their presence both in pathogen strains that are virulent and nonvirulent on a given plant brings into question their significance with respect to whole-plant disease resistance in the field.

Considerable study has been made of the mechanisms by which plants carrying certain disease resistance genes perceive the presence of a pathogen and invoke the HR. Classical genetic work earlier in this century established that a single plant *disease resistance gene* recognizes only those biotypes of a pathogen species that harbor a corresponding or complementary *avirulence gene* (Flor, 1942, 1956; Gabriel and Rolfe, 1990). This *gene-for-gene* relationship implies that the resistance gene modulates a plant recognitional system for some specific chemical feature of the avirulent pathogen biotype. Thus, it constitutes a simple plus/minus recognitional system superficially similar to antigen–antibody recognition in vertebrates. The *elicitor–receptor model* is commonly invoked to explain this relationship (Dixon and Lamb, 1990; Ebel and Grisebach, 1988; Keen and Dawson, 1992), and its occurrence has recently been supported by considerable biochemical and genetic evidence.

Avirulence genes have been cloned and characterized from viral, bacterial, and fungal pathogens (for reviews, see Gabriel and Rolfe, 1990; Keen, 1990). Thus far, all known avirulence genes are inherited as single, dominant alleles that encode single protein products. Recent work from several laboratories has shown that avirulence genes may result in production by the pathogen of specific signal molecules, called elicitors, which are perceived by plants carrying the complementary resistance genes to initiate the HR (for a review see Keen and Dawson, 1992). The production of these specific elicitors by the pathogen therefore requires the presence of a particular avirulence gene allele. For instance, in tobacco mosaic virus, Dawson and co-worker (e.g., Culver and Dawson, 1989) showed that the 17.5-kD coat protein of certain virus strains functions as a specific elicitor of the HR in tobacco plants carrying the N′ resistance gene. Similarly, an extracellular 28-member peptide from *Cladosporium fulvum* biotypes carrying avirulence gene 9 (van Kan et al., 1991; P. de Wit, personal communication) functions as a specific elicitor of the HR only in tomato plants carrying the complementary resistance gene, *Cf9*. A lactone from *Pseudomonas syringae* cells expressing avirulence gene D (Kobayashi et al., 1990) is a specific elicitor in soybean plants carrying the resistance gene, *Rpg4* (Keen and Buzzell, 1991; Keen et al., 1990; Sims, Midland, Keen, and Stayton, unpublished). Finally, the work of Ricci et al. (1989), Knogghe et al., (1991), and Wolpert and Macko (1989) indicates that three additional peptides of pathogen origin (two low molecular weight linear peptides and a cyclic peptide, respectively) probably function as elicitors and may interact with particular disease resistance genes. The structural diversity of all these elicitors suggests that additional pathogen elicitors recognized by plants carrying single disease resistance genes will differ considerably in structure. It is also probable that each plant is able to recognize dozens of different elicitors. There is precedence that a sufficiently diverse set of disease resistance genes could be maintained for recognition of a large number of elicitors because humans, for example, contain several hundred different receptor genes for olfactory perception alone (Buck and Axel, 1991).

Avirulence genes were initially thought to be involved only with race-specific resistance, but recent work has raised the possibility that they may also be involved in species-level, or nonhost resistance. This result was demonstrated by Kobayashi and Keen (1985) and Kobayashi et al. (1989) when they found that avirulence genes cloned from *P. syringae* pv. *tomato* functioned in the related bacterium, *P.s.* pv. *glycinea,* causing it to elicit the HR in certain cultivars of soybean. Whalen et al. (1988) also cloned an avirulence gene from *Xanthomonas campestris* pv. *vesicatoria* that, upon introduction into *X.c. phaseoli,* caused this bacterium to elicit a HR on its normal host plant, bean. The cloned avirulence gene also led to the HR in several plants when it was inserted into normally pathogenic *X. campestris* pathovars (pv. *glycines* on soybean, pv. *vignicola* on cowpea, pv. *alfalfae* on alfalfa, pv. *holcicola* on corn, and pv. *malvacearum* on cotton). Carney and Denny (1990) cloned an avirulence gene from a tomato strain of *Pseudomonas solanacearum* that elicited the HR on tobacco plants when introduced into a strain of the bacterium normally pathogenic on tobacco. Finally, Whalen et al. (1991) recently identified an avirulence gene from *P.s.* pv. *tomato* that is recognized by both soybean and *Arabidopsis.* The results above are important because they demonstrate that unrelated plant species contain functionally identical disease resistance genes. This finding fosters the speculation that resistance genes will function if transferred to unrelated plants.

4. HOW DO PLANT DISEASE RESISTANCE GENES WORK?

At the time of this writing, no disease resistance genes have been cloned from a higher plant. This deficiency greatly restricts our understanding of the genes function in conferring disease resistance. Considerable indirect evidence, however, suggests that primary or processed protein products of resistance genes may function as specific receptors for pathogen elicitors (Keen and Dawson, 1991; Gabriel and Rolfe, 1990; Dixon and Lamb, 1990; Scheel and Parker, 1990). Biochemical evidence also indicates the occurrence of specific binding sites in plant cells for certain elicitors (Blein et al., 1991; Cheong and Hahn, 1991; Cosio et al., 1988; Ebel et al., 1991; Wolpert and Macko, 1989). Proof of the involvement of receptors in the elicitor–receptor model, however, must await isolation and characterization of the receptors themselves and/or cloning of the disease resistance genes postulated to encode them.

Classical genetic experiments have supplied clues as to the nature of disease resistance genes. For instance, disease resistance genes frequently occur in tightly linked clusters or as true alleles (Flor, 1956; Pryor, 1987). By analogy with vertebrate immunoglobulin genes, this result implies that new resistance gene specificities may occur by mutation of a preexisting gene or by recombination of two genes. Resistance genes with altered recognitional specificities (i.e., a capability to recognize pathogen biotypes not possessed by either parent) have been recovered in the progeny of sexual crosses (Bennetzen et al., 1988, 1991; Islam et al., 1989; Pryor, 1987). Such alterations have thus far been observed for resistance genes occurring in tightly linked clusters and with allelic genes of different specificities. It is appealing, therefore, to speculate that such altered specificities result from recombination of closely linked or heterozygous alleles of resistance genes to generate different recognitional specificities. While it is not known if these events occur vegetatively or during meiosis, their occurrence is tantalizingly similar to the rearrangement mechanisms studied in vertebrate B cell antibody systems (Tonegawa, 1983; Yancopoulos and Alt, 1986).

Assuming that disease resistance genes encode specific receptors for pathogen-produced elicitors, we are left with the question of how elicitor–receptor interaction leads to activation of the defense response genes discussed earlier. Although this is also an unresolved area, several recent observations should be noted. First, there are clear indications that elicitor activity results in the specific phosphorylation or dephosphorylation of certain plant proteins (Conrath et al., 1991; Dietrich et al., 1990; Ebel et al., 1991; Farmer et al., 1989; Grab et al., 1989; Kauss, 1991). Of more interest are recent observations that chemical inhibitors of vertebrate protein kinases are potent inhibitors of elicitor activity in certain plants (Grosskopf et al., 1990; Shiraishi et al., 1990) but function as elicitors in others (Conrath et al., 1991; Sharma A, Murillo J, and Keen N, unpublished). These observations raise the possibility that protein kinases and/or protein phosphatases may be involved either as primary or secondary members of the signaling mechanism that results in derepression of defense response genes.

Rapid changes in plasma membrane integrity have been observed that may be related to signal transduction events following the inoculation of resistant plants with pathogens or treatment with elicitors (e.g., Keppler et al., 1989). Active oxygen species are also generated at the plasma membrane. These species include the superoxide anion and hydrogen peroxide (e.g., Apostol et al., 1989; Doke, 1985). There are some suggestions that these events may be related to signal transduction in the HR, but further experimentation is required to more fully understand their role.

5. CURRENT STATUS OF DISEASE AND PEST CONTROL USING DISEASE RESISTANCE GENES

Diseases of economically important plants have been controlled primarily by the use of disease resistance genes introduced into agronomic cultivars from other, sometimes wild, accessions of the same or closely related plant species. The con-

trol of insect pests, on the other hand, has tended to rely on pesticides and biological control agents, but genetic plant resistance has been less utilized. Table 4.1
summarizes the successes and problems associated with the use of genetic resistance for control of pests and pathogens. The major advantages revolve around
the simplicity and low cost of genetic resistance. Once a resistance gene(s) is
introduced into a commercial cultivar, no additional input is required as long as
the gene remains effective. This finding is a particularly important consideration
for third world agriculture, where expenditures for pesticides or biological control agents are not feasible. Norman Borlaug (1983), by realizing this fact, relied
on improved genetic disease resistance as one of the cornerstones of the cultivar
improvement programs that eventually resulted in his award of a Nobel prize.
Additional examples of the successful deployment of disease resistance genes are
plentiful. In wheat, for example, the most serious diseases in most parts of the
world, stem rust and leaf rust, have not caused serious epidemics for almost 40
years, due to the effective use of mixtures of disease resistance genes in prevail-
ing cultivars. In cucumber, J.C. Walker introduced single genes for resistance to
scab, caused by *Cladosporium cucumerinum,* and cucumber mosaic virus. Varieties carrying these genes (Walker, 1952) have remained disease resistant for over
40 years. Prior to their introduction, the diseases threatened to wipe out the cucumber industry in Wisconsin.

Successes notwithstanding, there have also been several problems in the utilization of genetic disease resistance during this century (Table 4.1). While some
disease resistance genes have proven stable for many years, cultivars carrying others were overcome relatively quickly by virulent pathogen races. An even more
serious problem has been the lack of resistance genes in the available germplasm
of several cultivated crop plant species. In some crops, it is difficult or impossible to introduce genes into agronomically important cultivars because of the difficulty or inability to perform conventional crosses. In other plants, crosses can
be made but generation times are extremely long, as for example, in tree crops.
These factors all limit the utility of genetic disease resistance. However, if a large
collection of disease resistance genes were molecularly cloned and deployed by
transforming them into desired plant cultivars, substantial improvements in disease and pest control could be quickly realized. Increased attention therefore needs
be given to strategies for efficiently cloning relatively large numbers of these genes
from a variety of crop plants.

6. IMPROVED DISEASE CONTROL WITH CLONED DISEASE RESISTANCE GENES

The availability of cloned disease resistance genes will permit studies of resistance gene structure and function. It should be possible to critically test the major

TABLE 4.1. Advantages and Problems Associated With the Use Of Resistance Genes to Control Plant Diseases and Possible New Approaches

Advantages

The resistance gene, once introduced, is inherited and therefore permanent at no extra cost.

The method is simple; no application or delivery systems are required and no attention need be given to populations of the pest, the presence of alternate hosts, and so on.

The method is safe and environmentally sound since no applied biological control agents or chemicals are involved.

Resistance is only expressed when needed, since the plant does not express resistance mechanisms until an infecting pathogen invades.

Genetic resistance is generally less affected by soil type, climate, or previous rotation crops than are most biocontrol agents.

The probability of an introduced disease resistance gene working in a new genetic background is high.

Current Problems

Virulent pathogen races may appear that overcome the effects of single plant disease resistance genes.

Disease resistance genes directed against particular pests or pathogens are unavailable in the gene pools of some plant species.

Resistance genes are difficult or impossible to deploy in crops that are refractory to crossing, in which generation times are long or vegetative propagation is routinely utilized.

Resistance genes are only detectable by inoculation with appropriate pathogen races; the introduction of multiple resistance genes by crossing thus requires complex progeny screening.

Introduction of resistance genes into elite cultivars by crossing necessitates extensive backcrossing to reestablish desired traits.

New Approaches If Several Cloned Resistance Genes Were Available

Resistance genes introduced into plants by transformation would avoid negative characteristics from crossing and eliminate the time for extensive backcrossing.

The availability of several cloned resistance genes against a pest or pathogen should permit deployment of only those that are difficult for the pathogen to overcome.

New combinations of resistance genes could be released periodically to avoid problems with the emergence of virulent pathogen races.

Resistance genes effective against a certain pest or pathogen of a particular plant might be obtained from unrelated plant species.

The availability of cloned resistance genes might permit the construction of chimeric or synthetic resistance genes encoding novel recognitional specificities.

predictions of the elicitor–receptor hypothesis. Do resistance gene protein products in fact function as receptors for corresponding specific elicitors from pathogens? Are the gene products localized in the plant plasma membrane? Do the gene products have distinct recognitional and effector domains? A number of other questions can also be explored: Do these genes have structural features in common with others of similar function? What other cell components do resistance gene products interact with? Do different resistance genes initiate similar signal transduction pathways?

Regardless of progress in basic biological understanding, the molecular cloning of disease resistance genes will have major impact on practical agriculture—providing that several different genes targeted against specific pathogens can be cloned. Table 4.1 summarizes the advantages that will accrue to practical agriculture when significant numbers of cloned disease resistance genes become available. Impact will be most obvious in cases where disease resistance has historically not been available, but it will also be significant where resistance genes have failed to afford long-term disease control. The potato late blight disease, caused by *Phytophthora infestans,* is a good example. Because potato is a difficult crop in which to breed and select acceptable new cultivars, disease resistance genes were historically introduced from the wild potato, *Solanum demissum* (Wastie and Solomon, 1988). These genes failed to provide long-term disease control, largely because only one gene was introduced at a time into commercial cultivars. This result provided strong selection for mutant *P. infestans* biotypes that had lost the complementary avirulence allele.

If six different cloned disease resistance genes were deployed against *P. infestans,* better results would be expected (e.g., Green and Campbell, 1979; Wheeler and Diachun, 1983). A deployment scheme could be used in which an elite potato cultivar was transformed with three of the genes and, following testing, released to growers. The same cultivar would subsequently be transformed with the other three disease resistance genes. After a predetermined period, the original line would be withdrawn and the second, near-isogenic, transformed line released to large scale culture. By alteration of resistant lines and the combination of resistance genes, the emergence of damaging virulent pathogen races should become a rare event.

As pointed out in Table 4.1, the use of genetic transformation to introduce resistance genes into desirable cultivars carries several other advantages. Resistance genes are often derived from germplasm sources that are inferior for many other traits. Transformation avoids introduction of these unlinked traits, and, therefore eliminates the need for repeated backcrossing and selection to recreate the elite traits. This method not only offers savings based on the generation time of a new cultivar but also offers economic savings for the labor required at each step of a backcross program. Transformation is all the more significant for floral plants, potato, cassava, and other crops that are generally propagated vegetatively, and also for trees and other plants with extremely long reproductive cycles. It is important to emphasize that several different cloned disease resistance genes could be introduced simultaneously into a plant by a single transformation.

The capability to introduce genetically determined traits by transformation eliminates the species boundaries that have traditionally limited germplasm sources. As noted earlier, research with pathogen avirulence genes has suggested that functionally identical disease resistance genes are shared between taxonomically diverse plants. This theory supports the prediction that disease resistance genes from a certain plant may function when introduced into an unrelated plant. If this prediction holds true, cloned disease resistance genes will have important uses against pathogens and pests that attack plant species with little or no currently available resistance. For example, *Phytophthora cinnamomi* is a serious pathogen on avocado trees, primarily because little naturally occurring resistance has been identified in *Persea americana,* the cultivated avocado (Zentmyer, 1980). However, several *Persea* species that cannot be intercrossed or successfully intergrafted with *P. americana* exhibit high level resistance to the fungus. If resistance genes could be cloned from these species and transformed into the cultivated avocado, resistance to *P. cinnamomi* might be obtained. Many resistance genes effective against *Phytophthora* sp. also occur in other plant species (e.g., soybean and peppers). Several resistance genes cloned from these plants might also confer resistance in avocado to *P. cinnamomi*. If such lateral introductions of disease resistance genes between taxonomically diverse plant species are effective, their value to society will be considerable.

Certain disease resistance genes are ''durable'' such that they continue to afford disease resistance despite the presence of virulent races in the pathogen population. These resistance genes, also called ''strong'' (Van der Plank, 1968), are assumed to remain effective because they confer a virulence or biological fitness penalty on the pathogen if mutations occur in complementary avirulence gene alleles (Crill, 1977). Kearney and Staskawicz (1990) recently demonstrated that a resistance gene in pepper plants, *Bs2*, and its complementary avirulence gene in *Xanthomonas campestris* pv. *vesicatoria* behave in this manner. Avirulence gene *avrBs2* was cloned from *X.c.* pv. *vesicatoria* and a marker exchange mutant strain was constructed that was identical to the wild type except that it carried a defective *avrBs2* gene. This strain was markedly less virulent than the wild-type bacterium on all tested cultivars of pepper plants, regardless of whether they carried the complementary disease resistance gene, *Bs2*. The experiment therefore indicated that *avrBs2* is important in the virulence of *X.c.* pv. *vesicatoria* and that the pepper disease resistance gene, *Bs2,* is a strong gene, which should not easily be overcome by a virulent pathogen race lacking *avrBs2*. It may be possible to identify particularly durable disease resistance genes by observing the relative virulence of pathogens that lack a given avirulence gene.

7. PROSPECTS FOR CLONING DISEASE RESISTANCE GENES

Isolation of the necessary large battery of cloned disease resistance genes effective against a wide range of pathogens will require new methods, particularly since no resistance genes have been cloned to date. How will such a large number of genes be cloned? Several projects are now underway (North, 1990), the popular approaches being chromosome walking and gene tagging (Ellis et al., 1988). With widespread attention being paid to technology development, these methods are becoming more manageable. Additional approaches, such as subtractive hybridization and functional cloning, are also being developed with plants.

7.1 Chromosome Walking

The development of chromosome walking techniques has allowed spectacular successes in the cloning of human genes, such as the cystic fibrosis gene (Rommens et al., 1989). This work exemplifies both the rapid technical advances that have occurred as well as the remaining practical pitfalls that characterize available methods for cloning eukaryotic genes of unknown molecular function. The initial challenge in chromosome walking is to identify genetically polymorphic DNA markers that are closely linked to disease resistance or other genes of interest. The frequency of polymorphism increases with the relative sequence divergence between parental lines, but this can vary greatly for different crops. Widespread polymorphism is present in cultivated varieties of U.S. seed corn, for instance, whereas crosses to wild relatives have been necessary in tomato (Bernatzky and Tanksley, 1986; Helentjaris et al., 1986). Linked markers have typically been obtained through isolation of restriction fragment length polymorphism (RFLP) clones, but this can be somewhat difficult. An exciting new technology for identification of linked markers uses polymerase chain reaction (PCR) with defined 10-nucleotide primers to amplify ''random'' genomic sequences (Williams et al., 1990). If side-by-side reactions are run on near-isogenic lines, amplified PCR products that show size polymorphisms will generally correspond to linked loci (Martin et al., 1991). Markers within a few centiMorgans (cM) can be identified by running and analyzing only a few hundred PCR reactions. An additional innovation, the use of pooled deoxyribonucleic acid (DNA) from 10 to 15 homozygous-resistant and homozygous-susceptible F2 individuals instead of extensively backcrossed lines, further extends the applicability of this approach.

While PCR refinements have greatly expedited the isolation of linked markers, a practical limit still exists after which further effort is unlikely to yield markers that are more tightly linked. A ''walk'' to identify overlapping clones that span the gap between markers flanking the resistance gene is initiated at this point. A reference population of region-specific plant recombinants is required for accurate "fine-structure" mapping of all clones—the more recombinants the better. Use of yeast artificial chromosome (YAC) libraries (e.g., Imai and Olson, 1990) has increased the distance spanned by 5–10 overlapping clones to upwards of 1 Mb, but construction of YAC libraries can be a time consuming process, and 1 Mb corresponds on average to less than 1 cM in many crop plants. Pulse-field gel electrophoresis (PFGE) is now available for the establishment of long-range restriction maps (e.g., Seidler and Graner, 1991). These maps can help to link up non-overlapping YAC clones, to define the maximum distance between markers that flank a resistance gene, and to localize recombination breakpoints and endpoints of large-scale genomic perturbations. Pulse-field gel electrophoresis can also be used to construct region-specific mini-libraries for isolation of more tightly linked RFLP markers. More importantly, PFGE bands with homology to linked markers can serve as the starting material for "chromosome jumps" to the ends of these bands, increasing the potential distance that can be spanned (Kandpal et al., 1990). Both YACs and PFGE minimize the potential problems posed by regions of repetitive DNA.

Once a tightly linked region has been cloned, gene identification can begin in earnest; complementary DNAs (cDNAs) encoded by the PFGE-band or YAC clone DNA represent obvious candidate genes, but the resistance gene message may be missing due to tissue-specific, pathogen-induced, or otherwise low levels of constitutive expression. If resistance gene alleles likely to have deletions or insertions larger than about 20 bp are available, genomic blots from these lines can serve as the substrate for identification of cDNA or other probes that show polymorphisms. Positive identification of a resistance gene, however, requires observation of appropriate phenotypic changes in transformed and regenerated plants. Thus, a primary consideration is always the number of transformants that can realistically be generated and assayed for resistance. Anything fewer than about 100 transformants will generally be too small because 1 Mb is spanned by about 50 overlapping or 100 random cosmids. Techniques for integratively introducing foreign genes into plants are improving rapidly, particularly through the use of transformation guns with embryos or embryoids (Christou et al., 1989, 1990; Finer and McMullen, 1991; Klein et al., 1987, 1988). Rapid seed and germline transformation protocols have been used with success in *Arabidopsis,* but their applicability remains untested for most plant species (Chang et al., 1990; Feldmann and Marks, 1987).

Gene cloning by chromosome walking, then, is clearly a significant undertaking that suffers from several problems. A great deal of time and effort is required to identify closely linked genetic markers, particularly with plants that have not been extensively studied genetically. In addition, it is only possible to clone one resistance gene locus in each walking exercise, such that another round of marker identification would be required to clone an additional resistance gene locus from the same plant. The approaches discussed above have minimized but not eliminated problems with highly repetitive DNA in higher plants, which frequently bar hybridization walking. Finally, after isolation of a DNA fragment carrying a putative resistance gene by walking, it will still be necessary to transform this DNA into a susceptible plant and inoculate with a pathogen to prove that the desired gene has in fact been cloned. Further improvements in technology are required before the resources available to plant biology will permit a large number of resistance genes to be cloned by walking.

7.2 Gene Tagging

Gene tagging exercises typically involve the use of transposable elements to inactivate, and at the same time molecularly tag, a gene of interest (Doring and Starlinger, 1986; Gierl et al., 1989). A plant line that carries a molecularly characterized transposable element and at least one scoreable resistance gene is established. Self-progeny are then screened with the pathogen to identify rare plants that have lost the resistance phenotype due to insertion of the transposable element. The DNA is then isolated from such a plant by hydridization with the cloned transposon, and genomic DNA fragments flanking the transposon sequence are isolated. Confirmation that these sequences encode a resistance gene requires cloning the wild-type locus from a nonmutagenized plant by hybridization with the flanking sequences. This DNA must then be transformed into a susceptible plant line, which is then tested for the acquisition of resistance following inoculation with the appropriate pathogen biotype.

One advantage of the gene tagging approach is the rapidity with which one can go from mutant plant to putative resistance gene sequences. "Inverted-PCR" methods have been successfully used for very rapid isolation of DNA sequences that flank transposons (Earp et al., 1990). Numerous practical constraints, however, limit the use of transposon tagging. Endogenous transposons are frequently present in multiple copies, complicating analysis of flanking DNA (Doring and Starlinger, 1986). Lack of suitable transposons temporarily limited tagging to use in maize and snapdragon (e.g., Federoff et al., 1984, Coen et al., 1990), but maize *Ac/Ds* and *En-1/Spm* elements have now been shown to transpose in entirely unrelated species, such as tobacco, tomato, *Arabidopsis,* and potato (Baker et al., 1986, Frey et al., 1989; Van Sluys et al., 1987; Yoder et al., 1988). If transposition is too frequent, a given locus can become "untagged" prior to incorporation in the germline. Nonautonomous elements and tissue-specific control of transposase activity are currently being used to maximize the chance that transposition events will be transmitted in the germ line.

Low frequency transposition has proven to be equally problematic. Screening methods must be developed that allow reliable inoculation and screening of many thousands of individuals. Aids for pathogen inoculation have included surfactants, abrasives, air guns, and bulk vacuum infiltration. Background mutation rates that are above the rate due to transposon mutagenesis will defeat the method, and these rates can vary from locus to locus (Bennetzen et al., 1988; Ellis et al., 1988). Transposon mutagenesis rates can be enhanced by starting with plant lines in which the transposon is closely linked to the gene of interest (Dooner and Belachew, 1989), and by using foreign promoters to elevate the level of transposase activity.

In *Arabidopsis,* a transposon-independent gene-tagging method has been developed. *Agrobacterium*-mediated T-DNA transformation is used to generate thousands of lines, each of which carries an independent T-DNA insertion (Feldmann and Marks, 1987). A number of loci have been cloned using this method (e.g., Marks and Feldmann, 1989, Yanofsky et al., 1990). More widespread application of this method awaits the development of rapid transformation–regeneration protocols for other plant species. The range of loci that can be isolated using T-DNA or transposons is always limited by the extent to which integration sites are random, and by the degree that mutations will cause complete gene knockouts rather than more subtly altered alleles. Nevertheless, gene tagging has been successfully used to isolate a large number of plant genes and thus shows considerable promise for the cloning of disease resistance genes.

7.3. Subtractive Hybridization Cloning

A new approach to cloning genes of unknown biochemical function has recently been devised (Straus and Ausubel, 1990, Wieland et al, 1990) and has been used to clone at least two genes from *Arabidopsis* (F. Ausubel, personal communication). This technique, known as genomic subtraction or subtractive hybridization, is used to isolate DNA sequences that are not present in a second organism. Thus a plant disease resistance gene could be isolated if all or part of that gene was missing in a second plant line. In subtractive hybridization, DNA from the plant lacking the gene of interest is linked to biotin and an excess of this DNA is then combined with the wild-type DNA, the wild-type strands are denatured and allowed to reanneal, and the biotin-linked DNA is removed. The DNA that remains is significantly enriched for sequences that were not present in the plant lacking the gene of interest. Following multiple rounds of association and subtraction, remaining sequences are amplified using PCR. The resulting products are then used to isolate corresponding genomic or cDNA clones. As for the other approaches, candidate clones must be tested for phenotypic function by transformation of a plant line lacking the gene of interest.

One important advantage of subtractive hybridization is its generality. Transposon systems are not required and it is not even necessary to start with deletion mutants. As long as the two DNA pools are primarily colinear, they can be derived not only from parental and deleted plant lines, but also from near-isogenic backcrossed lines or pools of homozygous-resistant and homozygous-susceptible F2 progeny in which the resistance gene has no counterpart in the susceptible parent. Unfortunately, if the susceptible parent contains a nonfunctional but highly homologous allele, the subtractive cloning approach will fail. The method is also presently limited by the unfavorable reassociation kinetics of extremely complex genomes (>500 Mb), but this barrier may be overcome by the use of dextran sulfate and other treatments that increase the effective concentration of DNA in solution (F. Ausubel, personal communication).

7.4 Cloning Directed by the Isolation of Resistance Gene Proteins

The availability of pathogen-produced elicitors permits a cloning strategy in which labeled elicitor is used to physically isolate the cognate plant receptor. If, as the elicitor–receptor model proposes, these receptors are the protein products of plant disease resistance genes, they should permit cloning these genes. Thus far, the only elicitor that appears to interact with a defined resistance gene and for which evidence of a plant receptor exists is victorin and the oat *Pc-2* disease resistance gene (Wolpert and Macko, 1989). The gene encoding the approximate 100-kD oat receptor protein has recently been cloned using antibody screening of a DNA library and is now being characterized (T. Wolpert, personal communication). It will be interesting to see if this gene is indeed the product of the *Pc-2* disease resistance gene.

The general strategy of using elicitors to clone disease resistance genes suffers from many of the same problems as the preceeding approaches. For example, each elicitor will presumably only yield one resistance gene, although other closely linked genes with different specificities may then be cloned by walking. This approach requires considerable effort to isolate the relevant elicitor, label it satisfactorily, and then devise conditions to solubilize and isolate the receptor. If the receptor protein is isolated, it is straightforward to clone the gene by antibody screening or by amino acid sequencing the protein and constructing oligonucleotide probes based upon it to approach cloning the gene from a DNA library.

7.5 Functional Cloning Approaches

A strategy with the potential to yield a substantial number of cloned resistance genes from a large number of plant species is functional cloning. In this method, also called expression cloning, a DNA library from an organism with the desired genes is first prepared in a suitable vector. This library is then introduced into a line lacking the gene and the resulting transformants are analyzed by phenotype in order to identify relevant clones. This approach has rarely been used in attempts to clone plant genes, in part because of the large size of many plant genomes. However, functional cloning is the method of choice with microorganisms and has also led to dramatic recent successes with vertebrates (Frech et al., 1989; Harada et al., 1990; Seed and Aruffo, 1987), particularly for genes with very low abundance messages, such as those encoding receptors.

The major problem facing the use of functional approaches to cloning plant disease resistance genes is establishment of the functional assay. Because disease resistance is generally only expressed with full fidelity in intact plants, functional screening requires that intact plant tissue be transformed. Thus, it is necessary to introduce library clones containing DNA from a resistant plant into intact cells of a genetically susceptible plant, to screen by inoculation with a pathogen, and to monitor for clones that cause the otherwise susceptible tissue to undergo an HR. Fortunately, DNA can be transformed into intact plant cells when coated onto microprojectiles that are accelerated to velocities sufficient to traverse the plant cell wall without killing or otherwise seriously damaging the cell (Christou et al., 1989, 1990; Klein et al., 1987, 1988).

How should functional methods be applied to the cloning of disease resistance genes? F factor (Leonardo and Sedivy, 1990), P1 (Sternberg, 1990), or conventional cosmid libraries could be constructed from genomic DNA of a cultivar carrying several plant disease resistance genes. High quality cDNA expression libraries could also be constructed (e.g., Lin et al., 1991) and the remaining procedure done in a similar manner to that previously used in vertebrates (e.g., Frech et al., 1989). Recently, our own laboratories modified a plant expression vector carrying a dual enhancer cauliflower mosaic virus 35S promoter with a viral leader sequence (Restrepo et al., 1990) to permit high level expression of cDNAs obtained from a soybean breeding line carrying at least six different disease resistance genes.

Pools of clones (50–1000, depending on the type of library and insert DNA size) of a cDNA expression library or genomic library prepared from a plant line carrying several resistance genes would be transformed into suitable tissues of a plant line lacking the resistance genes by using a transformation gun. Our own work, for example, has shown that intact juvenile soybean hypocotyls can be efficiently transformed with a ''flowing helium'' transformation gun, such that hundreds of transformed cells are obtained with negligible tissue damage (Takeuchi et al., 1992). After about 24 h to permit gene expression, the transformed tissue can be inoculated with an appropriate pathogen biotype or, alternatively, treated with a race specific elicitor, such as that specified by the *avrD* gene described earlier. In either event, the transformed and inoculated tissue would be observed visually and microscopically for signs of the HR. Most transformations should not yield a hypersensitive reaction, but a few clone pools should cause the transformed plant cells to react with a HR. By individually screening clones from a positive pool, it should be possible to identify one clone that contains the resistance gene. Inoculation of the transformed tissue with various pathogen races should also permit identification of the cloned resistance gene. Because most of the cells are only transiently transformed, it is important to emphasize that this strategy is designed only for the *selection* of clones containing resistance genes and not for the development of integratively transformed, resistant plants. Once a clone was selected that contained a putative disease resistance gene, it could be integratively transformed into a susceptible plant cultivar using a transformation gun (e.g., Finer and McMullen, 1991) or a conventional *Agrobacterium* Ti vector and the resulting transgenic plants tested for resistance. A major advantage of such a functional approach is the potential to isolate several different resistance genes against several different pathogens in the same experiment, even if the genes are unlinked in the donor plant.

Several major technical obstacles to such a functional cloning approach must be addressed before it becomes generally useful. Among these are the establishment of high quality libraries, the development of transformation gun techniques to obtain high efficiency transformation of cells in a suitable tissue on the recipient plant, and devising inoculation screening techniques with either elicitors or pathogens to provide highly sensitive detection of the HR in those recipient plant cells transformed with a resistance gene clone. Fortunately, most inoculation and screening requirements can be met in several plant–pathogen systems. Techniques are improving rapidly for construction of high quality genomic and cDNA expression libraries and transformation guns are being developed that provide transformation of large numbers of intact plant cells in the field bombarded. In principle, the tissue utilized for transformation can be any plant part that exhibits the correct host–pathogen specificity in response to inoculation with the pathogen or elicitor treatment, but in practice transformation frequencies are relatively low when dealing with highly differentiated tissues or cells that are highly vacuolated. The recent discovery of specific elicitors should serve as an impetus to the functional cloning of disease resistance genes, because their application results in dramatic, easily scoreable hypersensitive reactions on tissues of resistant but not susceptible cultivars (e.g., Keen et al., 1990; van Kan et al., 1991). Thus, there is reason to expect that recipient plant cells transformed with a resistance gene clone and treated with an elicitor specific for that gene should provide a readily scoreable necrotic hypersensitive reaction.

The functional approach to resistance gene cloning has many advantages for the objective of cloning a large number of disease resistance genes effective against a large number of pathogens from many plants. The major advantages of the approach are its speed and simplicity. The time required for screening library clones should be no more than 1 week from the time transformations are performed until the reaction phenotypes of the recipient plants are scored. Unlike cloning rationales based on chromosome tagging or walking, functional cloning does not require extensive mapping of the genome, large numbers of test crosses, or many Southern blots. Furthermore, the method would permit selection of several clones containing different resistance genes (and effective against several different pathogens) from the same library, even though the genes were not closely linked. Since functional cloning involves a phenotypic assay, any clones conferring a HR to the recipient plant cultivar must *a priori* contain a resistance gene. This situation is unlike the case in walking or tagging exercises in which putative clones must still be assayed by a functional method to establish whether they are meaningful or artifacts. Indeed, the intrinsic speed of the functional protocol indicates that it would constitute a rapid method for screening candidate resistance gene clones obtained by one of the rationales discussed earlier. Finally, in a functional cloning approach, the donor library can in principle be constructed from any plant species. That is, it need not utilize the same plant species as the recipient cultivar but can be prepared from any other species. As discussed earlier, research with pathogen avirulence genes has raised the probability that resistance genes from one plant species can be expected to function against the same or a related pathogen if introduced into a taxonomically diverse plant species. The functional cloning approach constitutes a method to isolate such resistance genes.

8. MODIFICATION OF THE RECOGNITIONAL SPECIFICITY OF CLONED DISEASE RESISTANCE GENES

Numerous widely divergent disease resistance genes enable plants to evoke the HR in response to pathogens that carry a complementary avirulence gene (see Gabriel and Rolfe, 1990). The HR, then, is a globally conserved plant resistance response. Different aspects of the HR are probably effective against different pathogens, and the timing and intensity of the HR induced in response to different resistance and avirulence genes vary even in the same plant–pathogen system (e.g., Keen and Buzzell, 1990; Tamaki et al., 1988). It is nonetheless reasonable to postulate that the HR induced by most resistance genes could produce overlapping resistance against many different pathogens.

If the protein products of resistance genes have separate recognition and activation domains, as predicted by the elicitor–receptor model, it may eventually be possible to design resistance genes encoding altered recognitional specificities. Dramatic recent advances have occurred in designing antibody genes with altered recognitional motifs (Roberts et al., 1990). One can envision using recombinant DNA technology to engineer receptors that are specific for pathogen molecules that are not recognized by existing resistance genes. A number of pathogen molecules could be considered for such targeting, but those conferring strong selective value to the pathogen would be preferred (e.g., virulence factors), because their loss would invoke a severe fitness penalty. For a similar reason, function of the target molecule should be dependent on a highly conserved structure, which may not be the case, for instance, with many polysaccharide virulence factors. It may also be necessary to select molecules produced by pathogens and not other epiphytic flora, so as to avoid superfluous induction of the HR. This result could, however, prove to be irrelevant if recognition only occurred in subepidermal tissues that are rarely exposed to large populations of nonpathogens. Cloned resistance genes are the logical starting material for the engineering of new specificities, because they already function to induce resistance. However, increased understanding of HR signal transduction pathways may eventually allow the use of entirely heterologous proteins for recognition and activation of plant defense responses.

9. RECAPITULATION

If a substantial number of disease resistance genes can be cloned from several plants, these genes will have great impact on practical disease control. The availability of large numbers of cloned resistance genes would permit the deployment of various combinations against a given pathogen. This technique would be expected to reduce the emergence of virulent pathogen and pest biotypes. Cloned resistance genes should also allow disease control to be extended to plant–pathogen combinations for which genetic resistance was not previously available. Furthermore, the introduction of resistance by DNA transformation would avoid simultaneous introduction of negative characters and minimize the time and labor required in backcrossing. To obtain these results, however, it will be necessary to develop practical and straightforward methods for the cloning of resistance genes. There is considerable potential in approaches such as functional cloning, gene tagging, and subtractive hybridization, but the successful cloning of a resistance gene has not yet been reported. Significant advances will therefore be required to achieve the goal of cloning large numbers of resistance genes.

When cloned resistance genes are available, progress on their characterization will be very rapid. We can indeed expect that many of the current uncertainties surrounding the mechanisms of the HR will be quickly clarified. The availability of cloned resistance genes will also permit more ambitious future goals. For instance, it should be possible to devise tailored receptor-signal transducer genes that can be programmed for a specific pathogen elicitor and will interface with the extant plant signal transduction mechanisms of the HR. In this way, entirely new recognitional specificities may be generated, permitting plants to detect organisms that are now refractory, such as mycoplasma like organisms, certain insects, and many viruses.

REFERENCES

Apostol I, Heinstein PF, Low PS (1989): Rapid stimulation of an oxidative burst during elicitation of cultured plant cells. Role in defense and signal transduction. Plant Physiol 90:109–116.

Baker B, Schell J, Lorz H, Fedoroff N (1986): Transposition of the maize controlling element 'activator' in tobacco. Proc Nat Acad Sci USA 83:4844–4848.

Bennetzen JL, Hulbert SH, Lyons PC (1991): Genetic fine structure analysis of a maize disease-resistance gene. In Patil SS et al.(ed): Molecular strategies of pathogens and host plants. New York, Springer-Verlag, pp 177–188.

Bennetzen JL, Qin M-M, Ingels S, Ellingboe AH (1988): Allele-specific and mutator-associated instability at the *Rp1* disease-resistance locus of maize. Nature (London) 332:369–370.

Bernatzky R, Tanksley SD (1986): Toward a saturated linkage map in tomato based on isozymes and random cDNA sequences. Genetics 112:887–898.

Blein J-P, Milat M-L, Ricci P (1991): Responses of cultured tobacco cells to cryptogein, a proteinaceous elicitor from *Phytophthora cryptogea*. Plant Physiol 95:486–491.

Borlaug NE (1983): Contributions of conventional plant breeding to food production. Science 219:689–693.

Bowles DJ (1990): Defense-related proteins in higher plants. Annu Rev Biochem 59:873–907.

Buck L, Axel R (1991): A novel multigene family may encode odorant receptors: a molecular basis for odor recognition. Cell 65:175–187.

Carney BF, Denny TP (1990): A cloned avirulence gene from *Pseudomonas solanacearum* determines incompatibility on *Nicotiana tabacum* at the host species level. J Bacteriol 172:4836–4843.

Chang S-S, Park S-K, Nam H-G (1990): Transformation of *Arabidopsis* by *Agrobacterium* inoculation on wounds. Abstr. Fourth Int. Conf. on *Arabidopsis* Research, June 2–5, Vienna.

Cheong JJ, Hahn MG (1991): A specific, high-affinity binding site for the hepta-β-glucoside elicitor exists in soybean membranes. Plant Cell 3:137–147.

Christou P, McCabe DE, Martinell BJ, Swain WF (1990): Soybean genetic engineering-commercial production of transgenic plants. Trends Biotechnol 8:145–151.

Christou P, Swain WF, Yang N-S, McCabe DE (1989): Inheritance and expression of foreign genes in transgenic soybean plants. Trends Biotechnol 8:145–151.

Coen ES, Romero JM, Doyle S, Elliott R, Murphy G, Carpenter R (1990): *floricaula,* a homeotic gene required for flower development in *Antirrhinum majus*. Cell 63:1311–1322.

Collinge DB, Slusarenko AJ (1987): Plant gene expression in response to pathogens. Plant Molec Biol 9:389–410.

Conrath U, Jeblick W, Kauss H (1991): The protein kinase inhibitor, K-252a, decreases elicitor-induced Ca^{2+} uptake and K^+ release, and increases coumarin synthesis in parsley cells. FEBS Lett 279:141–144.

Cosio EG, Popperl H, Schmidt WE, Ebel J (1988): High-affinity binding of fungal β-glucan fragments to soybean (*Glycine max* L.) microsomal fractions and protoplasts. Eur J Biochem 175:309–315.

Crill P (1977): An assessment of stabilizing selection in crop variety development. Annu Rev Phytopathol 15:185–202.

Culver JN, Dawson WO (1989): Tobacco mosaic virus coat protein: an elicitor of the hypersensitive reaction but not required for the development of mosaic symptoms in *Nicotiana sylvestris*. Virology 173:755–758.

Dickman MB, Podila GK, Kolattukudy PE (1989): Insertion of cutinase gene into a wound pathogen enables it to infect intact host. Nature (London) 342:446–448.

Dietrich A, Mayer JE, Hahlbrock K (1990): Fungal elicitor triggers rapid, transient, and specific protein phosphorylation in parsley cell suspension cultures. J Biol Chem 265:6360–6368.

Dixon RA, Lamb CJ (1990): Molecular communication in interactions between plants and microbial pathogens. Annu Rev Plant Physiol Plant Mol Biol 41:339–367.

Doke N (1985): NADPH-dependent O_2^- generation in membrane fractions isolated from wounded potato tubers inoculated with *Phytophthora infestans*. Physiol Plant Pathol 27:311–322.

Dooner HK, Belachew A (1989): Transposition pattern of the maize element *Ac* from the *bz-m2(Ac)* allele. Genetics 122:447–457.

Doring H-P, Starlinger P (1986): Molecular genetics of transposable elements in plants. Annu Rev Genet 20:175–200.

Earp DJ, Lowe B, Baker B (1990): Amplification of genomic sequences flanking transposable elements in host and heterologous plants: a tool for transposon tagging and genome characterization. Nucleic Acids Res 18:3271–3279.

Ebel J, Cosio EG, Frey T (1991): Perception of pathogen-derived elicitor and signal transduction in host defenses. In Hennecke H, Verma DPS (eds): Advances in molecular genetics of plant-microbe interactions. Vol. I. Dordrecht: Kluwer Academic Publishing, pp 421–427.

Ebel J, Grisebach H (1988): Defense strategies of soybean against the fungus *Phytophthora megasperma* f.sp. *glycinea*: a molecular analysis. TIBS 13:23–27.

Ellis JG, Lawrence GJ, Peacock WJ, Pryor AJ (1988): Approaches to cloning plant genes conferring resistance to fungal pathogens. Annu Rev Phytopathol 26:245–263.

Farmer EE, Ryan CA (1990): Interplant communication: airborne methyl jasmonate induces synthesis of proteinase inhibitors in plant leaves. Proc Natl Acad Sci USA 87:7713–7716.

Fedoroff NV, Furtek DB, Nelson OE (1984): Cloning of the *bronze* locus in maize by a simple and generalizable procedure using the transposable element (*Ac*). Proc Natl Acad Sci USA 81:3825–3839.

Feldmann KA, Marks DM (1987): *Agrobacterium*-mediated transformation of germinating seeds of *Arabidopsis thaliana*: a non-tissue culture approach. Mol Gen Genet 208:1–9.

Finer JJ, McMullen MD (1991): Transformation of soybean via particle bombardment of embryogenic suspension culture tissue. In Vitro Cellular & Devel Biol 278:175–182.

Flor HH (1942): Inheritance of pathogenicity in *Melampsora lini*. Phytopathology 32:653–669.

Flor HH (1956): The complementary genic systems in flax and flax rust. Adv Genet 8:29–54.

Frech GC, VanDongen AMJ, Schuster G, Brown AM, Joho RH (1989): A novel potassium channel with delayed rectifier properties isolated from rat brain by expression cloning. Nature (London) 340:642–645.

Frey M, Tavantzis SM, Saedler H (1989): The maize *En-1/Spm* element transposes in potato. Mol Gen Genet 217:172–177.

Gabriel DW, Rolfe BG (1990): Working models of specific recognition in plant–microbe interactions. Annu Rev Phytopathol 28:365–391.

Gierl A, Saedler H, Peterson PA (1989): Maize transposable elements. Annu Rev Genet 23:71–85.

Grab D, Feger M, Ebel J (1989): An endogenous factor from soybean (*Glycine max* L.) cell cultures activates phosphorylation of a protein which is dephosphorylated *in vivo* in elicitor-challenged cells. Planta 179:340–348.

Green GJ, Campbell AB (1979): Wheat cultivars resistant to *Puccinia graminis tritici* in western Canada: their development, performance, and economic value. Can J Plant Pathol 1:13–15.

Grosskopf DG, Felix G, Boller T (1990): K-252a inhibits the response of tomato cells to fungal elicitors in vivo and their microsomal protein kinase in vitro. FEBS Lett 275:177–180.

Hahlbrock K, Scheel D (1987): Biochemical responses of plants to pathogens. In Chet I (ed): Innovative approaches to plant disease control. New York: Wiley, pp 229–254.

Harada N, Castle BE, Gorman DM, Itoh N, Scheurs J, Barrett RL, Howard M, Miyajima A (1990): Expression cloning of a cDNA encoding the murine interleukin 4 receptor based on ligand binding. Proc Natl Acad Sci USA 87:857–861.

Helentjaris T, Slocum M, Wright S, Schaefer A, Nienhuis J (1986): Construction of linkage maps in maize and tomato using restriction fragment length polymorphisms. Theor Appl Genet 72:761–769.

Imai T, Olson MV (1990): Second-generation approach to the construction of yeast artificial-chromosome libraries. Genomics 8:297–303.

Islam MR, Shepherd KW, Mayo GME (1989): Recombination among genes at the L group in flax conferring resistance to rust. Theor Appl Genet 77:540–546.

Kandpal RP, Shukla H, Ward DC, Weissman SM (1990): A polymerase chain reaction approach for constructing jumping and linking libraries. Nucleic Acids Res 18:3081.

Kauss H (1991): Phosphoprotein-controlled changes in ion transport are common events in signal transduction for callose and phytoalexin induction. In Hennecke H and Verma DPS (eds): Advances in Molecular Genetics of plant–microbe interactions. Vol. I. Dordrecht: Kluwer Academic Publishers, pp 428–431.

Kearney B, Staskawicz BJ (1990): Widespread distribution and fitness contribution of *Xanthomonas campestris* avirulence gene *avrBs2*. Nature (London) 346:385–386.

Keen NT (1990): Gene-for-gene complementary in plant–pathogen interactions. Annu Rev Genet 24:447–463.

Keen NT, Buzzell RI (1991): New disease resistance genes in soybean against *Pseudomonas syringae* pv. *glycinea*: evidence that one of them interacts with a bacterial elicitor. Theor Appl Genet 81:133–138.

Keen NT, Dawson WO (1992): Pathogen avirulence genes and elicitors of plant defense. In Boller T, Meins F (eds): Genes involved in plant defense. Vol. 8, Plant Gene Research. New York: Springer-Verlag, pp. 76–103.

Keen NT, Tamaki S, Kobayashi DY, Gerhold D, Stayton M, Shen H, Gold S, Lorang J, Thordal-Christensen H, Dahlbeck D, Staskawicz B (1990): Bacteria expressing avirulence gene D produce a specific elicitor of the soybean hypersensitive reaction. Molec Plant Microbe Inter 3:112–121.

Keppler LD, Baker CJ, Atkinson MM (1989): Active oxygen production during a bacteria-induced hypersensitive reaction in tobacco suspension cells. Phytopathology 79:974–978.

Kiraly Z (1980): Defenses triggered by the invader: hypersensitivity. In Horstfall JG, Cowling EB (eds): Plant Disease, Vol. V, New York: Academic Press, pp 201–224.

Klein TM, Fromm M, Weissinger A, Tomes D, Schaaf S, Sletten M, Sanford JC (1988): Transfer of foreign genes into intact maize cells with high-velocity microprojectiles. Proc Natl Acad Sci USA 85:4305–4309.

Klein TM, Wolf ED, Wu R, Sanford JC (1987): High-velocity microprojectiles for delivering nucleic acids into living cells. Nature (London) 327:70–73.

Knogge W, Hahn M, Lehnackers H, Rupping E, Wevelsiep L (1991): Fungal signals involved in the specificity of the interaction between barley and *Rhynchosporium secalis*. In Hennecke H, Verma DPS (eds): Advances in molecular genetics of plant–microbe interactions, Vol. 1. Dordrecht: Kluwer Academic Publishing, pp 250–253.

Kobayashi DY, Keen NT (1985): Cloning of a factor from *Pseudomonas syringae* pv. *tomato* responsible for a hypersensitive response on soybean. Phytopathology 75:1355.

Kobayashi DY, Tamaki SJ, Keen NT (1989): Cloned avirulence genes from the tomato pathogen *Pseudomonas syringae* pv. *tomato* confer cultivar specificity on soybean. Proc Natl Acad Sci USA 86:157–161.

Leonardo ED, Sedivy JM (1990): A new vector for cloning large eukaryotic DNA segments in *Escherichia coli*. Bio/Technol 8:841–844.

Lin HY, Kaji EH, Winkel GK, Ives HE, Lodish HF (1991): Cloning and functional expression of a vascular smooth muscle endothelin 1 receptor. Proc Natl Acad Sci USA 88:3185–3189.

Marks MD, Feldmann KA (1989): Trichome development in *Arabidopsis thaliana*. I. T-DNA tagging of the GLABROUS1 gene. Plant Cell 1:1043–1050.

Martin GB, Williams JGK, Tanksley SD (1991): Rapid identification of markers linked to a *Pseudomonas* resistance gene in tomato by using random primers and near-isogenic lines. Proc Natl Acad Sci USA 88:2336–2340.

North G (1990): The race for resistance genes. Nature (London) 347:517.

Pryor AJ (1987): The origin and structure of fungal disease resistance genes in plants. Trends Genet 3:157–161.

Restrepo MA, Freed DD, Carrington JC (1990): Nuclear transport of plant potyviral proteins. Plant Cell 2:987–998.

Ricci P, Bonnet P, Huet J-C, Sallantin M, Beauvais-Cante F, Bruneteau M, Billard V, Michel G, Pernollet JC (1989): Structure and activity of proteins from pathogenic fungi *Phytophthora* eliciting necrosis and acquired resistance in tobacco. Eur J Biochem 183:555–563.

Roberts VA, Iverson BL, Iverson SA, Benkovic SJ, Lerner RA, Getzoff ED, Tainer JA (1990): Antibody remodeling: a general solution to the design of a metal-coordination site in an antibody binding pocket. Proc Natl Acad Sci USA 87:6654–6658.

Rommens JM, Iannuzzi MC, Kerem B-S, Drumm ML, Melmer G, Dean M, Rozmahel R, Cole JL, Kennedy D, Hidaka N, Zsiga M, Buchwald M, Riordan JR, Tsui L-C, Collins FS (1989): Identification of the cystic fibrosis gene: chromosome walking and jumping. Science 245: 1059–1065.

Scheel D, Parker JE (1990): Elicitor recognition and signal transduction in plant defense gene activation. Z Naturforsch 45c:569–575.

Seed B, Aruffo A (1987): Molecular cloning of the CD2 antigen, and T-cell erythrocyte receptor, by a rapid immunoselection procedure. Proc Natl Acad Sci USA 84:3365–3369.

Seidler H, Graner A (1991): Construction of physical maps of the *Hor*1 locus of two barley cultivars by pulsed field gel electrophoresis. Mol Gen Genet 226:177–181.

Shiraishi T, Hori N, Yamada T, Oku H (1990): Suppression of pisatin accumulation by an inhibitor of protein kinase. Annu Phytopath Soc Jpn 56:261–264.

Sternberg N (1990): Bacteriophage P1 cloning system for the isolation, amplification and recovery of DNA fragments as large as 100 kilobase pairs. Proc Natl Acad Sci USA 87:103–107.

Straus D, Ausubel FM (1990): Genomic subtractions for cloning DNA corresponding to deletion mutations. Proc Natl Acad Sci USA 87:1889–1893.

Takeuchi Y, Dotson M, Keen NT (1992): Plant transformation: a simple particle bombardment device based on flowing helium. Plant Molec Biol 18:835–839.

Tamaki S, Dahlbeck D, Staskawicz B, Keen NT (1988): Characterization and expression of two avirulence genes cloned from *Pseudomonas syringae* pv. *glycinea*. J Bacteriol 170: 4846–4854.

Tonegawa S (1983): Somatic generation of antibody diversity. Nature (London) 302:565–581.

Van der Plank JE (1968): Disease resistance in plants. New York: Academic Press.

VanEtten HD, Matthews DE, Matthews PS (1989): Phytoalexin detoxification: importance for pathogenicity and practical implications. Annu Rev Phytopathol 27:143–164.

van Kan JAL, van den Ackerveken GFJM, de Wit PJGM (1991): Cloning and characterization of cDNA of avirulence gene *avr9* of the fungal pathogen *Cladosporium fulvum*, causal agent of tomato leaf mold. Mol Plant Microbe Inter 4:52–59.

Van Sluys MA, Tempe J, Fedoroff N (1987): Studies on the introduction and mobility of the maize *Activator* element in *Arabidopsis thaliana* and *Daucus carota*. EMBO J 56:3881–3889.

Walker JC (1952): Diseases of vegetable crops. New York: McGraw-Hill.

Wastie R, Solomon RM (1988): The contribution and value of resistant cultivars to disease control in potatoes. In Clifford BC, Lester E (eds): Control of plant diseases: costs and benefits. Oxford: Blackwell Scientific Publications, pp 103–112.

Whalen MC, Innes RW, Bent AF, Staskawicz BJ (1991): Identification of *Pseudomonas syringae* pathogens of *Arabidopsis* and a bacterial locus determining avirulence on both *Arabidopsis* and soybean. Plant Cell 3:49–59.

Whalen MC, Stall RE, Staskawicz BJ (1988): Characterization of a gene from a tomato pathogen determining hypersensitive resistance in non-host species and genetic analysis of this resistance in bean. Proc Natl Acad Sci USA 85:6743–6747.

Wheeler H, Diachun S (1983): Mechanisms of pathogenesis. In Kommedahl T, Williams PH (eds): Challenging problems in plant health. St. Paul: American Phytopathological Society, pp 324–333.

Wieland I, Bolger G, Asouline G, Wigler M (1990): A method for difference cloning: gene amplification following subtractive hybridization. Proc Natl Acad Sci USA 87:2720–2724.

Williams JGK, Kubelik AR, Livak KJ, Rafalski JA, Tingey SV (1990): DNA polymorphisms amplified by arbitrary primers are useful as genetic markers. Nucleic Acids Res 18:6531–6535.

Wolpert TJ, Macko V (1989): Specific binding of victorin to a 100-kDa protein from oats. Proc Natl Acad Sci USA 86:4092–4096.

Yancopoulos GD, Alt FW (1986): Regulation of the assembly and expression of varible-region genes. Annu Rev Immunol 4:339–368.

Yanofsky MF, Ma H, Bowman JL, Drews GN, Feldmann KA, Meyerowitz FM (1990): The protein encoded by the *Arabidopsis* homeotic gene *agamous* resembles transcription factors. Nature (London) 346:35–39.

Yoder JI, Palys J, Alpert K, Lassner M (1988): *Ac* transposition in transgenic tomato plants. Mol Gen Genet 213:291–296.

Zentmyer GA (1980): *Phytophthora cinnamomi* and the diseases it causes. St. Paul: American Phytopathological Society.

5

VIRUS RESISTANCE THROUGH EXPRESSION OF COAT PROTEIN GENES

ROGER N. BEACHY

Division of Plant Biology, The Scripps Research Institute, La Jolla, California 92037

1. INTRODUCTION

During the past 20 years plant pathologists have identified an ever increasing number of diseases caused by plant viruses, until the current number exceeds 1400. Some, like rice tungro disease (Jones et al., 1991) are caused by two or more distinctly different viruses, and attempts to incorporate genes for resistance have not been successful. Some, like the disease caused by tomato mosaic virus (ToMV) in tomato, have been recognized for many years, and plant breeders have deployed a series of different genes to control the virus. Variant strains of the virus that eventually overcome the genes for resistance have been selected in a number of cases. The search for resistance genes has been fruitless for many viruses, yielding little or weak resistance, often multigenic, or providing tolerance that, while suppressing symptoms, permits virus replication. In none of the examples of disease resistance or tolerance have the cellular or molecular mechanisms of resistance been fully characterized, nor have the structural genes responsible for the resistance been identified. The isolation and characterization of resistance genes remains, however, a viable target for molecular biologists, who propose to transfer resistance genes to heterologous plant species to provide resistance to one or more virus diseases.

Biotechnology in Plant Disease Control, pages 89–104
© *1993 Wiley-Liss, Inc.*

As an alternative strategy to plant breeding for virus resistance, there are growing numbers of examples in which disease resistance or tolerance has been developed by expressing sequences derived from viral genomes in a variety of transgenic plants. Although the results of these studies have been mixed they have, nevertheless, led to innovative approaches to control plant virus infections. Equally as important is the fact that such studies have dramatically expanded our understanding of some of the molecular events of virus replication and pathogenicity. This knowledge will, in turn, lead to additional strategies for developing disease resistant plants. One should, perhaps, consider that today's technology is a forebearer of technologies to come, and that several strategies will be applied together, as needed, to control specific types of virus diseases.

The genetic engineering strategies that have been employed to date have included introduction into plants of transgenes that encode all or parts of a viral cistron (gene) to encode sequences that are sense or antisense relative to the viral genome [in the case of ribonucleic acid (RNA)-containing viruses] or to a viral transcript. Genes that lead to accumulation of the following types of molecules have been tested in transgenic plants. (1) Satellite RNAs (Harrison et al., 1987; Gerlach et al., 1987) that limit virus replication and/or otherwise decrease disease severity. Stanley et al. (1990) used a related approach to express a defective interfering deoxyribonucleic acid (DNA) of the African cassava mosaic geminivirus in transgenic plants resulting in amelioration of disease symptoms. (2) Antisense RNAs (i.e., RNAs that are complimentary in sequence to a target RNA) that should, in principle, decrease the frequencies of infection as well as virus replication. In general, however, the expression of antisense RNAs has not led to high levels of disease resistance (e.g., see Powell et al., 1989), although it had limited success against the tomato golden mosaic geminivirus (Day et al., 1991) and potato leafroll luteovirus (Kawchuck et al., 1991). (3) The whole or a portion of the viral replicase. Golemboski et al. (1990) reported very high levels of resistance against tobacco mosaic tobamovirus by expressing a gene encoding a portion of the tobacco mosaic virus (TMV) replicase. Hemenway (1991) expressed the entire replicase of potato virus X (PVX) and likewise achieved a high level of resistance against PVX. Similar strategies are currently being applied in attempts to limit a number of other viruses. (4) Viral coat (capsid) proteins to reduce the frequency of infection, virus replication, and disease severity. The number of examples of resistance due to capsid gene expression is rapidly increasing, and has proven to be effective in a number of field trials. This topic is discussed in this chapter. For a more complete review of this topic, the reader is referred to Beachy et al. (1990).

2. COAT PROTEIN-MEDIATED RESISTANCE

Coat protein-mediated resistance (CP-MR) describes resistance to virus infection and/or disease development in transgenic plants that express a gene(s) that encodes a viral CP. This functional definition of a type of resistance is substantially different from the resistance searched for by the more standard plant breeding approaches. Plant breeders traditionally search the plant germplasm pools to locate plants that are resistant to a target pathogen or group of pathogens, and devise strategies to either sexually or asexually (e.g., by protoplast fusion) transfer the gene(s) to the target plant. While this strategy has been highly successful in developing resistance in many types of crop plants, it is generally time consuming and, if the resistance trait is multigenic, it can be difficult to achieve the desired level of resistance.

In contrast to multigenic disease resistance, CP-MR is a relatively simple concept. It requires knowledge of the targeted viral pathogen, the capacity to isolate the viral gene or genes that encode the capsid or coat protein(s), and the ability to construct a chimeric gene that, when introduced into the target plant, is expressed at sufficiently high levels to confer disease resistance. The most difficult portion in the process is the capacity to regenerate whole, intact transgenic plants from the cells into which the gene encoding the CP is introduced. Nevertheless, CP-MR provides the scientist with a new source of genes for disease resistance, that is, the virus against which resistance is desired. Furthermore, since CP-MR is a dominant single-gene trait, transferring resistance from one transgenic plant to another by genetic crossing is relatively simple.

2.1. Developing Coat Protein-Mediated Resistance: Technical Aspects

The technical aspects of developing CP-MR are, on paper, relatively straightforward, but not necessarily easy. The isolation of the sequence containing the open reading frame for the CP involves the application of standard techniques of recombinant DNA. It also requires prior knowledge about the organization of the viral genome and the location of the CP coding sequence, as well as at least limited amounts of information about how the CP is produced during virus replication.

Like isolation of the CP coding sequence, the selection of an appropriate transcriptional promoter and polyadenylation regulatory sequences is relatively straightforward. In general, the expectation is that the greater the degree of gene expression and CP accumulation, the greater the likelihood of achieving CP-MR. While this is not always the result, it is the experimental goal in most studies. In most reported cases of CP-mediated resistance the promoter selected has been the nominally constitutive 35S promoter (P35S) from cauliflower mosaic virus (CaMV), or the enhanced (duplicated) 35S promoter. Some workers have chosen the CaMV P19S or a T-DNA promoter from the tumor inducing plasmid (Ti plasmid) of *Agrobacterium tumefaciens* and obtained CP-MR, although with difficulty in some cases. The polyadenylation signals used for construction of CP genes are those used to express other transgenes. There are as yet no reports that different polyadenylation signals affect the levels of CP-MR, but direct comparisons between polyadenylation signals have yet to be done.

2.2. Examples and Phenotypes of Coat Protein-Mediated Resistance

Since the first report of CP-MR against TMV in tobacco (Powell-Abel et al., 1986) there have been increasing numbers of examples described in the literature. A current summary is presented in Table 5.1 and shows that CP-MR is effective against pathogens in 12 different virus classification groups. Resistance has been observed in a number of different crop plants including tobacco, tomato, potato, sugarbeet, cucumber, cantaloupe (melon), papaya, alfalfa, and rice.

There are several phenotypes associated with CP-MR, some of which may be reflective of the cellular and molecular mechanisms of resistance. These phenotypes may also vary depending on the host and the environmental conditions of plant growth and testing, and the replication and disease strategies of the pathogen. Some of these differences will become apparent in subsequent portions of the text.

Coat protein-mediated resistance is characterized by one or more of the phenotypes listed in Table 5.2. Not all examples of CP-MR exhibit each of the phenotypes listed. For example, CP-MR in some cases leads to reduced severity of disease symptoms in individual plants in a population, yet the amount of virus that accumulates in transgenic plants that express the CP gene (CP+) may not be significantly different from the control plants (CP−). In such individuals CP-MR is similar to disease tolerance. From the standpoint of complete disease control, it would be desirable for CP-MR to encompass each of the traits listed in Table 5.2: This result occurs in some, but not all cases of CP-MR.

The phenotypes of CP-MR are the result of various types of molecular and cellular mechanisms that occur in the host as CP molecules interact with the virion and/or the replication of its genome. Certainly, a more complete understanding of the mechanism will lead to better resistance and to a better understanding of virus replication.

2.3. Molecular and Cellular Mechanisms of Coat Protein-Mediated Resistance

The phenotypes of CP-MR have been explored at fundamental levels in a limited number of examples, yet it is becoming clear that there are similarities as well as differences in different host–virus combinations. The most significant differences are, however, likely due to differences between viruses and their modes of infection and replication rather than in the host *per se*. The following sections

TABLE 5.1. Summary of Examples of Coat Protein-Mediated Resistance[a]

CP Gene	Virus Group	Virus	Challenge Group	CP-MR	Correlation CP-MR vs CP Levels	References
TMV	Tobamo	TMV-U1	Tobamo	Yes	Yes	Powell-Abel P et al., 1986
		TMV-P230	Tobamo	Yes	Yes	Nelson RS et al., 1988
		ToMV	Tobamo	Yes	Yes	Nelson RS et al., 1987
		TMGMV	Tobamo	Yes	Yes	Nejidat A and Beachy RN, 1990
		PPMV	Tobamo	Yes	Yes	Register III, JC and Beachy RN, 1988
		ORSV	Tobamo	Yes	Yes	
		RMV	Tobamo	No		
		SHMV	Tobamo	No		
		AlMV	Alfalfa mosaic	No		
AlMV	Alf Mosaic	AlMV	Alfalfa mosaic	Yes	Yes	Tumer NE et al., 1987
		TRV	Tobra	No		Loesch-Fries LS et al., 1987
		TSV	Ilar	No		van Dun CMP et al., 1988
		TMV	Tobamo	No		Tumer NE et al., 1991
						van Dun CMP et al., 1987
TSV	Ilar	TSV	Ilar	Yes	Yes	Tumer NE et al., 1991
		AlMV	Alfalfa mosaic	No		
		TMV	Tobamo	No		
TRV-TCM	Tobra	TRV-TCM	Tobra	Yes	Yes	van Dun CMP et al., 1988
		TRV-PLB	Tobra	No		
		PEBV	Tobra	Yes	Yes	
BNYVV	Furo	BNYVV	Furo	Yes		Kallerhoff J et al., 1990
CMV	Cucumo	CMV	Cucumo	Yes	No	Quemada HD et al., 1991
						Cuozzo M et al., 1988
						Namba S et al., 1991

(Continued)

TABLE 5.1. Summary of Examples of Coat Protein-Mediated Resistance[a] (Continued)

CP Gene	Virus Group	Virus	Challenge Group	CP-MR	Correlation CP-MR vs CP Levels	References
PVX	Potex	PVX	Potex	Yes	Yes	Hemenway C et al., 1988 / Hoekema A et al., 1989
SMV	Poty	PVY	Poty	Yes	No	Stark DM and Beachy RN, 1989
		TEV	Poty	Yes	No	
PRV	Poty	TEV	Poty	Yes	No	Ling K et al., 1991
		PeMV	Poty	Yes	No	
		PVY	Poty	Yes	No	
		CMV-C	Cucumo	No		
PPV	Poty	PPV	Poty	Yes	No	Regner F et al., 1992
PVS	Carla	PVS	Carla	Yes	No	MacKenzie DJ et al., 1990 / MacKenzie DJ et al., 1991
PLRV	Luteo	PLRV	Luteo	Yes	?	Kawchuk LM et al., 1990 / Kawchuk LM et al., 1991
TSWV	Tospo	TSWV	Tospo	Yes	No	Gielen JJL et al., 1991 / MacKenzie DJ and Ellis PJ, 1992
RSV	Tenui	RSV	Tenui	Yes	?	Hayakawa T and Schimamoto K, 1991

[a]List of abbreviations: AlMV, Alfalfa mosaic virus; BNYVV, Beet necrotic yellow vein; CaMV, Cauliflower mosaic virus; CMV, Cucumber mosaic virus; ORSV, Ondontoglossum ringspot virus; PEBV, Pea early browning virus; PeMV, Pepper mottle virus; PLRV, Potato leaf roll virus; PMMV, Pepper mild mottle virus; PPV, Plum pox virus; PRV, Papaya ringspot virus; PVS, Potato virus S; PVX, Potato virus X; PVY, Potato virus Y; RMV, Ribgrass mosaic virus; RSV, Rice stripe virus; SHMV, Sunn hemp mosaic virus; SMV, Soybean mosaic virus; TEV, Tobacco etch virus; TMGMV, Tobacco mild green mosaic virus; TMV, Tobacco mosaic virus; ToMV, Tomato mosaic virus; TRV, Tobacco rattle virus; TSV, Tobacco streak virus; TSWV, Tomato spotted wilt virus.

TABLE 5.2. Characteristics of Coat Protein Mediated Resistance

Reduced sites of infection following inoculation: fewer necrotic local lesions, starch lesions, or chlorotic zones on inoculated leaves

Fewer plants become systemically infected

Reduced severity of disease symptoms in plants that become infected

Reduced accumulation of virus in infected plants

summarize and interpret the results of studies of the mechanisms of CP-MR that can lead to better understanding of the phenotypes of CP-MR, and perhaps to improving the levels of disease resistance in transgenic plants. These studies also led to a greater understanding of virus replication and pathogenesis.

2.3.1. CP-MR Reduces the Numbers of Sites of Infection on Inoculated Leaves

Several early reports of CP-MR reported that the expression of CP genes resulted in reduced numbers of sites of infection in inoculated leaves. Studies by Nelson et al. (1987) on CP-MR against TMV reported fewer necrotic local lesions or chlorotic zones on CP(+) tobacco plants than on CP(−) plants. However, the size of the local lesions on chlorotic zones was not different in CP(+) and CP(−) plants. This result suggests that once infection occurs local spread of the infection was the same in the presence or absence of CP. Coat protein-mediated resistance against alfalfa mosaic virus (AlMV) in tobacco produced fewer necrotic spots on inoculated leaves of CP(+) plants than on CP(−) plants (Tumer et al., 1987). In a similar manner, CP-MR against PVX in tobacco resulted in fewer infection sites, detected as starch lesions, on CP(+) plants versus CP(−) plants (Hemenway et al., 1987). Each of these examples support the conclusion that CP-MR results in reduced numbers of sites of infection, and subsequent delay in disease development.

To better address the issues of reduced sites of infection in CP-MR, several groups have expanded their studies to include transgenic protoplasts: Studies in protoplasts make it possible to distinguish between events that occur in initially infected cells from those that require the involvement of multiple cells. Furthermore, protoplast studies enable one to distinguish between molecular events that affect infection and/or replication, apart from the movement of virus in the inoculated leaves and through the plant.

Coat protein-mediated resistance has been demonstrated in at least four protoplast systems: TMV-tobacco, AlMV-tobacco, tobacco rattle virus (TRV)-tobacco, and cucumber mosaic virus (CMV)-tobacco. The results of those studies have not been the same in each system. The level of CP-MR against TMV in protoplasts exceeded that in whole plants, considering the concentration of virus required to overcome CP-MR in protoplasts versus whole plants (Register and Beachy, 1988). While CP-MR in whole plants was overcome with virus inoculum concentrations of 1–10 μg/ml, protoplasts were resistant to concentrations greater than 1 mg/ml. The CP-MR was demonstrated against CMV at 100 μg/ml (Okuno et al., 1992); similar high levels of resistance in protoplasts was reported against AlMV (Loesch-Fries et al., 1987); and TRV (Angenent et al., 1990).

However, CP-MR in transgenic protoplasts does not necessarily mean that the plants from which the protoplasts are isolated are resistant. Clark et al. (1990) reported that leaf mesophyll protoplasts isolated from plants that express a gene comprising the promoter from *rbcs* (Prbcs) (the gene encoding the small subunit of ribulose bisphosphate carboxylase) and the TMV CP coding region were highly resistant to infection by TMV. However, intact plants that harbor this gene were not resistant to infection. The difference between the results of Clark et al. (1990) and Register and Beachy (1988) is that the latter study involved the 35S promoter P35S from CaMV rather than the Prbcs. This result indicates that whole plant resistance against TMV is related not only to the level of CP (see below) but also to the levels of CP in different types of cells. Whereas the level of CP that accumulates in mesophyll cells is about the same when either the P35S or the Prbcs promoters are used, the levels of CP are much lower in epidermal cells when the Prbcs is used than when the P35S is used. In the whole plant experiments TMV was mechanically inoculated to the upper leaf epidermis. These results indicates that it is essential to achieve high levels of CP in the inoculated cells in order to achieve high levels of whole plant resistance.

Such a situation may not be essential for all examples of CP-MR. Because there are other differences in the patterns of gene expression between the P35S and Prbcs promoters in addition to those in epidermal cells, studies should be extended to include other promoters that cause the protein to accumulate specifically in different cell types, and to viruses that are transmitted by insects.

2.3.2. Overcoming CP-MR by Inoculation With Viral RNA

Studies of the mechanisms of CP-MR included challenging resistant plant lines and their protoplasts with purified viral RNA rather than with virus. The rationale behind these studies is to determine whether resistance is manifested at the level of uncoating (release) of viral RNA from the virion, or after its release.

In whole plant studies, two types of results have been observed. In the case of TMV, CP-MR is largely overcome when CP(+) plant lines are inoculated with TMV–RNA (Nelson et al., 1987). Furthermore, treating TMV at elevated pH (pH 8.0, 4°C), which releases 60–100 subunits from the virus, also overcomes resistance (Register and Beachy, 1988). Release or removal of protein subunits is the first step in the infection process. Therefore, it was concluded that CP-MR against TMV acts at a step that precedes the uncoating of TMV–RNA.

Similar studies with other viruses resulted in conflicting results. Loesch-Fries et al. (1987) reported that inoculation of CP(+) plants with AlMV–RNA overcame CP-MR while Tumer et al. (1991) reported the opposite results. Tumer et al. (1991) attributed these differences in results to the higher levels of AlMV CP in the plant lines that they studied compared with those used by Loesch-Fries et al. (1987). In the case of CP-MR against PVX (Hemenway et al., 1987) and CMV (Okuno et al., 1992) inoculation with viral RNA did not overcome resistance. It is, therefore, clear that this aspect of CP-MR is different in different host–virus combinations.

To further clarify these differences, comparing CP-MR in protoplasts versus whole plants can be helpful. Studies of several viruses have shown that CP-MR is overcome in protoplasts following infection with viral RNA (TMV, Register and Beachy, 1988; AlMV, Loesch-Fries et al., 1987; CMV, Okuno et al., 1992; TRV, Antenent et al., 1990). Thus, CP-MR against CMV and AlMV is different in protoplasts and in whole plants. As discussed by Okuno et al. (1992) in the studies with CMV, whole plants may be resistant to CMV-RNA because CMV moves from cell to cell as virus particles. Therefore, the virus encounters CP-MR in each cell that it encounters as it spreads in CP(+) plants. In contrast, viruses such as TMV, which spread from cell to cell in the absence of CP (Dawson et al., 1988; Holt and Beachy, 1991) are able to overcome CP-MR following infection by TMV–RNA. In the case of TMV, there would be little resistance to local spread; resistance to long-distance spread is encountered because CP is required for long-distance movement of the infection (Dawson et al., 1988; Holt and Beachy, 1991). Studies of CP-MR with other viruses will provide additional evidence to lead to further understanding of overcoming CP-MR by inoculations with viral RNA.

2.3.3. Breadth of Protection Afforded by CP-MR

A very practical and important question in any type of disease resistance relates to the breadth of resistance conferred by the gene. If resistance is effective against one or several virus strains it may have less resilience or durability than a gene that is widely effective. Several types of studies suggest that CP-MR may be relatively broad in activity: Others suggest that it may be relatively narrow. In fact, the breadth of CP-MR may be related to a number of unknown variables.

It has been well documented that CP-MR is generally effective against the virus from which the CP sequence was obtained, and to related strains but not to other viruses. Table 5.1 presents a relatively complete summary of the results of studies in which CP-MR was tested against different viruses or strains of virus. In most of the cases cited, resistance was more effective against strains of viruses closely related to the CP sequence expressed in the host than to a more distant virus. Thus, the TMV-CP gene provided resistance (under laboratory conditions) to ToMV and pepper mild mottle virus, but not to Sunn hemp mosaic virus (Nejidat and Beachy, 1990). van Dun et al. (1988) reported that the CP of the TCM strain of TRV provided resistance against the TCM strain of TRV and against pea early browning virus, but not against the PLB strain of TRV. Pea early browning virus is more closely related to the TCM strain of TRV than is the PLB strain. In the case of CMV, two types of CP genes have given resistance in transgenic plants. The C strain of CMV gave resistance against the D and C strains of CMV (Cuozzo et al., 1988; Quemada et al., 1991) but not against the CMV strains in subgroup II (the C and D strains are in subgroup I). By contrast, the CP from the WL strain of CMV (a member of subgroup II) gives resistance to viruses in both subgroup I and II (Namba et al., 1991).

In the case of CP-MR against members of the potyvirus group, there is strong evidence that CP-MR is a broad type of resistance. Stark and Beachy (1989) expressed a CP gene of soybean mosaic potyvirus in tobacco plants. The CP(+) plants were resistant to the potyviruses tobacco etch virus (TEV), potato virus Y (PVY) (Stark and Beachy, 1989), and pepper mottle virus (PeMV) (C. Malpica and R.N. Beachy, unpublished). Likewise, Ling et al. (1991) found that the CP gene of papaya ringspot potyvirus gave resistance to TEV, PVY, and PeMV in CP(+) tobacco plants. These studies indicate that single genes encoding potyviruses can provide broad, multivirus CP-MR.

It should be noted that experimental results of CP-MR can be significantly different under greenhouse and field conditions. Nelson et al. (1988) reported that the TMV CP gene gives high levels of resistance in tomato to both TMV and ToMV. Under field conditions, however, these tomato plants proved significantly more resistant to TMV than to ToMV. When a ToMV CP gene was expressed in tomato plants, there was a high level of resistance to ToMV, and lesser resistance to TMV. Transgenic tomato lines that contained genes to both TMV and ToMV exhibited high levels of resistance to both viruses (Sanders et al., 1992). These types of studies demonstrate that some types of CP genes can give resistance to many strains and even different viruses, while others provide a more narrowly based resistance.

2.3.4. Levels of CP versus Levels of CP-MR

The first published reports of CP-MR demonstrated that plant lines with higher levels of CP exhibited higher levels of resistance. In some cases, no attempt was made to correlate resistance and levels of gene expression. In still other cases, however, there was a clear lack of correlation, and the levels of CP accumulation were not a guide to selecting virus resistant plant lines: In these cases resistance could be determined only by challenging CP(+) plants by virus inoculation. Table 5.1 provides a summary of the results of the attempts to find correlations between CP levels and CP-MR.

Perhaps the most puzzling cases of CP-MR are those in which the levels of resistance do not correlate with the levels of (extractable) CP. In these cases, it is particularly important to demonstrate that resistance is due to the accumulation of CP rather than CP messenger RNA (mRNA): This resistance is generally accomplished by introducing, as a control in CP-MR studies, a gene that produces a transcript that is incapable of translation.

Several groups have reported a lack of correlation between CP-MR and levels of potyvirus CP. Stark and Beachy (1989) first reported the results for the expression of the SMV CP gene in tobacco to develop resistance against TEV and PVY. Since this is considered a case of heterologous resistance, and since soybean mosaic virus (SMV) does not infect tobacco plants and is not closely related in amino acid sequence to either TEV or PVY, this example may have been unique. However, similar results were reported for CP-MR against PVY in potato lines that express the PVY CP gene (Lawson et al., 1990) and in tobacco plants that express the PRV CP gene (Quemada et al., 1991) and are resistant to TEV, PeMV, and PVY. It is evident that lack of correlation between CP levels and levels of resistance is not unique to heterologous versus homologous CP-MR with respect to the challenge virus.

To date, there is no clear data that addresses this lack of correlation. Several authors have suggested that it reflects an example in which a phenotype is governed by "when and where" the gene is expressed rather than simply "how much" CP accumulates. In such an hypothesis the precise pattern of expression of the CP gene is the critical factor that determines resistance. In each of the examples of CP-MR against potyvirus infection, the P35S promoter of CaMV was used. Although this promoter is nominally constitutive, that is, expressed in all or most cell types, the pattern of expression can be different in different plants.

In transgenic plants that contain the *gus*A reporter gene under the control of the P35S promoter, the gene was expressed to different degrees in different cell types in different plant lines (Robeff P, Lessard P, and Beachy RN, unpublished). For example, in some plant lines the *gus*A gene was expressed to high levels in epidermal cells, while in other lines there was little or no expression in epidermal cells. Such differences in gene expression have been attributed to "position" effects, that is, the chromosomal location of the target gene impacts the expression of the gene. Whether or not position affects control the efficacy of the CP to limit virus infection and disease in CP-MR in the case of the potyviruses (and others) remains to be determined. It is clear, however, that if this variable could be eliminated, it would increase the likelihood that CP-MR could be achieved following development of a few, rather than many, transgenic plant lines.

3. EFFICACY OF COAT PROTEIN-MEDIATED RESISTANCE UNDER FIELD CONDITIONS

Following the successful field testing of CP-MR against TMV in tomato plants (Nelson et al., 1988) additional testing of a number of plant lines has occurred.

In the United States all small-scale field tests for CP-MR follow the guidelines and protocols set forth by the U.S. Department of Agriculture and the Environmental Protection Agency. Other countries have developed or are currently developing guidelines to make it possible to carry out similar tests. There are unpublished reports of tests of CP-MR in China and Mexico, and it is anticipated that a number of other countries will grant approvals for limited field trials 1992 and 1993.

The first reported results of field tests of CP-MR against TMV in tomato plants (Nelson et al., 1988) was successful, and demonstrated that CP(+) tomato lines were nearly free of infection following intentional (primary) and inadvertent (secondary) infection by TMV. Crop yields in the CP(+) transgenic plants were not different from noninfected controls, while TMV infection reduced yields by 25–30% in control [i.e., CP(−)] plants. This study also demonstrated that expression of the CP gene had no effect on the agronomic characteristics of the transgenic tomato plants tested, while it provided protection against infection and disease development.

Since that first test in 1987, a number of other field trials have been done, most of which have yet to be reported: Others will not be reported until multiyear and multilocale tests are completed. Those reports published to date have confirmed that CP-MR is fully or partially effective under field situations.

Researchers at Monsanto Co. developed transgenic Russet-Burbank potato plants that express genes encoding the CPs of PVX and PVY (Lawson et al., 1990). Although a number of transgenic lines expressed both genes and were resistant to PVX, a very few were resistant to both viruses (Lawson et al., 1990). When these plant lines were vegetatively multiplied and tested under field conditions, several lines were found to be resistant to PVX. However, fewer lines were resistant to both PVX and PVY (Kaniewski et al., 1990). Lines with resistance to both viruses were highly productive and gave high tuber yields.

Researchers at Asgrow Seed Company (Upjohn), in conjunction with Dr. D. Gonsalves (Cornell University, and the NY State Experiment Station, Geneva, NY), tested R_1 transgenic cucumber plants that express the CP gene of cucumber mosaic virus. The source of inoculum was provided by planting CMV-infected plants; virus was spread by the indigenous aphid population. In this study 70% of the CP(−) control plants became infected, compared to about 24% of the R_1 plant population and 10% of the CMV-tolerant cultivar Marketmore (Slightom et al., 1990). In a 1990 field trial of the same plant lines, homozygous R_2, CP(+) plants were compared with Marketmore in terms of percentage infected plants, and plants that developed disease symptoms. From 0 to 5% of the plants of the most resistant CP(+) plant lines showed disease symptoms, compared with 1% for Marketmore, and 98% for the CP(−) controls. Furthermore, the number of plants that became infected with CMV [as determined by enzyme-linked immunosorbent assay (ELISA)] was not significantly different between the Marketmore variety and the

CP(+) plant lines (Gonsalves et al., 1991). These well-executed studies demonstrated that CP-MR against CMV in cucumbers provides a type of resistance to infection that is as effective as the tolerant commercial cultivar Marketmore. Additional studies of this type in different crops will ultimately determine how widely effective CP-MR will be to reduce disease losses due to virus infection. Because CP-MR is effective against virus infection via insect transmission, as well as mechanical transmission, there may be the added benefit of reducing the use of insecticides to control the vector.

4. SUMMARY

Coat protein-mediated resistance has been applied to develop resistance to virus infections in a number of crop plants. Small-scale field tests demonstrated that CP-MR is effective in controlling both mechanically transmitted as well as insect-vectored viruses, and, if future tests are equally successful, genetically modified plant materials could be ready for release by the mid-1990s. Meanwhile, research is ongoing to determine the cellular and molecular mechanisms that confer CP-MR: A better understanding may lead to second-generation CP genes that improve resistance beyond current levels, and extend its applications to still more viruses and crop plants.

REFERENCES

Angenent GC, van den Ouweland JMW, Bol JF (1990): Susceptibility to virus infection of transgenic tobacco plants expressing structural and nonstructural genes of tobacco rattle virus. Virology 175:191–198.

Beachy RN, Loesch-Fries, Tumer NE (1990): Coat protein-mediated resistance against virus infection. Annu Rev Phytopathol 28:451–474.

Clark LOG, Register III SC, Nejidat A, Eichholtz DA, Sanders PR, Fraley RT, Beachy RN (1990): Tissue-specific expression of the TMV coat protein in transgenic tobacco plants affects the level of coat protein-mediated virus protection. Virology 179:640–647.

Cuozzo M, O'Connell KM (1990): Kaniewski W, Fang RX, Chua N-H, Tumer NE (1988): Viral protection in transgenic tobacco plants expressing the cucumber mosaic virus coat protein or its antisense RNA. Bio/Technol 6:549–557.

Dawson WO, Burbrick P, Grantham GL (1988): Modifications of the tobacco mosaic virus coat protein gene affecting replication, movement, and symptomatology. Phytopathology 78:783–789.

Day AG, Bejarano ER, Buck KW, Burrell M, Lichtenstein CP (1991): Expression of an antisense viral gene in transgenic tobacco confers resistance to the DNA virus tomato golden mosaic virus. Proc Natl Acad Sci USA 88:6721–6725.

Gerlach WC, Llewellyn D, Haseloff J (1987): Construction of a plant disease resistance gene from the satellite RNA of tobacco ringspot virus. Nature (London) 328:802–805.

Gielen JJL, de Haan P, Kool AJ, Peters D, van Grinsven MQJM, Goldbach RW (1991): Engineered resistance to tomato spotted wilt virus, a negative-strand RNA virus. Bio/Technol 9: 1363–1367.

Golemboski DB, Lomonossoff GP, Zaitlin M (1990): Plants transformed with a tobacco mosaic virus nonstructural gene sequence are resistant to the virus. Proc Natl Acad Sci USA 87:6311–6315.

Gonsalves D, Chee P, Slightom JL, Provedentia R (1991): Field evaluation of transgenic cucumber plants expressing the coat protein gene of cucumber mosaic virus. Phytopathology (Suppl abstract) p 296.

Harrison BD, Mayo MA, Baulcombe DC (1987): Virus resistance in transgenic plants that express cucumber mosaic virus satellite RNA. Nature (London): 328:799–802.

Hayakawa T, Shimamoto K (1991): Genetically engineered rice resistant to rice stripe virus, an insect transmitted virus. Third International Congress of Plant Mol Biol., Tucson, AZ.

Hemenway C (1991): Presented at 3rd International Congress of Plant Molecular Biology, Tucson, AZ, October 1992.

Hemenway C, Fang RX, Kaniewski WK, Chua N-H, Tumer NE (1988): Analysis of the mechanism of protection in transgenic plants expressing the potato virus X coat protein or its antisense RNA. EMBO J 7:1273–1280.

Hemenway C, O'Connell K, Kaniewsi W, Tumer N, Fang R-X, Cuozzo M, Chua N-H (1987): Expression of coat protein genes in transgenic plants confers protection against AlMV, CMV and PVX. VII Intnl Congress of Virology, Edmonton, Canada.

Hoekema A, Huisman MJ, Molendijk L, Van Den Elzen PJM, Cornelissen BJC (1989): The genetic engineering of two commercial potato cultivars for resistance to potato virus X. Bio/Technol 7:273–278.

Holt CA, Beachy RN (1991): *In vivo* complementation of infectious transcripts from mutant tobacco mosaic virus cDNAs in transgenic plants. Virology 181:109–117.

Jones M, Gough K, Dasgupta I, Subba Rao BL, Cliffe J, Qu R, Shen P, Kaniewska MB, Davies JB, Beachy RN, Hull R (1991): Rice tungro disease is caused by an RNA and a DNA virus. J Gen Virol 72:757–761.

Kallerhoff J, Perez P, Bouzoubaa S, Ben Tahar S, Perret J (1990): Beet necrotic yellow vein virus coat protein-mediated protection in sugarbeet (*Beta vulgaris* L.) protoplasts. Plant Cell Rep 9:224–228.

Kaniewski W, Lawson C, Sammons B, Haley L, Hart J, Delannay X, Tumer NE (1990): Field resistance of transgenic Russet Burbank potato to effects of infection by potato virus X and potato virus Y. Bio/Technol 8:750–754.

Kawchuk LM, Martin RR, McPherson J (1990): Resistance in transgenic potato expressing the potato leafroll virus coat protein gene. Mol Plant Microbe Interact 3:301–307.

Kawchuk LM, Martin RR, McPherson J (1991): Sense and antisense RNA-mediated resistance to potato leafroll virus in Russet Burbank potato plants. Mol Plant Microbe Interact 4:247–253.

Lawson C, Kaniewski W, Haley L, Rozman R, Newell C, Sanders P, Tumer NE (1990): Engineering resistance to mixed virus infection in a commercial potato cultivar: resistance to potato virus X and potato virus Y in transgenic Russet Burbank. Bio/Technol 8:127–134.

Ling K, Namba S, Gonsalves C, Slightom JL, Gonsalves D (1991): Protection against detrimental effects of potyvirus infection in transgenic tobacco plants expressing the papaya ringspot virus coat protein gene. Bio/Technol 9:752–758.

Loesch-Fries LS, Merlo D, Zinnen T, Burhop L, Hill K, Krahn K, Jarvis N, Nelson S, Halk E (1987): Expression of alfalfa mosaic virus RNA 4 in transgenic plant confers virus resistance. EMBO J 6:1845–1851.

MacKenzie DJ, Ellis PJ (1992): Resistance to tomato spotted wilt virus infection in transgenic tobacco expressing the viral nucleocapsid gene. Mol Plant Microbe Interact 5:34–40.

MacKenzie DJ, Tremaine JH (1990): Transgenic *Nicotiana debneyii* expressing viral coat protein are resistant to potato virus S infection. J Gen Virol 71:2167–2170.

MacKenzie DJ, Tremaine JH, McPherson J (1991): Genetically engineered resistance to potato virus S in potato cultivar Russet Burbank. Mol Plant Microbe Interact 4:95–102.

Namba S, Ling K, Gonsalves C, Gonsalves D, Slightom JL (1991): Expression of the gene encoding the coat protein of cucumber mosaic virus (CMV) strain-WL appears to provide protection to tobacco plants against infection by several different CMV strains. Gene 107:181–188.

Nejidat A, Beachy RN (1990): Transgenic tobacco plants expressing a coat protein gene of tobacco mosaic virus are resistant to some other tobamoviruses. Mol Plant Microbe Interact 3:247–251.

Nelson RS, McCormick SM, Delannay X, Dubé P, Layton J, Anderson EJ, Kaniewska M, Proksch RK, Horsch RB, Rogers SG, Fraley RT, Beachy RN (1988): Virus tolerance, plant growth, and field performance of transgenic tomato plants expressing coat protein from tobacco mosaic virus. Bio/Technol 6:403–409.

Nelson RS, Powell-Abel P, Beachy RN (1987): Lesions and virus accumulation in inoculated transgenic tobacco plants expressing the coat protein gene of tobacco mosaic virus. Virology 158: 126–132.

Okuno T, Nakayama M, Yoshida S, Furusawa I, Komiya T (1992): Susceptibility to infection with virus and its RNA of transgenic tobacco plants and protoplasts expressing the coat protein gene of cucumber mosaic virus. Virology (submitted).

Powell-Abel P, Nelson RS, De B, Hoffmann N, Rogers SG, Fraley RT, Beachy RN (1986): Delay of disease development in transgenic plants that express the tobacco mosaic virus coat protein gene. Science 232:738–743.

Powell PA, Stark DM, Sanders PR, Beachy RN (1989): Protection against tobacco mosaic virus in transgenic plants that express tobacco mosaic virus antisense RNA. Proc Natl Acad Sci USA 86:6949–6952.

Quemada HD, Gonsalves D, Slightom JL (1991): Expression of coat protein from cucumber mosaic virus strain C in tobacco: protection against infection by CMV strains transmitted mechanically or by aphids. Phytopathology 81:794–802.

Register III JC, Beachy RN (1988): Resistance to TMV in transgenic plants results from interference with an early event in infection. Virology 166:524–532.

Register III JC, Powell PA, Beachy RN (1989): Genetic engineering of plants for viral disease resistance. In Staskowicz B, Ahlquist P, Yoder OC (eds): Molecular Biology of Plant–Pathogen Interactions. New York: Alan R Liss, pp 269–281.

Regner F, da Camara Machado A, da Camara Machado ML, Steinkellner H, Mattanovich D, Hanzer V, Weiss H, Katinger H (1992): Coat protein mediated resistance to plum pox virus in *Nicotiana clevelandii* and *benthamiana*. Plant Cell Rep 11:30–33.

Sanders PR, Kaniewski WK, Haley L, LaVallee B, Delannay X, Tumer N (1992): Field trials of transgenic tomatoes expressing the tobacco mosaic or tomato mosaic virus coat proteins. Phytopathology 82:683–690.

Sanders PR, Winter JA, Barnason AR, Rogers SG, Fraley RT (1987): Comparison of cauliflower mosaic virus 35S and nopaline synthase promoters in transgenic plants. Nucleic Acids Res 15:1543–1558.

Stanley J, Freschmuth T, Ellwood G (1990): Defective viral DNA ameliorates symptoms of geminivirus infection in transgenic plants. Proc Natl Acad Sci USA 87:6291–6295.

Stark DM, Beachy RN (1989): Protection against potyvirus infection in transgenic plants: evidence for broad spectrum resistance. Bio/Technol 7:1257–1262.

Tumer NE, Kaniewski W, Haley L, Gehrke L, Lodge JK, Sanders P (1991): The second amino acid of alfalfa mosaic virus coat protein is critical for coat protein-mediated protection. Proc Natl Acad Sci USA 88:2331–2335.

Tumer NE, O'Connell KM, Nelson RS, Sanders PR, Beachy RN, Fraley RT, Shah DM (1987): Expression of the alfalfa mosaic virus coat protein gene confers cross protection in transgenic tobacco and tomato plants. EMBO J 6:1181–1188.

van Dun CMP, Bol JF (1988) Transgenic tobacco plants accumulating tobacco rattle virus and pea early browning virus. Virology 167:649–652.

van Dun CMP, Bol JF, Van Vloten-Doting L (1987): Expression of alfalfa mosaic virus and tobacco rattle virus coat protein genes in transgenic tobacco plants. Virology 159:299–305.

van Dun CMP, Overduin B, Van Vloten-Doting L, Bol JF (1988): Transgenic tobacco expressing tobacco streak virus or mutated alfalfa mosaic virus coat protein does not cross-protect against alfalfa mosaic virus infection. Virology 164:383–389.

THE LOCAL LESION RESPONSE TO VIRUSES: POSSIBILITIES FOR ENGINEERING RESISTANT PLANTS

G. LOEBENSTEIN
A. GERA

Department of Virology, Agricultural Research Organization, The Volcani Center, Bet Dagan, 50250, Israel

1. INTRODUCTION

Viruses are intercellular pathogens closely integrated within the host's cellular machinery. Chemicals that affect the virus are also harmful to the plant. Therefore, as no chemical pesticides against viruses are available, the best way to overcome virus diseases is to breed resistant or tolerant plants. Major achievements in breeding virus resistant plants have been made, as, for example, cucumber varieties resistant to cucumber mosaic virus (CMV). Here resistance depends on three genes (Kooistra, 1969). However, traditional breeding procedures allow movement of resistance sources only between closely related species, and geneticists lack sufficient sources for their breeding programs. The use of DNA technology to manipulate and move resistance sources against plant pests has been widely advocated (Chet, 1987), but with little practical work in this direction. This research would be in addition to current approaches of inserting viral genes [coat protein (Nelson et al., 1990), satellite RNA (Gerlach et al., 1987), and antisense strategies (Hemenway et al., 1988)] to obtain resistant plants. The major obstacles to

Biotechnology in Plant Disease Control, pages 105–113
© *1993 Wiley-Liss, Inc.*

overcome in using natural resistance genes in DNA technology is choosing the right gene on one hand, and then its isolation on the other hand.

2. NATURAL RESISTANCES

Various resistances against plant viruses are known. We may distinguish between passive preformed defences prepared in advance of infections, and active defences that generally develop after infection. The former type may include examples such as mechanical barriers or lack of metabolites required for virus multiplication (Loebenstein et al., 1984). Another likely candidate for preformed resistance could be the inability of host- and virus-coded subunits to form a functional replicase.

Active host resistance generally develops after infection and probably requires induction of transcription of a gene or genes in the host. This development may lead to structural changes forming physical barriers to the spread of the virus, or to functional changes that interfere with viral replication or translocation (Loebenstein and Stein, 1985). Active resistance mechanisms are known in systemic infections, as in cucumbers resistant to CMV (Nachman et al., 1971; Levy et al., 1974) and in local lesion-responding plants (Loebenstein et al., 1984). Localization in most cases depends on one dominant gene (Fraser, 1987).

The local lesion response is probably the most notable active resistance phenomenon, whereby after inoculation the virus invades and multiplies in several hundred cells but does not spread to other tissues. The following types of localized infections are known (Loebenstein et al., 1982): necrotic lesions of the nonlimited type, self-limiting necrotic local lesions, chlorotic local lesions, ring-like patterns or ringspots, starch lesions, microlesions, and subliminal infections.

Resistance is also induced in uninoculated parts of a hypersensitive host by inoculating other parts of the plant with necrotic local lesion-producing viruses, fungi (Loebenstein, 1972; Kuc, 1983), and bacteria (Jenns et al., 1979). For example, inoculation of the lower leaves of tobacco plants containing the N gene for tobacco mosaic virus (TMV) resistance with TMV or with tobacco necrosis virus (TNV), induced resistance in the upper uninoculated leaves; inoculation of a primary leaf of pinto beans with TMV or TNV induced resistance in the opposite leaf. Lesions developing after challenge inoculation of the resistant tissue were consistently smaller, and usually fewer in number, than those formed on previously uninoculated control plants. In tobacco NN plants, induced resistance, as measured by reduction in number and size of lesions, was found to be closely correlated with virus concentration as determined by enzyme-linked immunosorbent assay (ELISA) (Stein et al., 1985). This indicates that virus replication, and not only the development of necrotic local lesions, was suppressed in the "resistant" tissues. However, there have been other reports indicating that virus multiplication is not suppressed in leaves with induced resistance (Fraser, 1979; Coutts and Wagih, 1983).

3. HYPOTHESES EXPLAINING LOCALIZATION

Several hypotheses have been raised to explain localization. Details have been discussed previously (Loebenstein et al., 1982; Loebenstein et al., 1984; Loebenstein and Stein, 1985), therefore only some major points are presented here. It has been suggested that the death of cells may localize or inactivate the virus. This explanation is not satisfactory even in necrotic local lesion hosts, and even less so in a chlorotic or starch-lesion host, as virus particles are found in apparently viable cells outside the necrotic area (Milne, 1966).

Several barrier substances such as lignin and callose, in cells surrounding the local lesion, have been implicated as possible factors preventing virus movement. However, callose depositions were also observed in infections that produce systemic necrosis, where the virus does not remain localized, as with tobacco infected with tomato spotted wilt and tobacco rattle virus (Shimomura and Dijkstra, 1975). On the other hand, no callose deposition was observed around TV-induced starch lesions on cucumber cotyledons, where the infection remains localized without necrosis (Cohen and Loebenstein, 1975). It seems, therefore, that barrier substances could be a response to necrotization and not necessarily responsible for localization.

Another possible mechanism of localization is the inactivation of a transport protein. Virus spread is thought to require the help of a virus-coded protein. For TMV, this factor has been identified as a 30-kD nonstructural protein (Leonard and Zaitlin, 1982). Using immunogold cytochemistry, it was shown that the 30-kD protein accumulated in plasmodesmata, an observation consistent with a role in the spread of virus (Tomenius et al., 1987). Similarly, immunogold labeling has shown the association of a nonstructural protein (P3) of alfalfa mosaic virus, considered to be involved in cell-to-cell spread, with the middle lamella of cell walls (Stussi-Garaud et al., 1987). Plasmodesmata have been identified as subcellular locations of the P1 protein product of cauliflower mosaic virus (CaMV) gene I, suggesting its involvement in the systemic spread of CaMV (Linstead et al., 1988). Recently, it was reported that the amount of the 30-kD protein in the cell wall fraction of TMV-infected Samsun NN tobacco plants decreased sharply as soon as necrosis was visible, compared with that in the systemic host Samsun nn. It was suggested that this may explain why TMV infection becomes localized (Moser et al., 1988). Whether the decrease of transport protein is responsible for localization or is a result of the necrosis remains to be seen, especially as virus particles are found outside the necrotic area (Milne, 1966). In this respect, it may be of interest to study the kinetics of the transport protein after transfer of plants with necrotic lesions to higher temperatures, where localization fails, as well as in chlorotic- or starch-lesion infections.

Finally, localization might be due to reduced virus multiplication caused by a substance or substances that inhibit virus multiplication. Indications of a host-mediated process that suppresses virus multiplication in a starch-lesion host were obtained from studies using antimetabolites (Loebenstein et al., 1969; Sela et al., 1969) and ultraviolet (UV) irradiation (Loebenstein et al., 1970). Virus concentration in cucumber cotyledons increased markedly when cotyledons were treated with actinomycin D, chloramphenicol, or UV 1 day after inoculation with TMV. The antimetabolites also stimulated the biosynthesis of TNV in *Chenopodium amaranticolor* (Faccioli and Rubies-Autonell, 1976). It was suggested that the localizing mechanism operates mainly by inhibition of virus multiplication (Loebenstein et al., 1984).

In summary, the local lesion response is probably the most notable active resistance phenomenon. It generally depends on one single dominant gene and may therefore be most suitable for isolation and transformation of plants.

4. STRATEGIES FOR ISOLATING GENES

The following strategies can be considered for the isolation of natural resistance genes.

1. *Gene tagging.* The recent isolation and characterization of plant transposable elements offer the possibility to mutagenize resistance genes. In cases where the resistance gene is dominant, the transposable element changes the phenotype from resistant to susceptible. Parallel to insertion, the resistance gene also gets ear-marked, and can thus be cloned molecularly, using the transposable element as a probe. This type of approach should allow the isolation of resistance genes in cases where resistance is conferred by a single dominant gene. Transposon tagging could be used to isolate the "N" gene from resistant tobaccos. Insertion of a transposon in this case should change the phenotypic expression from a local necrotic lesion response to TMV to a systemic infection. Another approach to tag genes by insertional mutagenesis is by using *Agrobacterium tumefaciens* T-DNA.

2. *Differential screening and substraction hybridization.* The isolation of a "resistance" gene might be achieved by obtaining messenger ribonucleic acid (mRNA) from the resistant line, which is then reverse transcribed into complimentary DNA (cDNA) using labeled nucleotides. This cDNA is then hybridized with a huge excess of mRNA from the susceptible, otherwise isogenic, line. Most of the mRNA sequences in the susceptible and resistant lines, are common, so most of the *resistant* cDNA will hybridize with the mRNA from the *susceptible*

line. Specific *resistant* cDNA, however, will remain single stranded, and can be separated from the cDNA–mRNA hybrids by chromatography on hydroxyapatite, which binds double-stranded molecules more avidly than single-stranded molecules. The labeled *resistant* cDNA can then be used to screen a *resistant* cDNA library to obtain a cloned copy of the *resistant* cDNA. By substraction hybridization it is possible to increase the concentration of the desired (resistance-associated) sequences (Sargent, 1987). Thus, mRNA from resistant plants is used as a template to produce radiolabeled cDNA. This cDNA is hybridized to an excess of mRNA isolated from the susceptible plant. The unhybridized cDNA from the resistant plant represents an enriched population of sequences expressed in the resistant tissue.

3. *Via the gene product.* Probably the best option to isolate a gene is by its product, providing that biochemical information on the product is available. In that case an oligonucleotide probe can be synthesized, based on the amino acid sequence of the protein product. If the N-terminal of the protein is not blocked, a sequence of 6–10 amino acid residues is sufficient for preparing a probe with at least 16–20 nucleotides. In case the N-terminal is blocked, partial digestion of the protein is necessary to obtain a polypeptide stretch. Another possibility is to prepare poly- or monoclonal antibodies as a probe. These probes are used to screen a recombinant DNA library, obtained by reverse transcription of poly(A) RNA from resistant plants. Recombinant DNA packaged, for example, in lambda gt 10 phage, plated on an appropriate *Escherichia coli* host, can be screened with the nucleic acid probe. When antibody probes are available for screening, lambda gt 11 is used, since it is an expression vector (Jendrisak et al., 1987). If a product of the gene responsible for localization could be identified, this would be the easiest way to isolate this gene. We suggest that the inhibitor of virus replication (IVR) might be this product.

5. INHIBITOR OF VIRUS REPLICATION

We have reported that a substance(s) IVR is released into the medium from TMV-infected protoplasts of Samsun NN, a cultivar in which infection in the intact plant is localized (Loebenstein and Gera, 1981). The IVR inhibited virus replication in protoplasts from both local-lesion-responding resistant Samsun NN and systemic responding susceptible Samsun plants, when applied up to 18 h after inoculation (virus being assayed by a local lesion assay, ELISA or a cDNA probe for TMV). This result indicates that IVR acts as an inhibitor of replication and not as an inhibitor of infection, or at a later stage, translocation. The IVR was also produced by *Nicotiana glutinosa* and Xanthi-nc protoplasts. It was not produced in protoplasts from susceptible plants or from noninoculated protoplasts of the resistant cultivar. The IVR was found to be sensitive to trypsin and chymotrypsin but not to RNAse, suggesting that it has a proteinaceous nature (Gera and Loebenstein, 1983). A specific protein band of an approximate molecular weight of 23 KD was regularly observed in polyacrylamide gel electrophoresis (PAGE) of crude preparations of IVR (Spiegel et al., 1989; Gera et al., 1990). Electroeluting protein from this band recovered a biologically active fraction with a 25-fold increase in specific activity over that of the crude preparation. This fraction revealed only one band at 23 KD in PAGE, providing evidence that the 23 KD band is IVR purified to a high degree. Polyclonal (Gera and Loebenstein, 1989) and monoclonal (Gera et al., 1989) antibodies were prepared, which neutralized IVRs antiviral activity and enabled its detection in immunoblots. The 23 KD band gave a specific reaction in immunoblots with IVR antiserum (Gera et al., 1990). In addition, it was found that IVR inhibited TMV replication in Samsun and tomato leaf tissue disks and inhibited replication of CMV, potato virus X (PVX) (Gera and Loebenstein, 1983), and potato virus Y (PVY) (Loebenstein and Gera, 1988) in different host tissues, thereby indicating that it is neither virus nor host specific. Inhibition rates by IVR in both protoplasts and leaf tissue disks were dose responsive (Gera et al., 1986). The IVR was also obtained from the intercellular fluid of Samsun NN tobacco infected with TMV and from induced resistant tissue (Spiegel et al., 1989).

Recently, we found that IVR is produced constitutively (without infection) in the amphidiploid of *N. glutinosa* × *N. debneyi* (Loebenstein et al., 1990). This hybrid is highly resistant to infection with TMV (Ahl and Gianinazzi, 1982). Preliminary results indicate that this hybrid is resistant to several viruses, which infect systemically other *Nicotiana* species (Loebenstein, Gera, Tam, unpublished).

Inhibitor of virus replication has a certain resemblance to a class of proteins termed b or "pathogenesis related" [PR(b) proteins]. Induction of PR(b) proteins in most *Nicotiana* species requires infection by either viruses, fungi, bacteria, or other external stimuli (Gianinazzi, 1984; van Loon, 1985). The characterized PR(b) proteins, however, occur in the plant in relatively large quantities, while only ngs of IVR can be obtained from 1 g of leaf tissue, Furthermore, IVR is sensitive to proteolytic enzymes (Gera and Loebenstein, 1983) in contrast to the stability of most of the characterized PR(b) proteins. The major difference between IVR and the PR(b) proteins is that no biological antiviral activity has been reported for the latter. As tested by Dr. S. Gianinazzi (INRA, Dijon, France, personal communication) no serological reaction was obtained between PR(b) proteins of Xanthi n.c. and IVR antiserum.

Recently, the N-terminal sequence for IVR was determined (in cooperation with Professor Y. Burstein, Weizmann Institute) and no homology was observed with the N-terminal sequences of all the PR-proteins that have been sequenced. It seems, therefore, that IVR is not related to those PR proteins that have been well characterized. Recently, it was reported that the constitutive expression of several PR-proteins in transformed tobacco had no effect on virus infection (Linthurst et al., 1989).

Poly(A)$^+$ RNA obtained from induced-resistant tissue of Samsun NN, where IVR is present, was translated in a rabbit reticolocyte cell-free system. Immuno-precipitation of the translation products with IVR antiserum yielded a specific product with an approximate molecular weight of 20 KD, similar to that of IVR. This product was not detectable in healthy control tissue of Samsun NN, or in inoculated susceptible tissue of Samsun (Aly R, unpublished).

The recovery of IVR from protoplasts and intact tissue of resistant cultivars, its purification, production of poly- and monoclonal antisera, and the presence of an IVR-like compound in the translation products of poly(A)$^+$ RNA from resistant tissue, are the base for isolation and cloning the gene coding for IVR.

Should plants transformed with the IVR gene become resistant to a spectrum of viruses (IVR inhibits replication of several viruses), this gene would broaden markedly the battery of resistance sources presently available to plant breeders.

ACKNOWLEDGMENTS

Contribution 3312-E, 1991 Series from the Agricultural Research Organization, The Volcani Center, Bet Dagan, 50250, Israel.

This work was supported in part by the United States—Israel Agricultural Research and Development Fund (BARD).

REFERENCES

Ahl P, Gianinazzi S (1982): b-Proteins as a constitutive component in highly (TMV) resistant inter-specific hybrids of *Nicotiana glutinosa* × *Nicotiana debneyi*. Plant Sci Lett 26:173–181.

Chet I (1987): Innovative Approaches to Plant Disease Control. New York: Wiley, p 372.

Cohen J, Loebenstein G (1975): An electron microscope study of starch lesions in cucumber cotyle-dons infected with tobacco mosaic virus. Phytopathology 65:32–39.

Coutts RHA, Wagih EE (1983): Induced resistance to viral infection and soluble protein alterations in cucumber and cowpea plants. Phytopathol Z 107:57–69.

Faccioli G, Rubies-Autonell C (1976): Action of antimetabolites on the biosynthesis of tobacco necro-sis virus in locally-infected cells of *Chenopodium amaranticolor*. Phytopathol Z 87:48–56.

Fraser RSS (1979): Systemic consequences of the local lesion reaction to tobacco mosaic virus in a tobacco variety lacking the N gene for hypersensitivity. Physiol Plant Pathol 14:383–394.

Fraser RSS (1987): Genetics of plant resistance to viruses. In Evered D, Harnett S (eds): Plant Resis-tance to Viruses. Ciba Foundation Symposium 133. Chichester: Wiley, pp 6–15.

Gera A, Loebenstein G (1983): Further studies of an inhibitor of virus replication from tobacco mosaic virus-infected protoplasts of a local lesion-responding tobacco cultivar. Phytopathology 73:111–115.

Gera A, Loebenstein G (1989): Evaluation of antisera to an inhibitor of virus replication. J Phytopathol 124:366–371.

Gera A, Loebenstein G, Salomon R, Franck A (1990): An inhibitor of virus replication (IVR) from protoplasts of a hypersensitive tobacco cultivar infected with tobacco mosaic virus is associated with a 23K protein species. Phytopathology 80:78–81.

Gera A, Loebenstein G, Shabtai S (1983): Enhanced tobacco mosaic virus production and suppressed synthesis of a virus inhibitor in protoplasts exposed to antibiotics. Virology 127:475–478.

Gera A, Sadka A, Spiegel S, Salomon R, Smorodinsky NI (1989): Use of monoclonal antibodies in the purification of an inhibitor of virus replication by affinity chromatography. J Gen Virol 70:1293–1296.

Gera A, Spiegel S, Loebenstein G (1986): Production, preparation and assay of an antiviral substance from plant cells. Methods Enzymol 119:729–733.

Gerlach WL, Llewellyn D, Haseloff J (1987): Construction of a plant disease resistance gene form the satellite RNA of tobacco ringspot virus. Nature (London) 328:802–805.

Gianinazzi S (1984): Genetic and molecular aspects of resistance induced by infections or chemicals. In Kosuge T, Nester EW (eds): Plant–Microbe Interactions. Molecular and Genetic Perspectives, Vol 1. New York: Macmillan, pp 321–342.

Hemenway C, Fang R, Kaniewski WK, Chua N, Tumer NE (1988): Analysis of the protein or its antisense RNA. EMBO J 7:1273–1280.

Jendrisak J, Young RA, Engel JD (1987): Cloning cDNA into λgt 10 and λgt 11. Methods Enzymol 152:359–371.

Jenns AE, Caruso FL, Kuc J (1979): Non-specific resistance to pathogens induced systemically by local infection of cucumber with tobacco necrosis virus, *Colletotrichum lagenarium* or *Pseudomonas lachrymans*. Phytopathol Mediterr 18:129–134.

Kooistra E (1969): The inheritance of resistance to Cucumis virus I in cucumber (*Cucumis sativus* L.) Euphytica 18:326–332.

Kuc J (1983): Induced systemic resistance in plants to diseases caused by fungi and bacteria. In Bailey JA, Deverall BJ (eds): Dynamics of Host Defense. Sydney: Academic Press, p 191.

Leonard DA, Zaitlin M (1982): A temperature sensitive strain of tobacco mosaic virus defective in cell-to-cell movement generates an altered viral-coded protein. Virology 117:416–424.

Levy A, Loebenstein G, Smookler M, Drori T (1974): Partial suppression by UV irradiation of the mechanism of resistance to cucumber mosaic virus in a resistant cucumber cultivar. Virology 60:37–44.

Linstead PF, Hills GJ, Plaskitt KA, Wilson IG, Harker CL, Maule AJ (1988): The subcellular location of the gene 1 product of cauliflower mosaic virus is consistent with a function associated with virus spread. J Gen Virol 69:1809–1818.

Linthurst HJM, Meuwissen RLJ, Kauffmann S, Bol JF (1989): Constitutive expression of pathogenesis-related proteins PR-1, GRP, and PR-S in tobacco has no effect on virus infection. Plant Cell 1:285–291.

Loebenstein G (1972): Localization and induced resistance in virus-infected plants. Annu Rev Phytopathol 10:177–208.

Loebenstein G, Chazan R, Eisenberg M (1970): Partial suppression of the localizing mechanism to tobacco mosaic virus by UV irradiation. Virology 41:373–376.

Loebenstein G, Gera A (1981): Inhibitor of virus replication released from tobacco mosaic virus infected protoplasts of a local lesion-responding tobacco cultivar. Virology 114:132–139.

Loebenstein G, Gera A (1988): Resistance responses of plants to virus infections. Acta Hortic 234:403–411.

Loebenstein G, Gera A, Gianinazzi S (1990): Constitutive production of an inhibitor of virus replication in the interspecific hybrid of *Nicotiana glutinosa* × *Nicotiana debneyi*. Phys Molec Plant Pathol 37:145–151.

Loebenstein G, Gera A, Stein A (1984): Plant defense mechanisms of viral infections. In Kurstak E (ed): Control of Virus Diseases. New York: Marcel Dekker, pp 375–391.

Loebenstein G, Sela B, van Praagh T (1969): Increase of tobacco mosaic local lesion size and virus multiplication in hypersensitive hosts in the presence of actinomycin D. Virology 37:42–48.

Loebenstein G, Spiegel S, Gera A (1982): Localized resistance and barrier substances. In Wood RKS (ed): Active Defense Mechanisms in Plants. New York: Plenum, pp 211–230.

Loebenstein G, Stein A (1985): Plant defense responses to viral infections. In Key JL, Kosuge T (eds): Cellular and Molecular Biology of Plant Stress. New York: Alan R. Liss, pp 413–433.

Milne RG (1966): Electron microscopy of tobacco mosaic virus in leaves of *Nicotiana glutinosa*. Virology 28:527–532.

Moser O, Gagey MJ, Godefroy-Colburn T, Stussi-Garaud C, Ellwart-Tschurtz M, Nitschko H, Mundry KW (1988): The fate of the transport protein of tobacco mosaic virus in systemic and hypersensitive tobacco hosts. J Gen Virol 69:1367–1373.

Nachman I, Loebenstein G, van Praagh T, Zelcer A (1971): Increased multiplication of cucumber mosaic virus in a resistant cucumber cultivar caused by actinomycin D. Physiol Plant Pathol 1:67–72.

Nelson RS, Powell PA, Beachy RN (1990): Coat protein-mediated protection against virus infection. In Fraser RSS (ed): Recognition and Response in Plant-Virus Interactions. Berlin: Springer-Verlag, pp 427–442.

Sargent TD (1987): Isolation of differentially expressed genes. Methods Enzymol 152:423–432.

Sela B, Loebenstein G, van Praagh T (1969): Increase of tobacco mosaic virus multiplication and lesion size in hypersensitive hosts in the presence of chloramphenicol. Virology 39:260–264.

Shimomura T, Dijkstra J. (1975): The occurrence of callose during the process of local lesion formation. Neth J Plant Pathol 81:107–121.

Spiegel S, Gera A, Salomon R, Ahl P, Harlap S, Loebenstein G (1989): Recovery of an inhibitor of virus replication from the intercellular fluid of hypersensitive tobacco infected with tobacco mosaic virus and from uninfected induced-resistant tissue. Phytopathology 79:258–262.

Stein A, Spiegel S, Loebenstein G (1985): Studies on induced resistance to tobacco mosaic virus in Samsun NN tobacco and changes in ribosomal fractions. Phytopathol Z 114:295–300.

Stussi-Garaud C, Garaud JR, Berna A, Godefroy-Colburn T (1987): In situ location of an alfalfa mosaic virus non-structural protein in plant cell walls: correlation with virus transport. J Gen Virol 68:1779–1784.

Tomenius K, Clapham D, Meshi T (1987): Localization by immunogold cytochemistry of the virus-coded 30K protein in plasmodesmata of leaves infected with tobacco mosaic virus. Virology 160:363–371.

van Loon LC (1985): Pathogenesis—related proteins. Plant Molec Biol 4:111–116.

TRANSGENIC PLANTS RESISTANT TO DISEASES BY THE DETOXIFICATION OF TOXINS

KATSUYOSHI YONEYAMA

Laboratory of Plant Pathology, Faculty of Agriculture, Meiji University, 1-1-1 Higashimita, Tama-ku, Kawasaki, Kanagawa 214, Japan

HIROYUKI ANZAI

Pharmaceutical Research Center, Meiji Seika Kaisha, Ltd., Morooka-cho, Kohoku-ku, Yokohama, Kanagawa 222, Japan

1. INTRODUCTION

Plant diseases are a major cause of decreased yield of foods both agriculturally and horticulturally. These losses are an important factor when taking into consideration the stability and constant supply of staple foods for humans. A great effort to achieve disease protection (i.e., breeding of disease-resistant crops; improvement of cultivation systems, and development of chemical and biological control agents) has led to considerable progress in the increase of food production. However, disease control is still very important in agriculture. New methods of intensive agriculture include increased fertilization, use of genetically uniform and high-yielding varieties, and increased irrigation. One of the most effective and environmentally safe means of combating disease is, no doubt, the

Biotechnology in Plant Disease Control, pages 115–137
© 1993 Wiley-Liss, Inc.

use of resistant plant varieties, since the adverse side effects sometimes resulting from pesticides can be avoided.

Recent advances in genetic engineering have provided significant potential to
modify and improve agricultural plants at molecular levels. This molecular breeding has already allowed construction of transgenic plants resistant to herbicides,
insects, and plant viruses. Such a strategy could be used to achieve plant protection against bacterial, fungal, and viral diseases. In plant parasitic microorganisms, some pathogens produce phytotoxic substances, (phytotoxins or pathogenic
toxins), which play an important role in causing a disease to infect plants. The
pathogenic toxins might be one of the most attractive targets for plant genetic
engineering of disease resistance, since the resistance can be conferred to plants
by introducing a single dominant gene as demonstrated in many cases of herbicides (Comai et al., 1985; DeBlock et al., 1987; Lee et al., 1988), fungicides
(Kamakura et al., 1990), and insect toxins of *Bacillus thuringiensis* (Vaeck et al.,
1987; Fischhoff et al., 1987; Hilder et al., 1987). Thus, our attention has been
attracted to the tobacco wildfire disease caused by *Pseudomonas syringae* pv.
tabaci, which produces a phytotoxic tabtoxin. A tabtoxin resistance gene (*ttr*)
was cloned as a self-resistance gene from the wildfire pathogen itself (Anzai et
al., 1990), and introduced into tobacco plants by *Agrobacterium*-mediated transformation. The transgenic tobacco plants expressing the *ttr* gene have been demonstrated to show a high resistance to tabtoxin and to infection of wildfire disease
(Anzai et al., 1989).

In this chapter, we describe a genetic engineering approach in the design of
disease resistance transgenic plants by introducing the detoxifying enzyme gene
of pathogenic toxins. The results will be compared to those obtained with the
wildfire-resistant tobacco plants.

2. PATHOGENICITY AND TOXIGENICITY IN PLANT PATHOGENS

2.1. Pathogenicity and Disease Protection

Because of our increasing knowledge about the mechanisms of pathogenicity
on pathogens, more progress is expected to be made regarding the development
of new types of disease protectants, as well as in genetic engineering of resistant
plant varieties. The pathogenicity that determines the ability of a pathogen to cause
disease is generally defined as an integrated ability of both aggressiveness and
virulence of a pathogen to the host plant. The aggressiveness of the pathogen is
its ability to be exerted in the process of invasion, overcoming host resistance,
and becoming established in nutritional affiliation with its host. On the other hand,
virulence is its ability to develop disease symptoms by attacking plant cells using
biologically active substances, such as enzymes, toxins, and phytohormones. There-

fore, when either one of the ability of aggressiveness or virulence of a pathogen is completely interfered with by an appropriate means, it would be possible to protect the host plant from attack by the pathogen.

In the case of chemical controls, as observed by the use of nonfungicidal chemicals for rice blast (Watanabe et al., 1979; Woloshuk et al., 1980; Chida et al., 1982; Yamaguchi et al., 1983), it has been demonstrated that specific inhibitors of pathogen aggressiveness have an excellent effect on disease protection, even though the chemicals do not kill the pathogens. As a result, disease control using fungicidal chemicals has been replaced with the use of nonfungicidal protectants, which in turn provide a higher degree of safety for animals and environmental toxicity due to the high specificity to a particular pathogen. At present, disease protection based on the pathogenicity of a pathogen is becoming more important. As for the use of chemical protectants as disease controls, no specific inhibitors blocking the virulence of pathogens have been developed. This lack of success suggests a difficulty or a limitation in specifically targeting the virulence of a pathogen by chemical means.

In contrast to the use of chemical controls, genetic engineering of plants may well be a potent and precise tool as a means of specifically blocking the virulence of bacterial and fungal parasites. It is known that some pathogenic toxins are directly involved in the development of disease symptoms, either as a primary determinant or a main damaging factor of a pathogenicity. Bearing this in mind, the most practical target for disease control seems to be the pathogenic toxins that are produced by plant parasitic microorganisms. Although lytic enzymes or plant hormones play an important role as virulence factors of some pathogens, some new and innovative approaches are needed to block the activity of these factors by means of genetic engineering. In the case of pathogenic toxins, a promising technique is to degrade or inactivate the toxin. If the detoxifying-enzyme gene of the pathogenic toxin is introduced into the host plants, the genetically engineered host plants could be protected from the corresponding disease.

2.2. Role of Toxins in Pathogenicity

In most plant diseases caused by microbial attack, development of disease symptoms may result from a direct or indirect effect of the toxic metabolites produced by pathogens. The importance of the role of toxins in pathogenicity has been pointed out by Gäumann (1954), who reported the first isolation of the *Fusarium* wilt toxin, lycomarasmin (Crouson-Kaas et al., 1944). Meeham and Murphy (1947) indicated that the development of symptoms by *Helminthosporium victoriae* on oat plants was closely related to the productivity of toxin by the fungus. This work brought more attention to pathogenic toxins and probably has had a greater impact on furthering research on pathogenic toxins. To date, a number of toxins have been isolated and identified from phytopathogenic bacteria and fungi. Table 7.1 represents main phytopathogenic toxins in plant diseases.

TABLE 7.1. Main Phytopathogenic Toxins in Plant Diseases

Pathogens	Toxins	Host Plants	Action or Target Sites
Host specific toxins			
Fungi			
Alternaria alternata			
apple pathotype	AM-toxin	Apple	Membrane
Japanese pear pathotype	AK-toxin	Japanese pear	Membrane
strawberry pathotype	AF-toxin	Strawberry	Membrane
tobacco pathotype	AT-toxin	Tobacco	Mitochondoria
tomato pathotype	AL-toxin	Tomato	Mitochondoria
rough lemon pathotype	ACR-toxin	Citrus	Mitochondoria
Helminthosporium maydis	HMT-toxin	Maize	Mitochondoria
H. carbonum	HC-toxin	Maize	Membrane
H. sacchari	HS-toxin	Sugarcane	Membrane
H. victoriae	HV-toxin	Oat	Membrane
Periconia circinata	PC-toxin	Sorghum	Membrane
Phyllostica maydis	PM-toxin	Maize	Mitochondoria
Nonhost specific toxins			
Fungi			
Alternaria solani	Tentoxin	Potato	Chloroplasts
Fusarium oxysporum	Fusaric acid	Vegetables	Water permeability
Gibberella fujikuroi	Gibberellin	Rice	Plant hormone
Helminthosporium sativum	Helminthosporol	Barley	Respiration
Ophiobolus miyabeanus	Ophiobolin	Rice	Membrane
Bacteria			
Agrobacterium tumefaciens	Indolacetic acid cis-Zeatin	Dicot plants	Plant hormone
Pseudomonas syringae pv. *tabaci*	Tabtoxin	Tobacco	Glutamine synthesis
P. syringae pv. *phaseolicola*	Phaseolotoxin	Bean	Arginine synthesis
P. syringae pv. *syringae*	Syringomycin	Peach	Membrane
P. syringae pv. *atropurpurea*	Coronatine	Italian ryegrass	Membrane
Rhizobium japoni cus	Rhizobitoxine	Soybean	Methionine metabolism

The toxins produced by plant pathogens are classified into two categories: host specific and nonhost specific, on the basis of selective phytotoxicity to compatible and noncompatible hosts. A host-specific toxin selectively damages only those plant varieties that are susceptible to the pathogens, so that the potential virulence of the pathogen is dependent on the amounts of toxin produced. Also, the level of disease resistance in a plant variety is correlated with that of toxin resistance. All the host-specific toxins isolated so far are produced by filamentous fungi, especially those belonging to two genera of *Alternaria* and *Helminthosporium,* representing an AK-toxin produced by *Alternaria alternata* Japanese pear pathotype, AM-toxin by *A. alternata* apple pathotype, HMT-toxin by

Helminthosporium maydis and HV-toxin by *H. victoriae*. It is also known that the primary sites of action of their host-specific toxins are related to cell functions such as cell membrane, mitochondria, or chloroplasts.

On the other hand, nonhost-specific toxins have a toxicity not only to host plants but also to a wide variety of nonhost plants and living organisms. In this case, the toxins are not a primary determinant of pathogenicity in the pathogens, but in several cases they may play an important role as a virulence factor causing disease development or metabolic disturbance of plant cells. Such fungal toxins are represented by an ophiobolin produced by *Helminthosporium oryzae* (the causal agent of rice leaf spot disease) and a gibberellin (well known as a phytohormone) produced by *Gibberella fujikuroi*, which causes Bakanae disease in rice. All the plant bacterial toxins isolated so far fall into nonhost-specific toxins, such as a tabtoxin produced by *P. syringae* pv. *tabaci*, a phaseolotoxin by *P. syringae* pv. *phaseolicola* that causes halo blight of bean, and a coronatine by *P. syringae* pv. *atropurpurea*, the causal agent of halo blight in ryegrass. These toxins have a close relationship between the toxigenicity of pathogen and the symptom development of plants. The bacterial toxins have various actions on plant cells: for example, the phytohormones produced by *Agrobacterium tumefaciens* and *P. syringae* pv. *savastanoi* induce abnormal proliferation of plant cells, both tabtoxin and phaseolotoxin inhibit amino acid metabolism, and coronatine acts on cell membranes to destroy cellular organelles.

3. PATHOGENIC TOXINS BY WILDFIRE BACTERIA

Diseased tobacco leaves infected with wildfire bacteria show a small necrotic spot containing bacterial cells, surrounded by a chlorotic zone that is free from bacteria. The wildfire bacterium *P. syringae* pv. *tabaci* produces a pathogenic toxin called tabtoxin, as mentioned earlier. When a tobacco leaf was treated with tabtoxin, it caused chlorotic spots similar to the halos on leaves infected by the wildfire bacteria. It has been suggested that the formation of disease symptom by the attack of *P. syringae* pv. *tabaci* is directly correlated to the inhibitory effect of tabtoxin on plant cells (Braun, 1955).

The first isolation of tabtoxin was reported by Woolley et al. (1952) and an acceptable structure was published by Stewart (1971). As shown in Figure 7.1, the structure of tabtoxin is composed of tabtoxinine-β-lactam[2-amino-4-(3-hydroxy-2-oxoazacyclobutan-3-yl) butanoic acid] and threonine. Another type of tabtoxin is also produced in a small amount as a derivative containing a serine molecule in the place of threonine, called 2-serine tabtoxin (Taylor et al., 1972). Although both types of tabtoxin are synthesized in a biologically inactive form, they are readily converted to the active moiety of tabtoxinine-β-lactam by cleavage of threonine or serine with some aminopeptidases present in either the bacte-

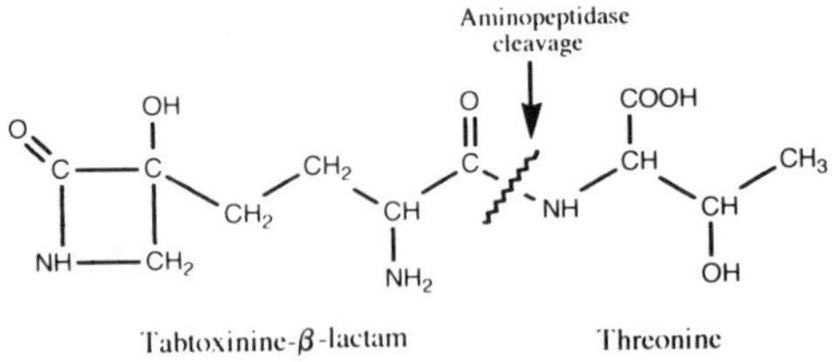

Fig. 7.1. Structure of tabtoxin.

ria or the plant. The active moiety tabtoxinine-β-lactam inhibits the target enzyme glutamine synthetase, which catalyzes the synthesis from glutamic acid to glutamine in amino acid metabolism. This inhibition results in the abnormal accumulation in tobacco cells of ammonia causing the characteristic chlorosis (Sinden and Durbin, 1968).

In addition to tobacco wildfire bacteria, some plant pathogenic bacteria, such as *P. syringae* pv. *coronafaciens,* causing halo blight of oats; *P. syringae* pv. *garcae,* causing bacterial scorch of coffee plants; and *P. syringae* sp., causing wildfire of soybean, are also known to produce tabtoxin (Mitchell, 1981). These bacteria also elicit chlorosis of leaf tissues similar to the wildfire disease of tobacco plants.

4. CLONING AND ANALYSIS OF TABTOXIN RESISTANCE GENES

4.1. Self-Resistance to Toxic Metabolites in Microorganisms

Since tabtoxin secreted by wildfire bacteria is a nonhost specific toxin, it shows a toxic activity on a wide range of living organisms from higher plants to algae, bacteria, and animals. However, it is interesting that tabtoxin has no toxic effect on the producer *P. syringae* pv. *tabaci* itself. This fact suggests that the wildfire bacterium has a particular mechanism of resistance to tabtoxin, such as inactivation of the toxin, insensitivity of the target enzyme to the toxin, or impermeability of the toxin through cell membrane, thereby protecting the bacterium from the inhibitory action of tabtoxin. This kind of self-resistance is well known in the microbial producers of physiologically active substances, especially in *Actinomycetes* of antibiotic producers. Representative examples are the acetylating enzyme of kanamycin-producing microorganism (Hotta et al., 1981), the phosphorylating enzyme of streptomycin producer (Cella and Vining, 1975), or the acetylating enzyme of bialaphos producer (Kumada et al., 1988).

The nonselective herbicide bialaphos is produced by *Streptomyces hygroscopicus* SF 1293, and its herbicidal activity is exerted by inhibiting the glutamine synthetase in a similar mechanism as an action of tabtoxin. Several intermediates in the latter processes of bialaphos biosynthesis in *S. hygroscopicus* also have an inhibitory activity on glutamine synthetase. The bialaphos producer defends itself from the toxic metabolites by converting the active intermediates to the acetylated inactive forms, then advances the biosynthetic steps, and finally secretes bialaphos as a deacetylated active product outside the cells (Imai et al., 1985; Kumada et al., 1988). In this case, the acetylating enzyme, which functions in the biosynthesis of bialaphos in *S. hygroscopicus,* is considered as a self-resistance factor protecting itself from the toxic metabolites. Murakami et al. (1986) and Thompson et al. (1987) succeeded in cloning the gene of bialaphos acetyltransferase, designated *bar* gene, from *S. hygroscopicus.* Recently, De Block et al. (1987) introduced the *bar* gene into tobacco plants, and created the transgenic tobacco showing high resistance to bialaphos.

Another mechanism of self-resistance is exemplified by the plant pathogenic bacterium *P. syringae* pv. *phaseolicola,* which produces the nonhost-specific toxin phaseolotoxin (Mitchell, 1976). In this case, the phaseolotoxin specifically inhibits the enzyme ornithine carbamoyltransferase (OCTase) to accumulate several hundredfold of ornithine in infected leaves (Turner and Mitchell, 1985). Since the biosynthesis of arginine in *P. syringae* pv. *phaseolicola* is presumed to follow through ornithine cycles in other prokaryotes (Cunin et al., 1986), inhibition of OCTase by phaseolotoxin can lead to autotoxicity. However, the bacterium does not require exogenous arginine for growth in the presence of phaseolotoxin. It suggests that a self-resistance mechanism must be functioning in the cells. Further works suggested that the bacterium possesses two OCTase activities, one sensitive and another insensitive to phaseolotoxin *in vitro* (Staskawicz et al., 1980). The insensitive form of the enzyme was found only in toxin-producing strains, though the sensitive form was present in toxin-nonproducing as well as toxin-producing strains. Therefore, the toxin-insensitive OCTase is biologically significant to the self-resistance mechanism *in vivo.* The gene encoding a phaseolotoxin-insensitive target enzyme has already been cloned from *P. syringae* pv. *phaseolicola* (Peet et al., 1986).

Because the presence of self-resistance enzymes have been found in many microorganisms that produce physiologically active secondary metabolites, it is likely that plant pathogens producing the nonhost-specific toxins, including tabtoxin, may have a certain kind of enzyme for self-resistance to the toxins they produce. Therefore, a quicker way to isolate toxin resistance genes, such as toxin-detoxifying enzyme genes, should be by screening the chromosomal or extrachromosomal deoxyribonucleic acid (DNA) from toxin-producing pathogen itself.

4.2. Cloning of Tabtoxin Resistance Genes

The previous examples on self-resistance suggest that a defense mechanism to tabtoxin might be present in the producer *P. syringae* pv. *tabaci.* Thus an attempt was made to isolate the tabtoxin resistance genes from the pathogen itself. The genomic DNA of *P. syringae* pv. *tabaci* MAFF03-01075 was isolated and digested with the restriction enzyme *Bcl*I, then cloned to a *Bam* HI site of the plasmid vector pUC13. The genomic DNA library obtained was transformed into *Escherichia coli* DH1 sensitive to tabtoxin, and the transformants were selected on the ampicillin-supplemented Luria-Bertani (LB) medium, followed by reselecting on the M9 minimum medium containing tabtoxin (Fig. 7.2). The several clones isolated were analyzed for the recombinant plasmid to identify the insert of genomic DNA. The results of restriction enzyme fragments revealed that the cloned plasmids of two different kinds of insert derived from genomic DNA were obtained; that is, one plasmid (pARK 10) contains a 2-kb insert, and the other plasmid (pARK 11) has a 11-kb insert.

4.3. Functions of Tabtoxin Resistance Genes

In tabtoxin-resistant transformants of *E. coli,* several mechanisms are possible for this resistance. These possibilities include: overproduction of glutamine syn-

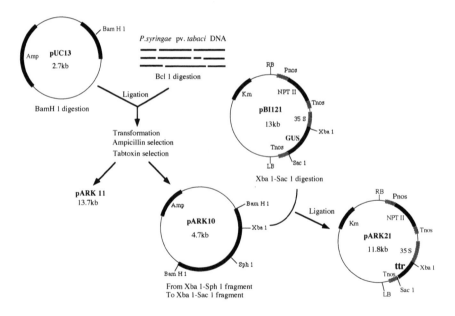

Fig. 7.2. Cloning strategy of tabtoxin resistance gene (*ttr*) and construction of the plasmid pARK21. Amp, ampicillin resistance gene; Km, kanamycin resistance gene; NPTII, neomycin phosphotransferase II gene; GUS, β-glucuronidase gene; Pnos and Tnos, promoter and terminator of nopaline synthase gene in pTiC58 respectively; 35S, cauliflower mosaic virus 35S promoter; RB (right border) and LB (left border) sequences of T-DNA in Ti plasmid, respectively.

TABLE 7.2. Characteristics of Tabtoxin-Resistant *E. coli* to Glutamine Synthetase Inhibitors

	Transformants	Tabtoxin[a]	Bialaphos[a]	MSO[b]
E. coli	none	S	S	S
	pARK10	R	S	S
	pARK11	R	R	S

[a]R = resistant and S = sensitive.
[b]Methionine sulfoximine is MSO.

thetase; production of tabtoxin-resistant glutamine synthetase; impermeability of tabtoxin through bacterial cell membrane; and the production of tabtoxin-detoxifying enzyme. To examine these possibilities, the *E. coli* tabtoxin-resistant clones including pARK 10 or pARK 11 were grown on medium containing each of the following inhibitors of glutamine synthetase: tabtoxin, bialaphos, and methionine sulfoximine. As shown in Table 7.2, the results showed that *E. coli* with pARK 10 was specifically resistant to tabtoxin only but sensitive to both bialaphos and methionine sulfoximine, whereas *E. coli* with pARK 11 was resistant to both tabtoxin and bialaphos. When the recombinant plasmids were transformed into the *E. coli* mutant YM11 lacking the activity of glutamine synthetase, the auxotrophic transformants with pARK 11 were able to grow on the medium without glutamine by complementation of nutrient requirement, while no growth was observed in the auxotrophic transformants with pARK 10 (Fig. 7.3). In addition, another experiment on *E. coli* with pARK 11 showed an increased activity of glutamine synthetase. This evidence suggests that the gene inserted in the plasmid pARK11 codes for the glutamine synthetase of *P. syringae* pv. *tabaci,* indicating that the mechanism of resistance to both tabtoxin and bialaphos is due to overproduction of the target enzyme, because the glutamine synthetase from wildfire bacterium is sensitive to tabtoxin.

On the other hand, *E. coli* with pARK 10 showed a specific resistance only to tabtoxin but not to the other two chemicals (Table 7.2). This suggests the possibility that the resistance might be due to action of a tabtoxin-detoxifying enzyme, which may catalyze acetylation of the amino group of tabtoxin or tabtoxinine-β-lactam. Due to its chemical structure, it is probable that the same mechanism is involved when bialaphos is acetylated at its amino group. To examine this possibility, tabtoxin was incubated in the presence of ^{14}C-acetyl coenzyme A (CoA) with cell extracts of *E. coli* or *P. syringae* pv. *tabaci.* The reactants were chromatographed on a silica gel thin-layer plate, and detected by radioautography. As seen in Figure 7.4, the radioautograph showed the new radioactive bands corresponding to the acetylated compounds of tabtoxin in both extracts from *E. coli* containing pARK 10 and *P. syringae* pv. *tabaci,* but these new bands were not

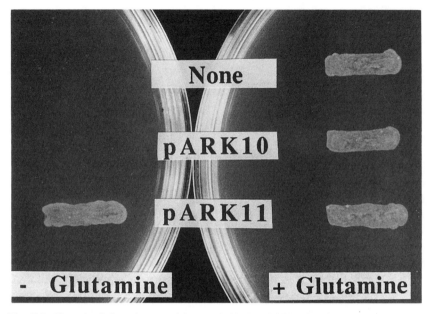

Fig. 7.3. Glutamine-independent growth in auxotrophic *E. coli* YM11 transformed with the plasmid pARK11. Glutamine-requiring *E. coli* mutant YM11 (None), with pARK10 and with pARK11 were inoculated on the M9 minimum medium in the absence (−Glutamine) or presence (+Glutamine) of glutamine. The glutamine auxotrophic *E. coli* mutant with pARK 11 grew on the medium without glutamine by complementation of nutrient requirement.

detected in extract from the original strain of *E. coli*. These results suggest that the gene inserted in pARK 10 encodes the enzyme acetyltransferase specific to tabtoxin and/or tabtoxinine-β-lactam. The acetylation probably occurs at an amino group of tabtoxinine-β-lactam in a mechanism similar to the inactivation of bialaphos, as shown in Figure 7.5. In conclusion, the cloned gene, named the tabtoxin resistance gene (*ttr*), was identified as a acetyltransferase gene, which specifies the tabtoxin-acetylating enzyme but not catalytic enzymes for bialaphos and methionine sulfoximine. This gene was used for further experiments.

4.4. Structure of Tabtoxin-Acetylating Enzyme Gene

Restriction enzyme analysis of a 2-kb insert in pARK 10 identified unique sites for *Bgl*I, *Xba*I, *Bal*I, *Sph*I, *Pst*I, and *Sal*I. Two *Hind*III sites and two *Eco*RV sites were also identified. The restriction enzyme DNA fragments were used for

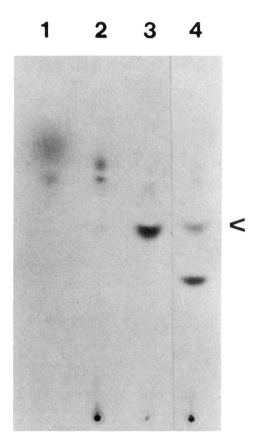

Fig. 7.4. Chromatograms of the radioactive tabtoxin derivatives produced by incubation of tabtoxin with ^{14}C-acetyl CoA in cell extracts of *E. coli* or *P. syringae* pv. *tabaci*. The mark < shows the position of an acetylated derivative of tabtoxin. Lane 1. ^{14}C-acetyl CoA; lane 2. *E. coli* DH1; lane 3. *E. coli* DH1 with pARK10; lane 4. *P. syringae* pv. *tabaci*.

subcloning the minimum DNA region responsible for tabtoxin resistance and for the determination of the direction of gene transcription. As seen in Figure 7.6, the experiments showed that a 700 bp fragment from *Xba*I to *Sph*I was enough to confer resistance to tabtoxin. When the *ttr* gene was inserted downstream into the *lac* Z promoter on the pUC vector, tabtoxin resistance was expressed only in the plasmid fused at the *Xba*I site of the *Xba*I-*Sph*I fragment to *lac* Z promoter, indicating that transcription of the *ttr* gene is directed from the *Xba*I to *Sph*I site.

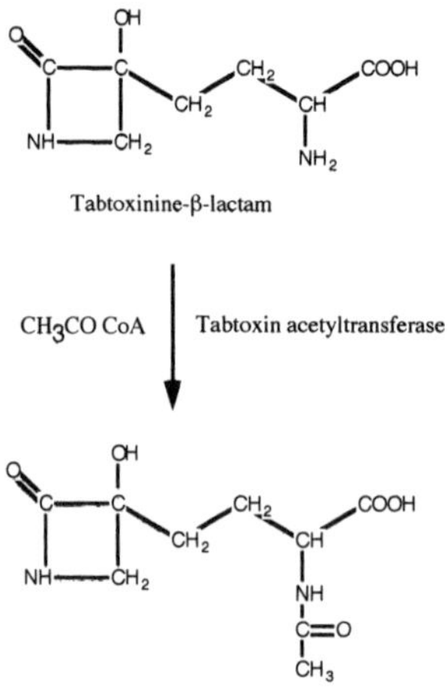

Tabtoxinine-β-lactam

CH₃CO CoA | Tabtoxin acetyltransferase

Fig. 7.5. Hypothetical structure for tabtoxin acetylation.

Nucleotide sequence of the *ttr* gene of about 700 bp of the *Xba*I–*Sph*I fragment was determined by the dideoxy method using M13 phage system. As shown in Figure 7.7, the open reading frame of 531 bp encodes a protein of 177 amino acid residues, which has a molecular weight of about 19,200 (Anzai et al., 1990). In terms of enzymatic functions and molecular weights, tabtoxin acetyltransferase is also quite similar to bialaphos acetyltransferase, which is composed of 185 amino acid residues corresponding to a protein with a molecular weight of 20,600 (Thomp-

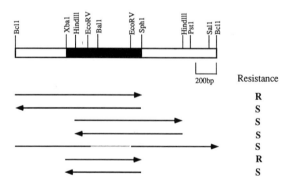

Fig. 7.6. Subcloning and direction of tabtoxin resistance gene. Arrows show the direction of transcription in restriction enzyme DNA fragments. The dotted line indicates the deletion of a fragment between *Eco*RV–*Eco*RV sites. Both R and S are resistant and sensitive to tabtoxin in the *E. coli* with a given plasmid, respectively.

```
  1 TCTAGAGAGCACGCCATGAACCATGCACAACTGCGACGTGTTACCGCTGAAAGCTTCGCC
                M  N  H  A  Q  L  R  R  V  T  A  E  S  F  A

 61 CATTATCGCCATGGTCTGGCGCAGTTGCTGTTTGAAACGGTTCACGGCGGCGCATCGGTG
     H  Y  R  H  G  L  A  Q  L  L  F  E  T  V  H  G  G  A  S  V

121 GGGTTCATGGCTGATCTGGATATGCAGCAGGCGTACGCTTGGTGCGATGGCCTGAAGGCC
     G  F  M  A  D  L  D  M  Q  Q  A  Y  A  W  C  D  G  L  K  A

181 GATATCGCCGCAGGCAGTCTGCTGCTGTGGGTGGTGGCCGAAGACGATAATGTGCTGGCC
     D  I  A  A  G  S  L  L  L  W  V  V  A  E  D  D  N  V  L  A

241 AGCGCGCAGTTGTCACTGTGCCAGAAGCCCAATGGCCTGAACCGCGCCGAAGTGCAGAAA
     S  A  Q  L  S  L  C  Q  K  P  N  G  L  N  R  A  E  V  Q  K

301 CTGATGGTGCTGCCCAGCGCCCGAGGTCGTGGTCTGGGCAGGCAACTGATGGACGAAGTC
     L  M  V  L  P  S  A  R  G  R  G  L  G  R  Q  L  M  D  E  V

361 GAGCAGGTTGCAGTAAAACACAAACGCGGGCTGCTGCACCTGGACACCGAAGCGGGCTCG
     E  Q  V  A  V  K  H  K  R  G  L  L  H  L  D  T  E  A  G  S

421 GTGGCTGAAGCGTTCTATAGCGCGCTGGCCTACACCCGCGTCGGCGAACTGCCAGGCTAC
     V  A  E  A  F  Y  S  A  L  A  Y  T  R  V  G  E  L  P  G  Y

481 TGCGCCACACCGGACGGTCGCCTTCATCCGACCGCCATTTATTTCAAAACCTTGGGGCAA
     C  A  T  P  D  G  R  L  H  P  T  A  I  Y  F  K  T  L  G  Q

541 CCGACATGATCCCTGGTCAATACCAGATCCAGCCCGGCGATATCGAACTCAACGCCGGAC
     P  T  ***

601 GCCGTACCCTCAGCCTGACGGTTGCCAACAGCGGCGACCGGCCGATTCAGGTGGGCTCGC

661 ATTTCCACTTTTTCGAAACCAATGACGCACTGACCTTTGATCGCGCCGCCAGCCGCGGCA

721 TGC
```

Fig. 7.7. Nucleotide sequence of the *ttr* gene determined by dideoxy method and the putative amino acid sequence for its gene product.

son et al., 1987). However, a computer search showed low homology in nucleotide sequences between the *ttr* gene for tabtoxin resistance and the *bar* gene for bialaphos resistance (Yoneyama et al., unpublished).

5. INTRODUCTION OF THE *TTR* GENE INTO PLANTS

5.1. Gene Transfer Into Plants

Dramatic progress has been made in the development of gene transfer systems for higher plants. The techniques used to introduce foreign genes into plant cells, tissues or to regenerate viable, fertile plants, have provided an ideal opportunity for modifying and improving crop plants. The first transgenic plants expressing engineered foreign genes were produced by the use of *A. tumefaciens* vectors (Horsch, 1984; DeBlock et al., 1984). The transformation was confirmed by the presence of foreign DNA sequences in primary transformants and their progeny, as well as by an antibiotic resistance conferred by a chimeric neomycin phosphotransferase gene. This *Agrobacterium*-mediated gene transfer system has been proved to be a highly efficient and versatile vehicle by which subsequent development of methods, based on regenerable explants, such as leaves, stems, and roots, are used today for many dicotyledonous plant species.

Another type of gene transfer method, generally designated as the direct DNA transfer method, which includes the following: electroporation (Fromm et al., 1985); chemical-mediated transfer by polyethyleneglycol (Lörz et al., 1985) or phosphate calcium (Potrykus et al., 1985); and particle gun technique (Klein et al., 1988), has also been developed for the transformation of monocotyledonous plants (corn, rice, and wheat). Apart from the recently developed particle gun method, and the tumor inducing (Ti) plasmid vector systems, most direct gene transfer methods use plant protoplasts as recipient cells and, therefore, more skillful techniques are required for manipulation. Plant regeneration systems must be established in each objective plant before transformation. Thus, direct gene transfer methods are not readily available for dicotyledonous plants that are relatively easy to regenerate from tissues, such as solanacearum plants, apart from use for special experimental purposes. However, these methods are essential for the transformation of monocotyledonous plants, including cereal plants where it is difficult to use the *Agrobacterium*-mediated transformation system.

Currently, in the *Agrobacterium*-mediated transformation system, two species of the *Agrobacterium* genus are used; one is *A. tumefaciens*, the causal agent of crown gall disease, and the other is *A. rhizogenes*, the causal agent of hairy root disease in dicotyledonous plants. *Agrobacterium tumefaciens* is generally used for transformation in the plants that regenerate easily from explanted leaves such as tobacco plants. *Agrobacterium rhizogenes* is convenient for gene transfer into plants that are difficult to regenerate from stems and leaves. It is also convenient for transformation of the plant cells aimed at the production of secondary metabolites. Both *Agrobacterium* species, even though they cause different types of disease symptoms, are similar in their fundamental mechanisms leading to disease development in plants. Although the chromosomal virulence genes are necessary for attachment of the bacterium to plant cells in the case of *A. tumefaciens*, the pathogenicity is associated with about 200 kb of a giant plasmid, Ti plasmid, which is present as an extra chromosome in bacterial cells in the case of *A. tumefaciens*. The pathogen causes crown gall disease through the transfer of a part of the Ti plasmid, called T-DNA (transferred-DNA) containing tumorigenic genes. Ti plasmid has two genetic elements essential for the transfer of T-DNA into plant cells: *vir* genes that set in motion the molecular events leading to T-DNA transfer, and the border sequences of 25 bp direct repeats that result in the generation of site-specific nicks at both sides of T-DNA.

Plant cells targeted by a pathogen produce phenolic compounds, such as acetosyringone, which plays a role in the bacterium–host cell recognition mechanism (Stachel and Zambryski, 1986). Such compounds activate the *vir* gene expression on Ti plasmids through the action of protein kinase (Jin et al., 1990), thereby inducing *vir* functions leading to nicks in the bottom strands of the T-DNA border sequences, and thus generating free, linear and single-stranded copies of (T-strand) T-DNA, which transfer into plants from the bacterial cells. Finally, the T-strands are incorporated into plant chromosomes, and cause a tumor formation in the expression of pathogenic genes in T-DNA. Using the same mechanism as that of *A. tumefaciens,* the T-DNA of root-inducing (Ri) plasmid in *A. rhizogenes* is transferred into the chromosomal DNA of plant cells where it induces hairy root disease on infected plants. Using such a natural gene transfer system, early experiments demonstrated that the foreign DNA within the T-DNA of Ti plasmids was physically integrated into the genome of transformed plant cells. Efficient plant transformation systems were constructed by removing the tumorigenic genes from the T-DNA region, thereby eliminating the ability of the bacteria to induce abnormal cell proliferation.

Modern plant transformation vectors have been modified to be capable of replicating in *E. coli* cells as well as in *Agrobacterium* cells, allowing for convenient gene manipulations such as binary vector systems. The general features of binary vector systems are based on the principle that the T-DNA region does not necessarily have to be physically linked to the *vir* genes of Ti plasmids, provided both the T-DNA- and *vir* region-containing plasmids coexist *in trans* in the same cell of *Agrobacterium* (Hoekema et al., 1983; Bevan, 1984). In highly developed binary vectors, a T-DNA-deficient Ti plasmid, where the whole T-DNA region, including the border sequences, has been removed by deletion, provides trans-acting *vir* gene products in *Agrobacterium* cells. A broad host range plasmid flanked in proper polarity by right and left 25 bp T-DNA border sequences is used as a plant expression vector, which contains the multiple cloning site, the plant marker genes for selection of transformed plants, the appropriate bacterial marker genes, and the mobilization functions for bacterial conjugation. Interest genes are introduced into a multiple cloning site between the 25 bp repeats of the broad host range plasmid, and moved to *Agrobacterium* cells with the Ti plasmid lacking T-DNA region, by conjugational transfer with the aid of a helper plasmid. The genes in the vectors become part of the T-DNA unit and are transferred from *Agrobacterium* hosts to plant genomes (Schell, 1987; Gasser and Fraley, 1989).

5.2. Construction of Binary Vector With *TTR* Gene

The plasmid pBI121, a binary vector commercially used, is a convenient and excellent leading plasmid for gene manipulation. The plasmid pBI121 possesses two genes as plant selection makers between the right and left border sequences of T-DNA; neomycin phosphotransferase II (NPT II) gene, coding for the NPTII enzyme, conferring resistance to aminoglycoside antibiotics such as kanamycin, fused behind the promoter of nopaline synthase (NOS) deriving from the Ti plasmid pTiC58, and β-glucuronidase (GUS) gene, enabling the expression to be measured by fluorometric assay, flanked by the 35S promoter of cauliflower mosaic virus (CaMV) and by NOS polyadenylation terminator. This plasmid pBI121 was used for the construction of the plant expression vector with the *ttr* gene. The 700 bp *Xba*I–*Sph*I fragment of the minimum DNA region conferring resistance to tabtoxin was recloned into the same restriction sites of the plasmid pTZ18R, and from the resulting plasmid an 800 bp *Xba*I–*Pvu*II fragment containing the 700 bp insert was reinserted between the *Xba*I and *Sma*I sites of the plasmid pUC19 to obtain a fragment with *Xba*I–*Sac*I restriction sites. As shown in Fig. 7.2, the *Xba*I–*Sac*I fragment replaced the fragment containing the GUS gene with the same restriction sites on the plasmid pBI121, so that the *ttr* gene was located between the 35S promoter of CaMV and the NOS polyadenylation signal of pTiC58. This plant expression vector with the *ttr* gene was named pARK21. Because pARK21 still has the NPTII gene, it could confer resistance to kanamycin in plants.

For plant transformation, the pARK21 was transformed and amplified in *E. coli* DH1, and then mobilized into disarmed and streptomycin-resistant *A. tumefaciens* LBA4404 with the Ti plasmid lacking the T-DNA region, by the triparental mating culture of the donor *E. coli* with pARK21, the helper *E. coli* with plasmid pRK2013, and the recipient *A. tumefaciens* LBA4404 (Ditta et al., 1980). The transconjugant *Agrobacterium* cells with pARK21 were obtained by selecting in the presence of both kanamycin and streptomycin antibiotics.

5.3 Introduction of the *TTR* Gene Into Plants

For infection or transformation of plant cells by *A. tumefaciens,* there are some techniques such as (a) cocultivation of growing protoplast-derived cells with the bacteria (Marton et al., 1979), (b) inoculation of the bacteria to cultured plant cells (An, 1985), and (c) direct inoculation to plant leaf disks (Horsch et al., 1985). In tobacco plants, the leaf disk inoculation method is extensively used for transformation, because the mesophyll cells in leaf explants provide a large number of transformed shoots in short periods through simple tissue culture manipulations. We carried out our experiments using this method to introduce the *ttr* gene into the tobacco variety *Nicotiana tabacum* bv. Petit Havana SR1.

Leaf disks were punched from axenically grown tobacco leaves, and dipped for a few minutes in an overnight culture of *A. tumefaciens* carrying the plasmid pARK21. Then, an excess amount of bacterial suspension on the leaf disks was removed with sterilized filter papers, and the disks were placed upside down on the leaf surface on Murashige–Skoog (MS) medium supplemented with 1 mg/l of benzyladenine (BA) and 0.1 mg/l of indole acetic acid (IAA). After infection with bacteria for 2 days the leaf disks were transferred to the same medium containing 100 μg/ml of kanamycin in order to select the transformed plant cells and 500 μg/ml of cefotaxim to kill the bacterial cells. The tobacco shoots can be differentiated within 3 weeks. Putatively transformed shoots were excised and planted for rooting and reselection on the hormone-free MS medium containing the above two antibiotics. At this stage, it should be confirmed that kanamycin-resistant shoots are free from infection of *Agrobacterium*. The young transformed tobacco plants were maintained by successive cultures on the medium consisting of one-half concentration of MS base without the antibiotics, or by transplanting on soils through acclimatization.

5.4. Selection of Tabtoxin-Resistant Plants

To examine whether the kanamycin-resistant tobacco plants can also express the *ttr* gene, leaf segments from the transformed axenic tobacco were tested for tabtoxin resistance by callus induction on the MS medium containing 10 μg/ml of tabtoxin, 2 mg/l of naphthalene acetic acid (NAA), and 0.2 μg/l of BA. Among 15 independent clones of the transformants that were examined, the leaf cells of the 11 clones proliferated actively and induced calli even in the presence of tabtoxin, while no callus was formed in the leaf segments from nontransformed tobacco plants. The results indicate that the *ttr* gene can express and produce the tabtoxin-acetylating enzyme, which confers tabtoxin resistance to tobacco cells.

The five typical transgenic tobaccos showing high resistance to tabtoxin were chosen for biochemical analysis. The isolated genomic DNAs were digested with the restriction enzymes *Eco*RI and *Xba*I. Samples of the digested DNA fragments were electrophoresed on an agarose gel, transferred to a nylon membrane, and followed by hybridizing with the ^{32}P-labeled *ttr* DNA as a probe. This Southern blotting revealed that all of the five tobacco clones resistant to tabtoxin contained the *ttr* gene, which was integrated in the genomic DNA. In addition, northern blotting was made to detect for the *ttr* transcripts by using the total RNA isolated from control and transgenic plant leaves. All five transgenic plants allowed transcription of the *ttr* gene to mRNA, although the transcriptional level was varied among the individual plants (Fig. 7.8). Two of these clones, named TAB3 and TAB7, showed much higher expression at levels of transcription. In nontransgenic tobacco, however, no *ttr* transcript was detectable.

6. RESISTANCE TO WILDFIRE DISEASE IN TRANSGENIC PLANTS

Because the transgenic tobacco plants expressing the *ttr* gene are resistant to tabtoxin produced by the wildfire bacterium, our interest was attracted to the resistance of the transgenic tobaccos against wildfire disease. Experiments were carried out using TAB3 and TAB7 transgenic clones, which showed a higher transcriptional level of the *ttr* gene. The transgenic and control tobacco plants were inoculated with a concentration of 10^7 cells/ml of *P. syringae* pv. *tabaci* at two points per leaf by the multiple needle method, and kept at 25°C for 6 days in a humid box to develop the disease. As expected, none of the two transgenic

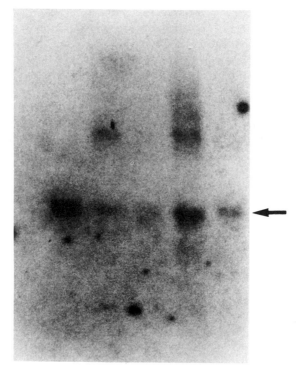

Fig. 7.8. Northern blotting of the *ttr* gene in transgenic tobacco plants. Total RNAs isolated from transgenic tobaccos were electrophoresed on a 1.5% agarose gel containing formaldehyde, transferred to nylon membrane, and hybridized using as a probe the ^{32}P-labeled 700 bp *XbaI–SphI* fragment. Lane 1, untransformed tobacco; lane 2–6, transgenic tobacco clones TAB3–TAB5, TAB7, and TAB9, respectively. The transgenics TAB3 and 7 show strong expression of the *ttr* gene, whereas no *ttr* transcript is detectable in untransformed tobacco.

plants formed any chlorotic halo at the inoculation sites on the leaves. However, nontransgenic controls produced the chlorotic lesions typical of wildfire disease at the infected sites, as shown in Fig. 7.9. Even when inoculated at a higher inoculum of 10^9 cells/ml, the transgenic shows no chlorosis in most cases, though in some cases the restricted minute yellow spots appeared at the inoculation sites. In the case of high concentration of inoculum, it is believed that when a large amount of bacterial cells are injected by force into plant leaf tissues by wounding, the plant cells around the inoculation sites may sometimes be affected by the pathogenic toxin escaping from the detoxifying enzyme. That is, extremely large bacterial populations may rapidly produce an excess amount of toxins over the detoxi-

Fig. 7.9. Disease development on the leaves of tobaccos inoculated with *P. syringae* pv. *tabaci*. (*Left*) Untransformed tobacco plant with chlorotic symptoms at the inoculation sites on the leaves. (*Right*) Transgenic tobacco plant (TAB7) without any chlorotic halo at the inoculation sites on the leaves.

fying ability of the transgenic plant cells. However, this case is not necessarily a natural case of bacterial infection. In natural circumstances, bacterial populations at primary infection sites might be extremely low in cell density, so that the transgenics could be resistant enough to the attacks by wildfire bacteria in the field. Therefore, it is concluded that the transgenic tobaccos expressing the *ttr* gene become resistant not only to tabtoxin, but also to the bacterial attack by *P. syringae* pv. *tabaci*.

7. CONCLUSIONS

We described a simple and useful strategy for molecular breeding of disease-resistant plants by detoxification of the pathogenic toxins as a typical example of tobacco wildfire disease. The resistance to wildfire disease was designed by introducing the tabtoxin-detoxifying enzyme gene derived from *P. syringae* pv. *tabaci* into tobacco plants. This case might be the first success of the transgenic plants resistant to fungal and bacterial diseases using a genetic engineering technique, although numbers of virus-resistant transgenic plants have been created by introducing a gene of viral coat proteins, anti-sense RNAs, or a cDNA of satellite RNAs (Gasser and Fraley, 1989).

In a genetic engineering approach for disease resistance, a major problem is how to search for the detoxifying-enzyme genes of pathogenic toxins. A successful application for wildfire disease is a typical case that utilized a toxin-inactivating enzyme gene present in the pathogen itself. A similar way to isolate the objective genes may be used in other plant pathogens, which produce especially nonhost-specific toxins. When plant pathogens do not possess any toxin-detoxifying enzyme genes, another biological target for gene cloning would be focused on toxin-degrading enzymes present in soil microorganisms, plants, algae, and other living organisms. Moreover, if the detoxifying enzyme gene corresponding to a toxin of each pathogen is found, our genetic engineering strategy for disease resistance could be widely applied to fungal and bacterial diseases caused by pathogenesis related toxins. In some cases, however, gene manipulation must be careful, depending on the mechanisms of action of pathogenic toxins, because some toxins interfere with the functions in the cell membranes of plants, such as AK- and AM-toxins, coronatine, and syringomycin. One possible genetic trick to detoxify the toxins outside the cells could be possible by fusion of a signal peptide gene to the structure gene of an enzyme that leads to translocation of the enzyme to the outside of plant cells.

The first step toward disease resistance using genetic engineering has been taken by detoxification of pathogenic toxins that play an important role as a virulence factor of plant pathogens. Selections for toxin resistance of plant tissues and cells have also been used to isolate plants resistant to diseases. In this case, genetic variability of cultured cells, calli, or protoplasts of plants has provided a chance to isolate toxin-resistant plants, such as eyespot disease of sugarcane (Daub, 1986), late blight of potato (Shepard et al., 1980), and leaf blight of corn (Gengenbach et al., 1977). Since such selections are based on a somaclonal variation of plant cells that may result in the course of cell culture, it sometimes takes a disadvantage of inducing the undesirable mutation of genetic characters together with the desirable disease resistance in the regenerated plants. Also, toxin-selected plant varieties must be refined in most cases by repetitive cross mating for some generations to obtain the desirable varieties. As a result, a genetic engineering technique provides significant benefits, which significantly reduce the period of breeding in plants and probably improves plants without alteration of their beneficial genetic characters.

The potential for improvement of plants through genetic engineering seems vast, even in the area of disease resistance. Although only disease resistance to pathogenic toxins was discussed in this chapter, it is apparent that many other ideas will be tried to design disease resistance, for example, using the defense genes of plants induced by pathogen attack. Genetic engineering will certainly stimulate a new direction of disease protection research including biological and chemical controls, as considered in the recent situation of the appearance of fungicide-resistant pathogens, the emergence of new fungal and bacterial races attacking the resistant plant varieties, and the environmental or toxicological problems of agricultural pesticides. Such progress will also contribute to our basic understanding of plant diseases and the relationship between symptom development and pathogenic toxins, as well as on the interaction between pathogens and plants.

REFERENCES

An G (1985): High efficiency transformation of cultured tobacco cells. Plant Physiol 79:568–570.

Anzai H, Yoneyama K, Yamaguchi I (1989): Transgenic tobacco resistant to a bacterial disease by the detoxification of a pathogenic toxin. Mol Gen Genet 219:492–494.

Anzai H, Yoneyama K, Yamaguchi I (1990): The nucleotide sequence of tabtoxin resistance gene (*ttr*) of *Pseudomonas syringae* pv. *tabaci*. Nucleic Acids Res 18:1890.

Bevan M (1984): Binary *Agrobacterium* vectors for plant transformation. Nucleic Acids Res 12:8711–8721.

Braun AC (1955): A study on the mode of action of the wildfire toxin. Phytopathology 45:659–664.

Cella R, Vining LC (1975) Resistance to streptomycin in a producing strain of *Streptomyces griceus*. Can J Microbiol 21:463–472.

Chida T, Uekita T, Satake K, Hirano K, Aoki K, Noguchi T (1982): Effect of fthalide on infection process of *Pyricularia oryzae* with special observation of penetration site of appressoria. Annu Phytopath Soc Jpn 48:58–63.

Clauson-Kaas N, Plattner PA, Gäumann E (1944): Uber ein Welkeerzeugendes Stoffwechselprodukt von *Fusarium lycopersici* Sacc. Ber Schweiz Bot Ges 54:523–528.

Comai L, Facciotti D, Haitt WR, Thompson G, Rose RE, Stalker DM (1985): Expression in plants of a mutant *aro*A gene from *Salmonella typhimurium* confers tolerance to glyphosate. Nature (London) 317:741–744.

Cunin R, Glandsorffn, Pierard A, Stalon V (1986): Biosynthesis and metabolism of arginine in bacteria. Microbial Rev 50:314–352.

Daub ME (1986): Tissue culture and the selection of resistance to pathogens. Annu Rev Phytopathol 24:159–186.

DeBlock M, Bottermann J, Vandewiele M, Dockx J, Thoen C, Gossele V, Movva NR, Thompson CJ, Van Montagu M, Leemans J (1987): Engineering herbicide resistance in plants by expression of a detoxifying enzyme. EMBO J 6:2513–1518.

DeBlock M, Herrera-Estrella L, Van Montagu M, Schell J, Zambryski P (1984): Expression of foreign genes in regenerated plants and their progeny. EMBO J 3:1681–1689.

Ditta G, Stanfield S, Corbin D, Helinski DR (1980): Broad host range DNA cloning system for Gram-negative bacteria: Construction of a gene bank of *Rhizobium meliloti*. Proc Natl Sci USA 77:7347–7351.

Fischhoff DA, Bowdish KS, Perlak FJ, Marrone PG, McCormick SM, Niedermeyer JG, Dean DA, Kusano-Kretzmer K, Mayer EJ, Rochester DE, Rogers SG, Fraley RT (1987): Insect tolerant transgenic tomato plants. Bio/Technol 5:807–812.

Fromm M, Taylor LP, Walbot V (1985): Expression of genes transferred into monocot and dicot plant cells by electroporation. Proc Natl Acad Sci USA 82:5824–5828.

Gasser CS, Fraley RT (1989): Genetically engineering plants for crop improvement. Science 244:1293–1299.

Gäumann E (1954): Toxins and plant diseases. Endeavour 13:198–204.

Gengenbach BG, Green CE, Donovan CM (1977): Inheritance of selected pathotoxin resistance in maize plants regenerated from cell cultures. Proc Natl Acad Sci USA 74:5113–5117.

Hilder VA, Gatehouse AMR, Sheerman SE, Barker RF, Boulter D (1987): A novel mechanism of insect resistance engineered into tobacco. Nature (London) 300:160–163.

Hoekema A, Hirsch PR, Hooykaas PJJ, Schilperoort RA (1983): A binary plant vector strategy based on separation of vir- and T-region of the *Agrobacterium tumefaciens* Ti-plasmid. Nature (London) 303:179–180.

Horsch RB, Fraley RT, Rogers SG, Sanders PR, Lioyd A, Hoffmann N (1984): Inheritance of functional foreign genes in plants. Science 223:496–498.

Horsch RB, Fry JE, Hoffmann NL, Eichholtz D, Rogers SG, Fraley RT (1985): A simple and general method for transferring genes into plants. Science 227:1229–1231.

Hotta K, Yamamoto H, Okami Y, Umezawa H (1981): Resistance mechanisms of kanamycin-, neomycin-, and streptomycin-producing *Streptomycetes* to aminoglycoside antibiotics. J Antibiot 34:1175–1182.

Imai S, Seto H, Sasaki T, Tsuruoka T, Ogawa H, Satoh A, Inouye S, Niida T, Otake N (1985): Studies on the biosynthesis of bialaphos(SF-1293). 6. Production of *N*-acetyldemethylphosphinothricin and *N*-acetylbialaphos by blocked mutants of *Streptomyces hygroscopicus* SF-1293 and their roles in the biosynthesis of bialaphos. J Antibiot 38:687–690.

Jin S, Prusti RK, Roitsch T, Ankenbauer RG, Nester EW (1990): Phosphorylation of the *vir*A protein: Essential role in biological activity of virG. J Bacteriol 172:4945–4950.

Kamakura T, Yoneyama K, Yamaguchi I (1990): Expression of the blasticidin S deaminase gene (bsr) in tobacco: Fungicide tolerance and a new selective marker for transgenic plants. Mol Gen Genet 223:332–334.

Klein TM, Harper EC, Svab Z, Sanford JC, Fromm ME, Maliga P (1988): Stable genetic transformation of intact *Nicotiana* cells by the particle bombardment process. Proc Natl Sci USA 85:8502–8505.

Kumada Y, Anzai H, Takano E, Murakami T, Hara O, Itoh R, Imai S, Satoh A, Nagaoka K (1988): The bialaphos resistance gene (*bar*) plays a role in both self-resistance and bialaphos production in *Streptomyces hygroscopicus*. J Antibiot 41:1838–1845.

Lee KY, Townsend J, Teppermann J, Black M, Chui CF, Mazur B, Dunsmuir P, Bedbrook J (1988): The molecular basis of sulfonylurea herbicide resistance in tobacco. EMBO J 7:1241–1248.

Lörz H, Baker B, Schell J (1985): Gene transfer to cereal cells mediated by protoplast transformation. Mol Gen Genet 199:178–182.

Marton L, Wullems JG, Molendijk L, Schilperoort R (1979): In vitro transformation of cultured cells from *Nicotiana tabacum* by *Agrobacterium tumefaciens*. Nature (London) 277:129–131.

Meeham F, Murphy HC (1947): Differential phytotoxicity of metabolic by-products of *Helminthosporium victoriae*. Science 106:270–271.

Mitchell RE (1976): Isolation and structure of a chlorosis-inducing toxin of *Pseudomonas phaseolicola*. Phytochemistry 15:1941–1947.

Mitchell RE (1981): Structure: bacterial. In Durbin RD: Toxins in plant disease. New York: Academic Press, pp 259–293.

Murakami T, Anzai H, Imai S, Satoh A, Nagaoka K, Thompson CJ (1986): The bialaphos biosynthetic genes of *Streptomyces hygroscopicus*: Molecular cloning and characterization of the gene cluster. Mol Gen Genet 199:183–188.

Peet RC, Lindgren PB, Willis DK, Panopoulos NJ (1986): Identification and cloning of genes involved in phaseolotoxin production by *Pseudomonas syringae* pv. *phaseolicola*. J Bacteriol 166:1096–1105.

Potrykus I, Saul MW, Petruska J, Paszkowski J, Schillito RD (1985): Direct gene transfer to cells of a graminaceous monocot. Mol Gen Genet 199:183–188.

Schell J (1987): Transgenic plants as tools to study the molecular organization of plant genes. Science 237:1176–1183.

Shepard JF, Bidney D, Shahin E (1980): Potato protoplasts in crop improvement. Science 208:17–24.

Sinden SL, Durbin RD (1968): Glutamine synthetase inhibition: Possible mode of action of wildfire toxin from *Pseudomonas tabaci*. Nature (London) 219:379–380.

Stachel SE, Zambryski PC (1986): *Agrobacterium tumefaciens* and the susceptible plant cell: a novel adaptation of extracellular recognition and DNA conjugation. Cell 47:155–157.

Staskawicz BJ, Panopoulos NJ, Hoogenraad NJ (1980): Phaseolotoxin-insensitive ornithine carbamoyl-transferase of *Pseudomonas syringae* pv. *phaseolicola*: a basis for immunity to phaseolotoxin. J Bacteriol 142:720–723.

Stewart W (1971): Isolation and proof of structure of wildfire toxin. Nature (London) 229:174–178.

Taylor PA, Schnoes HK, Durbin RD (1972): Characterization of chlorosis-inducing toxins from a plant pathogenic *Pseudomonas* sp. Biochim Biophys Acta 286:107–117.

Thompson CJ, Movva NR, Tizard R, Crameri R, Davies JE, Lauwereys M, Botterman J (1987): Characterization of herbicide-resistance gene *bar* from *Streptomyces hygroscopicus*. EMBO J 6:2519–2523.

Turner JG, Mitchell RE (1985): Association between symptom development and inhibition of ornithine carbamoyltransferase in bean leaves treated with phaseolotoxin. Plant Physiol 79:468–473.

Vaeck M, Reynaerts A, Hofte H, Jandens S, DeBeuckeleer M, Dean C, Zabeau M, Van Montagu M, Leemans J (1987): Transgenic plants protected from insect attack. Nature (London) 328:33–37.

Watanabe T, Sekizawa Y, Shimura M, Suzuki Y, Matsumoto K, Iwata M, Mase S (1979): Effects of probenazole (Oryzemate) on rice plants with reference to controlling rice blast. J Pesticide Sci 4:53–59.

Woloshuk CP, Sisler HD, Tokousbalides MR, Dutky SR (1980): Melanin biosynthesis in *Pyricularia oryzae*: Site of tricyclazole inhibition and pathogenicity of melanin-deficient mutants. Pestic Biochem Physiol 14:256–264.

Woolley DW, Pringle RB, Braun AC (1952): Isolation of the phytopathogenic toxin of *Pseudomonas tabaci,* an antagonist of methionine. J Biol Chem 197:409–417.

Yamaguchi I, Sekido S, Misato T (1983): Inhibition of appressorial melanization in *Pyricularia oryzae* by non-fungicidal anti-blast chemicals. J Pesticide Sci 8:229–232.

8

THE ROLE OF CELL WALL DEGRADING ENZYMES IN FUNGAL DISEASE RESISTANCE

KAREN BROGLIE
RICHARD BROGLIE
E.I. du Pont de Nemours & Co. Agricultural Products, Experimental Station, Wilmington, Delaware 19880

NICOLE BENHAMOU
Department of Phytology, Laval University, Quebec, Canada

ILAN CHET
The Hebrew University, Otto Warburg Center for Agricultural Biotechnology, Faculty of Agriculture, Rehovot 76100, Israel

1. INTRODUCTION

Plants react to pathogen attack by the activation of a variety of defense mechanisms (Hahlbrock and Scheel, 1987; Bowles, 1990) that culminate in a number of physical and biochemical changes in the host plant. Among the physical changes are those that directly affect the properties of the plant cell wall and are designed to strengthen this structure in order to inhibit entry of the invading pathogen. Examples include (1) the accumulation of hydroxyproline-rich glycoproteins (Esquerre-Tugaye et al., 1979; Showalter et al., 1985), (2) lignification and suberization (Vance et al., 1980; Espelie et al., 1986), (3) the deposition of callose (Ride, 1975, 1983; Bonhoff et al., 1987), and (4) the accumulation of phenolic compounds (Matta et al., 1969; Hunter, 1974). Among the major biochemical changes are (1) the biosynthesis and accumulation of phytoalexins, secondary metabolites that are toxic to bacteria and fungi (Darvill and Albersheim, 1984; Hahlbrock and Grieseback, 1979; Dixon et al., 1983), (2) the accumulation of

Biotechnology in Plant Disease Control, pages 139–156
© *1993 Wiley-Liss, Inc.*

protease inhibitors (Ryan, 1973; Peng and Black, 1976) and the release of oligo-saccharide elicitors of plant origin (Nothnagel et al., 1983).

Infection of plants by potentially pathogenic microorganisms has been shown to result in the accumulation of a novel class of proteins termed "pathogenesis-related proteins," or PR-proteins (Van Loon 1985). Although the role of some PR proteins in the defense response of plants is not known, their presence is often correlated with disease resistance (Van Loon, 1985; Gianinazzi, 1984). Recently, several classes of PR proteins have been shown to correspond to the hydrolytic enzymes, chitinase and β-1,3-glucanase. These enzymes have received increasing attention as important components in the arsenal of plant defense proteins, since they catalyze the hydrolysis of the main carbohydrate components of most fungal cell walls, chitin and β-1,3-glucan (Chet et al., 1967; Bartnick-Garcia, 1968; Wessels and Sietsma, 1981; Sivan and Chet, 1989). In this chapter, we will concentrate on the role of lytic enzymes in the defense response of plants with special emphasis on experiments designed to modify chitinase gene expression in transgenic plants. The production and ultrastructural characterization of transgenic plants with enhanced resistance to attack by the fungal pathogen *Rhizoctonia solani* will be described.

2. ACTIVATION OF LYTIC ENZYMES DURING FUNGAL ATTACK

Support for the hypothesis that chitinases and β-1,3-glucanases play an active role in the defense response of plants has been obtained largely from indirect evidence. Both chitinase and β-1,3-glucanase are known to be induced during fungal infection (Schlumbaum et al., 1986; Carr and Klessig, 1989). The purified enzymes are capable of degrading isolated fungal cell walls (Young and Pegg, 1981; Boller et al., 1983). While no endogenous substrate for chitinase is known to exist in plants, β-1,3-glucans are known to be present in several plant structures including the wall of pollen tubes, endosperm cell walls, and the walls of sieve elements (Fincher and Stone, 1981). Although this latter observation necessitates complex regulation of glucanase by hormonal and developmental factors, the coordinate regulation of chitinase and glucanase expression during pathogen attack implies that the two enzymes share a common secondary function in the plant defense response (Boller, 1987).

The enzyme chitinase [poly[1,4-(N-acetyl-β-D-glucosaminade)] glycanohydrolase, EC 3.2.1.1.4.] catalyzes the hydrolysis of chitin, a β-1,4-linked homopolymer of N-acetyl-D-glucosamine. Endo-type chitinase activities have been observed in many plant species including bean, soybean, tomato, melon, cotton, wheat, barley, and tobacco (Boller et al., 1983; Shinshi et al., 1987; Leah et al., 1987; Nasser et al., 1990; Jacobsen et al., 1989; Ride and Barber, 1990; Kragh et al., 1991). Multiple, acidic and basic forms of the protein are known to be present in certain species including bean, tobacco, cucumber, and arabidopsis (Legrand et al., 1987; Awade et al., 1989). In some cases, the genes encoding these hydrolytic enzymes have been isolated and characterized (Broglie et al., 1986; Shinshi et al., 1987; Metraux et al., 1989; Swegle et al., 1989; Samac et al., 1990).

Assays of enzyme activity, as well as measurements of steady-state messenger ribonucleic acid (mRNA) levels, indicate that the expression of plant chitinases may be induced by fungal infection, wounding, treatment with fungal elicitors, exposure to exogenous ethylene, and in the case of cucumber and tobacco, infection with viral agents. In melon plants, the level of translatable chitinase mRNAs was found to increase following infection by the fungus *Colletotrichum lagenarium* (Roby and Esquerre-Tugaye, 1987). When melon hypocotyls were locally infected with this pathogen, chitinase activity increased throughout the plant. Treatment with fungal elicitor also gave rise to elevated levels of enzyme activity. When these elicitor-treated plants were then subsequently exposed to *C. lagenarium,* an increased resistance to infection was observed. While disease symptoms appeared in these plants, necrosis was more limited than in infected, untreated plants. These results suggested that the elevated level of chitinase enzyme activity was responsible, at least in part, for the induced systemic resistance against *C. lagenarium* (Roby et al., 1988). In addition, a similar, systemic induction of chitinase activity has been observed in cucumber plants. In this case, abiotic agents were effective inducers (albeit weaker so) in addition to bacterial, fungal, and viral pathogens (Metraux and Boller, 1986). Tobacco infected with the pathogens, *Pseudomonas tabaci* and *Phytophthora parasitica* showed only localized induction of chitinase and β-1,3-glucanase enzyme activities (Meins and Ahl, 1989). In contrast, tobacco mosaic virus (TMV) infection led to systemic induction of the two enzymes (Vogeli-Lange et al., 1988).

Numerous studies have led to the assignment of a function for several pathogenesis-related proteins. The results of these studies have further suggested compartmentalization of the plant defense response with expression of intracellular and extracellular forms of chitinase and β-1,3-glucanase. Infection with various plant pathogens or treatment with salicylic acid leads to the accumulation of PR proteins. These proteins generally have an extracellular location, are of relatively low molecular weight, and often possess extreme isoelectric points. In tobacco, within the PR-3 group, two acidic proteins (PR-P and PR-Q) and two basic proteins (Ch.32 and Ch.34) have been purified and shown to possess chitinase activity. While they differ in size and charge, the acidic and basic forms share common antigenic determinants (Legrand et al., 1987). Within the PR-2 group, three basic and four acidic proteins were found to display β-1,3-glucanase activity. The acidic and basic glucanases exhibit 50–60% homology at the amino acid level. The basic isoforms were localized primarily in the central vacuole of the cell while the acidic forms (PR-2, PR-N, PR-O, and PR-Q') were extracellular (Van den Bulcke et al., 1989; Linthorst et al., 1990). No transport of the vacuolar forms to the extracellular space was discerned following infection of tobacco plants with *Pseudomonas syringae* (Van den Bulke et al., 1989). Kombrink et al. (1988) identified two distinct β-1,3-glucanases and six chitinases that accumulate following infection of potato with *Phytophthora infestans* or treatment of detached potato leaves with a *P. infestans* derived elicitor. In contrast to tobacco, where the basic forms reside in the vacuole and the acidic proteins reside in the intercellular fluid, all of the potato glucanases and chitinases were found to have basic isoelectric points and to be localized predominantly in the intercellular fluid (Kombrink et al., 1988).

Infection of tomato with the foliar pathogen, *Cladosporium fulvum* results in the accumulation of several pathogenesis-related proteins in the apoplast. Two of the predominant proteins were purified and one of these was shown to exhibit β-1,3-glucanase activity, while the other displayed chitinase activity. Western blot analysis of apoplastic fluid from infected leaf tissue indicated the presence of one additional glucanase and three additional chitinases. Analysis of the kinetics of induction indicated that the increase in chitinase and glucanase activity occurred more rapidly, and to a greater extent, in the incompatible interaction rather than in the compatible interaction (Joosten and De Wit, 1989). An ultrastructural investigation of *Fusarium oxysporum* infected tomato root tissues similarly showed an earlier induction of chitinase in the incompatible interaction. In both the compatible and incompatible systems, the presence of chitinase was found to be highly correlated with the distribution of pathogen. The preferential association of the enzyme with altered segments of the fungal cell wall suggested that the action of chitinase may be preceded by, or coincident with, that of other hydrolytic enzymes such as β-1,3-glucanase (Benhamou et al., 1990).

Mauch et al. (1988) investigated the effect of chitinase and glucanase on the *in vitro* growth of several fungi by utilizing protein extracts from pea pods infected with *Fusarium solani* f.sp. *phaseoli,* as well as purified enzyme preparations. Of the 18 fungi tested, 15 were found to be sensitive to protein extracts prepared from infected pods. When the purified lytic enzymes were added separately, only the growth of *Trichoderma viride* was inhibited by chitinase and only the growth of *F. solani* f.sp. *pisi* was inhibited by β-1,3-glucanase. Combinations of the two purified proteins provided effective inhibition of growth for 8 of the tested fungi. Microscopic investigation of treated mycelium indicated that the inhibitory effect is presumably derived from enzyme catalyzed hydrolysis of the chitin and glucan polymers in the cell walls of the growing hyphal tips (Mauch et al., 1988).

In bean plants, the major chitinase activity is known to be associated with a basic, 30-kD protein that is localized in the vacuole (Boller et al., 1983; Boller and Vogeli, 1984). In addition to the basic enzyme, two acidic chitinases have been purified from alfalfa mosaic virus infected bean plants. In contrast to the basic protein, these were found to have an extracellular location. Although they share 65–70% homology in amino acid sequence, the acidic proteins do not cross react with antiserum raised against the basic bean chitinase (Awade et al., 1989). A full length complimentary deoxyribonucleic acid (cDNA) clone encoding basic bean chitinase has been synthesized from ethylene-induced poly(A) RNA and used as a hybridization probe to isolate the corresponding nuclear genes (Broglie et al., 1986). The expression of two of the three genes encoding the basic bean chitinase is known to be regulated by the phytohormone, ethylene (Broglie et al., 1986; Broglie et al., 1989). The promoter of the chitinase 5B gene has also been shown to be activated in bean protoplasts by the addition of purified fungal elicitors (Roby et al., 1991). The DNA sequences contributing to the ethylene and elicitor response have been identified in the CH5B upstream region (Broglie et al., 1989; Roby et al., 1991).

3. ACTIVATION OF THE BEAN CHITINASE 5B PROMOTER BY FUNGAL INFECTION

The *E. coli* β-glucuronidase (GUS) gene has been used as a reporter gene to monitor induction of the bean chitinase 5B promoter in transgenic tobacco plants. A 1.7-kb fragment bearing 5B upstream sequences was fused to the coding region of GUS, followed by the termination signals of the nopaline synthase (NOS) gene. This chimeric gene, contained on a binary plasmid vector, was introduced into tobacco plants by *Agrobacterium*-mediated transformation. When the transgenic plants were treated with exogenous ethylene, an induction of GUS enzyme activity was observed, indicating that the sequence information necessary for ethylene-regulated expression of the chitinase 5B gene was contained within this DNA segment (Broglie et al., 1989). Infection of chitinase-GUS plants with fungal pathogens also gave rise to a significant induction of β-glucuronidase activity that was readily apparent upon histochemical staining of the infected tissue. When the transgenic plants were inoculated on one-half of a leaf with the foliar pathogen, *Botrytis cinerea,* positive staining could be found within and immediately around the necrotic areas (Fig. 8.1). Assays of the infected tissue indicated that although GUS activity was preferentially localized at the site of inoculation, a weaker induction of the chitinase promoter was evident 3–6 mm away from the lesion as well as on the uninoculated one-half of the leaf. No significant β-glucuronidase activity could be detected in leaves above or below that which had been inoculated with the fungal spore suspension. When the infected tissue was fixed, stained with fluorescein isothiocyanate-conjugated wheat germ agglutinin, and then subjected to microscopic investigation, fungal mycelia were found to be concentrated within the necrotic lesion. Significantly fewer mycelia were found at the periphery of the lesion in the region that exhibited a blue staining reaction with the chromogenic GUS substrate. These appeared to be confined to the first few cell layers on the surface of the leaf. Beyond this zone, essentially no fungal structures were apparent. Thus, while the distribution of the fungus was similar to the distribution of GUS enzymatic activity, activation of the bean chitinase promoter was clearly found in regions of the leaf that had not been exposed to the fungus.

The induction pattern observed for the bean chitinase promoter in the *Botrytis*-infected plants was not restricted to this pathogen. Qualitatively similar results

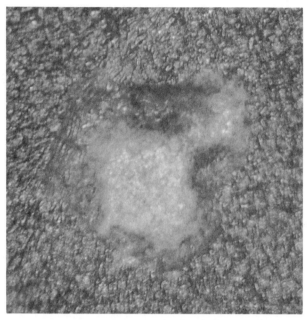

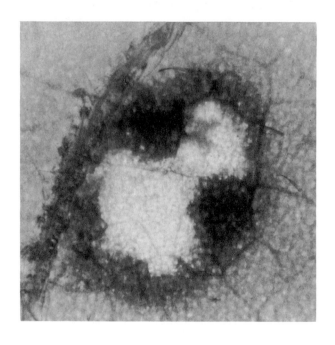

were obtained when the transgenic tobacco plants were infected with two soil-borne pathogens, *R. solani* and *Sclerotium rolfsii*. In both cases, GUS activity was highest in the vicinity of the lesion and declined as the distance from the lesion increased. While inoculation of tobacco plants with *Sclerotium* was achieved by placing an agar plug containing fungus on the surface of the soil near the plant, for *Rhizoctonia* seedlings were transplanted into fungal-infested soil. In the latter case, little if any induction of GUS activity was evident in the roots of the plant indicating that activation of the chitinase 5B promoter is dependent not only on the presence of the fungus but also on colonization and damage to the plant tissue (Roby et al., 1990).

4. MODIFICATION OF CHITINASE GENE EXPRESSION IN PLANTS

Previous studies have shown that purified chitinases are capable of hydrolyzing isolated fungal cell walls (Young and Pegg, 1982; Boller et al., 1983). The enzyme from pea, alone and in combination with β-1,3-glucanase, is able to inhibit the growth of certain pathogenic fungi in culture (Mauch et al., 1988). In addition, the activation profile of the 5B promoter during fungal infection has indicated that chitinase expression is intimately associated with the response of the plant to pathogen invasion (Roby et al., 1990). While these and other *in vitro* findings have attested to the participation of chitinase in the defense response of the plant, little corroborating evidence had been accumulated *in planta*. In order to evaluate the role that chitinase plays during attack by pathogenic organisms, we decided to modify the pattern of bean chitinase expression in transgenic plants. Because the timing of the defense response appears to contribute to the outcome of the interaction between host and pathogen (Bell et al., 1986; Joosten and De Wit, 1989; Benhamou et al., 1990), we have eliminated the temporal factor in chitinase expression. We have found that constitutive expression of a bean endochitinase in transgenic tobacco and canola plants affords increased protection against disease caused by the fungal pathogen *R. solani*.

A chimeric chitinase gene was constructed by replacing the 5' regulatory region of the bean chitinase 5B gene with the promoter region of the cauliflower mosaic virus (CaMV) 35S transcript. The 35S promoter has been shown to be active in various plant tissues and has been used to obtain high level, constitutive expression of a number of chimeric genes in transgenic plants (Abel et al., 1986; Shah et al., 1986; Fischhoff et al., 1987). The 35S-chitinase gene was introduced into

---

Fig. 8.1. Histochemical staining of β-glucuronidase activity in transgenic tobacco leaves infected with *B. cinerea*. (**Top**): Typical appearance of lesions produced by *B. cinerea* infection of transgenic tobacco plants. The lesions pictured are 2 days postinoculation; (**Bottom**): Histochemical staining of infected tissue for GUS activity (magnification 12 ×).

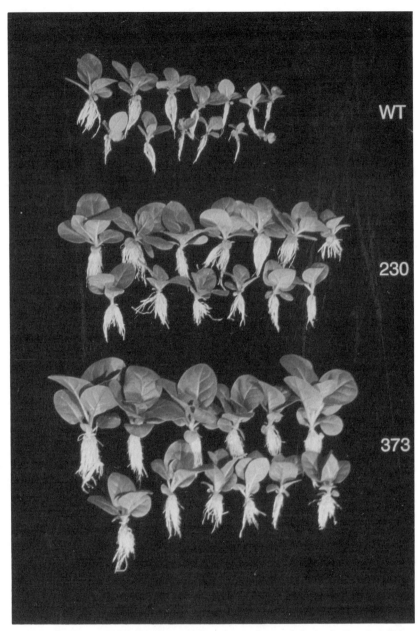

Fig. 8.2. Resistance of 35S-chitinase containing tobacco plants to infection by *R. solani*. Tobacco seedlings, 18 days old, were transplanted into *R. solani* infested quartz sand. *Rhizoctonia solani* inoculum was prepared by growing the fungus on a medium consisting of 500 ml of quartz sand, 40 ml of cream of wheat, 40 ml of corn meal, and 75 ml of water. Plants were maintained under the growth conditions defined above and nutrients supplied by watering with Hoagland's solution. Plants were uprooted 18 days postinoculation and thoroughly washed with tap water for visual comparison of roots and aerial portions of the plants. The figure shows a comparison of control (wt = #548) and 35S-chitinase plants (#230 and #373).

Agrobacterium tumefaciens by bacterial conjugation and cells carrying the cointegrate plasmids were used to infect leaf disks of *Nicotiana tabacum*. Twenty-two primary transformants were screened for expression of the bean chitinase polypeptide by western blot analysis. All were found to contain a 30-kD protein band that was immunoreactive with antibodies raised to bean chitinase. Assays of protein extracts indicated an increased chitinase activity in the 35S-chitinase transformed plants.

The 35S-chitinase tobacco plants were assayed for fungal resistance using the phytopathogen, *R. solani*. This phytopathogen is an endemic, chitinous, soil-borne fungus that infects numerous plant species, including corn and soybean, and produces severe stem and root rotting symptoms. Seeds planted in heavily infested fields typically have problems with stand and early season growth. Damping-off, seedling blight, and brown girdling root-rot are examples of important diseases that are attributable to *Rhizoctonia* infection. Homozygous progeny of tobacco plants were grown in the presence of the soil-borne pathogen in order to determine their susceptibility to infection by this fungus. When 18-day-old tobacco seedlings were transplanted into *R. solani*-infested soil, a decrease in seedling mortality was observed for plants harboring the 35S-chitinase gene relative to control plants that lacked the 35S-chitinase gene (Fig. 8.2). For the four different transformants tested, seedling mortality ranged from 23 to 37% in comparison to 53% for the control. When the surviving plants were compared, a significant difference was evident in the growth and development of control tobacco plants versus plants harboring the 35S-chitinase gene. Control plants were noticeably stunted and displayed poor development of the root system. In contrast, the 35S-chitinase plants were larger, hardier, and showed near normal root growth and morphology. The enhanced resistance of 35S-chitinase tobacco plants to *R. solani* infection appeared to be correlated with the level of bean chitinase expression: Plants containing higher levels of the bean polypeptide displayed a greater survival rate in fungal-infested soil. When the 35S-chitinase plants were grown in the presence of a pathogen, *Pythium aphanidermatum,* that lacks a chitin-containing cell wall, no difference in survival was detected compared to control plants.

The chimeric 35S-chitinase gene has also been introduced into canola, *Brassica napus* cv. Westar. Consistent with the results observed in transgenic tobacco, canola plants carrying the 35S-chitinase gene exhibited increased resistance to infection by the root and stem rot pathogen, *R. solani*. In both systems, seedlings constitutively expressing the bean chitinase polypeptide were delayed in the development of and progression of disease symptoms. Fifteen days after transplanting into *Rhizoctonia*-inoculated soil, 53% of the transgenic canola seedlings survived while only 24% of the wild-type plants remained (Broglie et al., 1991).

5. ANTIFUNGAL ACTIVITY OF BEAN CHITINASE

The addition of purified bean chitinase to actively growing cultures of *R. solani* produces zones of inhibition near the point of application of the protein (Broglie et al., 1991). The inhibition of fungal growth increases with increasing concentrations of the lytic enzyme. No inhibition can be detected in the presence of either boiled enzyme or bovine serum albumin (BSA). Previous studies would suggest that the observed growth inhibitory effect arises from enzyme-catalyzed hydrolysis of newly formed chitin and resultant disruption of the growing fungal hyphal tips (Molano et al., 1979). Light microscopic examination of *R. solani* that has been incubated with purified bean chitinase shows the presence of both

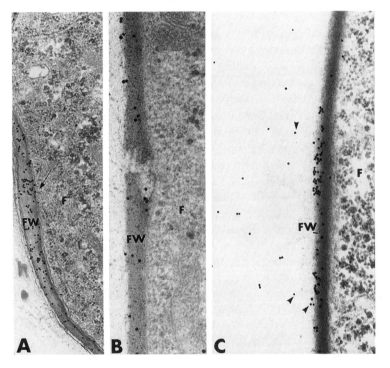

Fig. 8.3. Transmission electron microscope (TEM) photographs of *R. solani* hyphae after exposure to purified bean chitinase. **(A):** After 30 min of exposure to the enzyme, a marked swelling of the hyphal wall is observed (arrow). A decrease in the electron opacity of the wall is also noticeable (magnification 36,000 ×). **(B):** Some wall alterations are visible after 30 min of exposure to the enzyme. A decrease in wall labeling is observed following incubation of sections with WGA/ovomucoid–gold complex (magnification 72,000 ×). **(C):** Some gold particles are apparently released from a fungal cell wall (arrowheads) (magnification 54,000 ×).

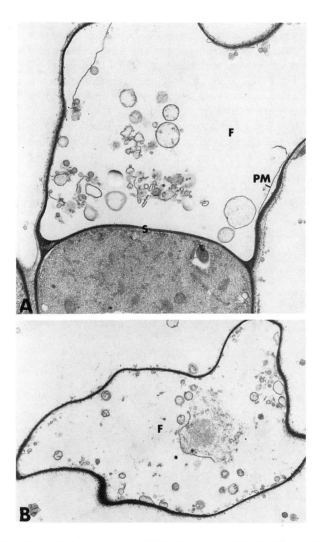

Fig. 8.4. Transmission electron microscope (TEM) photographs of *R. solani* hyphae after exposure to purified bean chitinase. (**A**) After 60 min of exposure to the enzyme, one cell portion is apparently preserved, while the other shows marked damage (magification 15,000 ×). (**B**) Two to three hours after exposure to the enzyme, the fungus is severely damaged as judged by the presence of an altered protoplasm reduced to only a few vesicles (magnification 15,000 ×). F, fungal cell; PM, plasma membrane; S, septum.

swollen and lysed hyphal tip cells and thus provides support for this explanation (Benhamou et al., in press). Ultrastructural and cytochemical investigation of chitin distribution using wheat germ agglutinin (WGA)/ovomucoid–gold labeling (Benhamou, 1991) following chitinase treatment revealed noticeable hyphal wall alterations (Fig. 8.3 **A** and **B**). The main alterations after 30 min of exposure to the enzyme were a marked swelling and a decrease in electron opacity of some wall areas. These changes were accompanied by a decrease in gold labeling intensity over the affected areas and a release of gold particles into the surrounding environment (Fig. 8.3**C**). Upon prolonged exposure to bean chitinase, severe structural modifications including plasmalemma retraction, wall disruption, and protoplasm leakage were commonly observed (Fig. 8.4**A**). Most fungal cells were severely damaged and suffered from pronounced distortion (Fig. 8.4**B**). A marked reduction or absence of gold labeling over these areas indicated an extensive break-

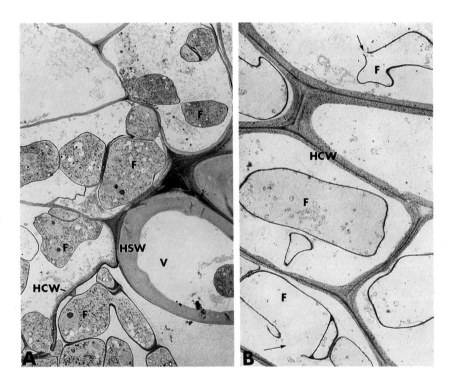

Fig. 8.5. Transmission electron microscope micrographs of canola root tissues infected with *R. solani*. (**A**): Wild-type canola plants. Numerous fungal cells are abundantly colonizing the vascular parenchyma of infected root tissues (magnification 2700 ×). (**B**): Transgenic canola plants expressing the 35S-chitinase gene. Invading fungal cells are markedly altered and some are disrupted (arrows). Nearly all hyphae exhibit a highly damaged protoplasm (magnification 3000 ×). F, fungal cell; HCW, host cell wall; HSW, host secondary wall; V, vacuole.

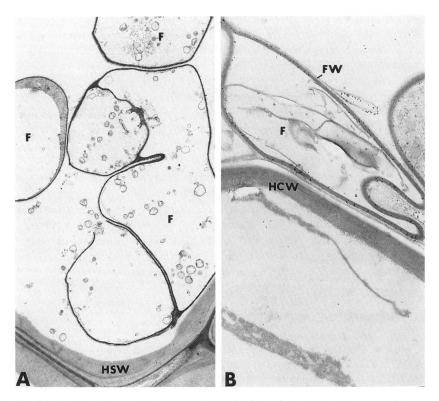

Fig. 8.6. Transmission electron microscope micrographs of roots from transgenic canola plants infected with *R. solani*. (**A** and **B**): Fungal cell wall breakdown and protoplasm leakage are evident. After treatment with WGA/ovomucoid–gold complex, little labeling is observed in the fungal wall. **A:** magnification 7200 ×; **B:** magnification 21,800 ×. F, fungal cell; HCW, host cell wall; HSW, host secondary wall; FW, fungal wall.

down of chitin. These data indicate that chitinase alone is sufficient to cause lysis of *R. solani* hyphae and that the chitin in this fungus is readily accessible to enzymatic attack presumably due to its peripheral location.

To determine whether the lytic effects of purified bean chitinase on *R. solani* seen *in vitro* are responsible for the enhanced resistance of 35S-chitinase transgenic plants a similar ultrastructural and cytochemical analysis was performed on infected control and transgenic canola plants. Analysis of root tissues of infected wild-type canola plants revealed that *R. solani* is capable of extensive colonization of all root tissues including the xylem vessels (Fig. 8.5A). Pathogen ingress was associated with marked host cell wall alterations, such as disruption of the middle lamella matrices. In contrast, the fungal hyphae do not appear to suffer significant damage as judged by their typical morphological features (Fig. 8.5A). In infected transgenic canola plants, however, the pattern of fungal colonization was different than that observed in wild-type plants. Penetration of the host cuticle and invasion of the epidermis was frequently observed, but fungal colonization was usually restricted to the cortex (in a few cases, some fungal cells were found to reach the xylem vessels) (Fig. 8.5**B**). However, in all samples examined, severe hyphal alterations ranging from increased vacuolization to cell lysis were seen. Those hyphae found in the xylem vessels were severely altered and often reduced to highly convoluted wall fragments (Fig. 8.6**A**). Labeling studies with WGA/ovomucoid revealed that the structural alterations of *R. solani* cell walls were accompanied by chitin degradation (Fig. 8.6**B**).

Thus, reduction in fungal biomass, increase in hyphal alterations leading to fungal lysis, and chitin breakdown are typical features observed in fungal resistant transgenic canola plants. Because these features were not seen in fungal infected wild-type canola plants, it is likely that constitutive expression of the bean chitinase gene is, at least in part, responsible for the enhanced protection against fungal attack seen in these plants. It is not known, however, if other components of the host defense response contribute to the resistance phenotype.

6. CONCLUSIONS AND FUTURE PROSPECTS

In experiments summarized here, we have shown that transgenic plants containing high, constitutive levels of bean endochitinase are more resistant to infection by the soil-borne pathogen, *R. solani* than are wild-type or control plants that lack the chimeric chitinase gene. The resistance of these plants to fungal infection was found to vary with the level of fungal inoculum and is indicative of a quantitative resistance. Both the delay in the onset of disease symptoms as well as the lower incidence of disease may enable young seedlings to survive the critical period in the field during stand establishment when they are most susceptible to attack by soil-borne pathogens.

The resistance of the 35S-chitinase plants to *Rhizoctonia* infection presumably results from the enzyme-catalyzed hydrolysis of chitin in the fungal cell wall. In agreement with this explanation, purified bean chitinase was found to inhibit the growth of *Rhizoctonia* on solid media and to promote disruption and lysis of growing hyphal tips (Broglie et al., 1991; Benhamou et al., submitted). Other *in vitro* studies have led to similar conclusions regarding the suggested participation of chitinase in the defense response of the plant (Young and Pegg, 1982; Boller et al., 1983; Schlumbaum et al., 1986; Mauch et al., 1988; Roberts and Selitrennikoff, 1988). In the studies summarized here, we have been able to demonstrate *in planta* that chitinase plays an important role in protecting plants against potentially pathogenic organisms. This protective effect, manifested by increased survival, was reinforced upon ultrastructural investigation of control and 35S-chitinase plants that had been grown in fungal-infested soil.

Through the generation and analysis of transgenic plants harboring the 35S-chitinase gene, we have shown that natural host defense mechanisms may be manipulated to produce fungal resistant plants. It is not known, however, if modification of chitinase expression alone is sufficient to provide protection against a wide range of fungal pathogens. While the chitin polymer is an important component of the fungal cell wall, the often predominant carbohydrate constituent is β-1,3-glucan with branches of β-1,6-linked units (Wessels and Sietsma, 1981). Mauch et al. (1988) found that in some instances, purified chitinase and β-1,3-glucanase alone are ineffective in inhibiting fungal growth *in vitro* and that instead, combinations of the two enzymes are required. Other work suggests that the antifungal action of chitinase may be enhanced synergistically in the presence of chitin-binding lectins (Broekaert et al., 1989; Chrispeels and Raikhel, 1991). It thus may be possible to enhance and/or extend the resistance response by integrating constitutive chitinase expression with the expression in transgenic plants of these and other available antifungal polypeptides.

REFERENCES

Abel PP, Nelson RS, De B, Hoffman N, Rogers SG, Fraley RT, Beachy RN (1986): Delay of disease development in transgenic plants that express the tobacco mosaic virus coat protein gene. Science 232:738–743.

Awade A, De Tapia M, Didierjean L, Burkard G (1989): Biological function of bean pathogenesis-related (PR 3 and PR 4) proteins. Plant Sci 63:121–130.

Bartnicki-Garcia S (1968): Cell wall chemistry, morphogenesis, taxonomy of fungi. Annu Rev Microbiol 22:87–108.

Bell JN, Ryder TB, Wingate VPM, Bailey JA, Lamb CJ (1986): Differential accumulation of plant defense gene transcripts in a compatible plant–pathogen interaction. Mol Cell Biol 6:1615–1623.

Benhamou N (1991): Electron microscopic localization of polysaccharides in fungal cell walls. In Latge JP, Boucias D (eds): Fungal Cell Wall and Immune Response, Vol. 53. NATO Advanced Study Institute Series. Berlin: Springer-Verlag, pp 205–218.

Benhamou N, Joosten MHA, De Wit PJGM (1990): Subcellular localization of chitinase and of its potential substrate in tomato root tissues infected by Fusarium oxysporum f. sp. radicis-lycopersici. Plant Physiol 92:1109–1120.

Boller T (1987): Hydrolytic enzymes in plant disease resistance. In Kosuge T, Nester EW (eds): Plant Microbe Interaction, Vol 2. New York: Macmillan Publishing, pp 385–414.

Boller T, Gehri A, Mauch F, Vogeli U (1983): Chitinase in bean leaves: induction by ethylene, purification, properties, and possible function. Planta 157:22–31.

Boller T, Vogeli U (1984): Vacuolar localization of ethylene-induced chitinase in bean leaves. Plant Physiol 74:442–444.

Bonhoff A, Reith B, Golecki J, Grisebach H (1987): Race cultivar-specific differences in callose deposition in soybean roots following infection with phytophthora megasperma f.sp. glycinea. Planta 172:101–105.

Bowles DJ (1990): Defense-related proteins in higher plants. Annu Rev Biochem 59:873–907.

Broekaert WF, Van Parijs J, Leyns F, Joos H, Peumans WJ (1989): A chitin-binding lectin from stinging nettle rhizomes with antifungal properties. Science 245:1100–1102.

Broglie KE, Biddle P, Cressman R, Broglie R (1989): Functional analysis of DNA sequences responsible for ethylene regulation of a bean chitinase gene in transgenic tobacco. Plant Cell 1:599–607.

Broglie KE, Chet I, Holliday M, Cressman R, Biddle P, Knowlton S, Mauvais CJ, Broglie R (1991): Transgenic plants with enhanced resistance to the fungal pathogen *Rhizoctonia solani*. Science 254:1194–1197.

Broglie KE, Gaynor JJ, Broglie RM (1986): Ethylene-regulated gene expression: Molecular cloning of the genes encoding an endochitinase from *Phaseolus vulgaris*. Proc Natl Acad Sci USA 83:6820–6824.

Carr JP, Klessig DF (1989): The pathogenesis-related proteins of plants. In Setlow JK (ed): Genetic Engineering, Principles and Methods, Vol II. New York: Plenum Press, pp 65–89.

Chet I, Henis Y, Mitchell R (1967): Chemical composition of hyphal and sclerotial walls of *Sclerotium rolfsii* Sacc. Can J Microbiol 13:137–141.

Christpeels MJ, Raikhel NV (1991): Lectins, lectin genes, and their role in plant defense. Plant Cell 3:1–9.

Darvill AG, Albersheim P (1984): Phytoalexins and their elicitors—A defense against microbial infection in plants. Annu Rev Plant Physiol 35:243–275.

Dixon RA, Day PM, Lamb CJ (1983): Phytoalexins: Enzymology and molecular biology. Adv Enzymol 55:1–136.

Espelie KE, Francheschi VR, Kolattukudy PE (1986): Immunocytochemical localization and time course of appearance of an anionic peroxidase associated with suberization in wound-healing potato tuber tissue. Plant Physiol. 81:487–492.

Esquerre-Tugaye MT, Lafitte C, Mazau D, Toppan A, Touze A (1979): Cell surfaces in plant-microorganism interactions II. Evidence for the accumulation of hydroxyproline-rich glycoproteins in the cell wall of diseased plants as a defense mechanism. Plant Physiol 64:320–326.

Fincher GB, Stone BA (1981): Metabolism of non-cellulosic polysaccharides. Encycl Plant Physiol New Series 13B:68–132.

Fischhoff DA, Bowdish KS, Perlak FJ, Marrone PG, McCormick SM, Niedermeyer JG, Dean DA, Kusano-Kretzmer K, Mayer EJ, Rochester DE, Rogers SG, Fraley RT (1987): Biotechnology 5:807–813.

Gianinazzi S (1984): Genetic and molecular aspects of resistance induced by infections or chemicals. In Nester EW, Kosuge T (eds): Plant–microbe interactions, molecular and genetic perspectives. Vol I. New York: Macmillan Publishing, pp 321–342.

Hadar Y, Chet I, Henis Y (1979): Biological control of *Rhizoctonia solani* damping-off with wheat bran culture of *Trichoderma harzianum*. Phytopathology 69:64–68.

Hahlbrock K, Grisebach H (1979): Enzymic controls in the biosynthesis of lignin and flavonoids. Annu Rev Plant Physiol 30:105–130.

Hahlbrock K, Scheel D (1987): Biochemical responses of plants to pathogens. In Chet I (ed): Innovative Approaches to Plant Disease Control. New York: Wiley, pp 229–252.

Hunter RE (1974): Inactivation of pectic enzymes by polyphenols in cotton seedlings of different ages infected with *Rhizoctonia solani*. Physiol Plant Pathol 4:151–159.

Jacobsen S, Mikkelsen JD, Hejgaard J (1989): Characterization of two antifungal endochitinases from barley grain. Physiol Plant 79:554–562.

Joosten MHAJ, DeWit PJM (1989): Identification of several pathogenesis-related proteins in tomato leaves inoculated with *Cladosporium fulvum* as 1,3-β-glucanases and chitinases. Plant Physiol 89:945–951.

Kombrink E, Schroder M, Hahlbrock K (1988): Several "pathogenesis-related" proteins in potato are 1,3-β-glucanases and chitinases. Proc Natl Acad Sci USA 85:782–786.

Kragh KM, Jacobsen S, Mikkelsen JD, Nielsen KA (1991): Purification and characterization of three chitinases and one β-1,3-glucanase accumulating in the medium of cell suspension cultures of barley (*Hordeum vulgare* L.). Plant Sci 76:65–77.

Leah R, Mikkelsen JD, Mundy J, Svendsen I (1987): Identification of a 28,000 dalton endochitinase in barley endosperm. Carlsberg Res Commun 52:31–37.

Legrand M, Kauffmann S, Geoffroy P, Fritig B (1987): Biological function of pathogenesis-related proteins: Four tobacco pathogenesis-related proteins are chitinases. Proc Natl Acad Sci USA 84:6750–6754.

Linthorst HJM, Melchers LS, Mayer A, Van Roekel JSC, Cornelissen BJC, Bol JF (1990): Analysis of gene families encoding acidic and basic β-1,3-glucanases of tobacco. Proc Natl Acad Sci USA 87:8756–8760.

Matta A, Gentile U, Giai I (1969): Accumulation of phenols in tomato plants infected by different forms of *Fusarium oxysporum*. Phytopathology 59:512–513.

Mauch F, Mauch-Mani B, Boller T (1988): Antifungal hydrolases in pea tissue. Plant Physiol 88:936–942.

Meins F Jr, Ahl P (1989): Induction of chitinase and β-1,3-glucanase in tobacco plants infected with *Pseudomonas tabaci* and *Phytophthora parasitica* var. *Nicotianae*. Plant Sci 61:155–161.

Metraux JP, Boller T (1986): Local and systemic induction of chitinase in cucumber plants in response to viral, bacterial and fungal infections. Physiol Mol Plant Pathol 28:161–169.

Metraux JP, Burkhart W, Moyer M, Dincher S, Middlesteadt W, Williams S, Payne G, Carnes M, Ryals J (1989): Isolation of a complementary DNA encoding a chitinase with structural homology to a bifunctional lysozyme/chitinase. Proc Natl Acad Sci USA 86:896–900.

Molano J, Polacheck I, Duran A, Cabib E (1979): An endochitinase from wheat germ. J Biol Chem 254:4901–4907.

Nasser W, De Tapia M, Burkard G (1990): Maize pathogenesis-related proteins: characterization and cellular distribution of 1,3-β-glucanases and chitinases induced by brome mosaic virus infection or mercuric chloride treatment. Physiol Plant Pathol 36:1–14.

Nothnagel EA, McNeil M, Albersheim P, Dell A (1983): Host-pathogen interactions. XXII. A galacturonic acid oligosaccharide from plant cell walls elicits phytoalexins. Plant Physiol 71:916–926.

Peng FH, Black LL (1976): Increased proteinase inhibitor activity in response to infection of resistant tomato plants by *Phytophthora infestans*. Phytopathology 66:958–963.

Ride JP (1975): Lignification in wounded leaves in response to fungi and its possible role in resistance. Physiol Plant Path 5:125–134.

Ride JP (1983): Cell walls and other structural barriers in defense. In Callow JA (ed): Biochemical Plant Pathology. New York: Wiley, pp 215–236.

Ride JP, Barber MS (1990): Purification and characterization of multiple forms of endochitinase from wheat leaves. Plant Sci 71:185–197.

Roberts WK, Selitrennikoff CP (1988): Plant and bacterial chitinases differ in antifungal activity. J Gen Micro 134:169–176.

Roby D, Broglie KE, Cressman R, Biddle P, Chet I, Broglie R (1990): Activation of a bean chitinase promoter in transgenic tobacco plants by phytopathogenic fungi. Plant Cell 2:999–1007.

Roby D, Broglie KE, Gaynor J, Broglie R (1991): Regulation of a chitinase gene promoter by ethylene and fungal elicitor in bean protoplasts. Plant Physiol 97:433–439.

Roby D, Esquerre-Tugaye M (1987): Induction of chitinases and of translatable mRNA for these enzymes in melon plants infected with *Colletotrichum lagenarium*. Plant 52:175–185.

Roby D, Toppan A, Esquerre-Tugaye MT (1988): Systemic induction of chitinase activity and resistance in melon plants upon fungal infection or elicitor treatment. Physiol Mol Plant Pathol 33:409–417.

Ryan CA (1973): Proteolytic enzymes and their inhibitors in plants. Annu Rev Plant Physiol 24:173–196.

Samac DA, Hironaka CM, Yallaly PE, Shah DM (1990): Plant Physiol 93:907–914.

Schlumbaum A, Mauch F, Vogeli U, Boller T (1986): Plant chitinases are potent inhibitors of fungal growth. Nature (London) 324:365–367.

Shah DM, Horsch RB, Klee HJ, Kishore GM, Winter JA, Tumer NE, Hironaka CM, Sanders PR, Gasser CS, Aykent S, Siegel NR, Rogers SG, Fraley RT (1986): Engineering herbicide tolerance in plants. Science 233:478–481.

Shinshi H, Mohnen D, Meins F Jr (1987): Regulation of plant pathogenesis-related enzyme: Inhibition of chitinase and chitinase mRNA accumulation in cultured tobacco tissues by auxin and cytokinin. Proc Natl Acad Sci USA 84:89–93.

Showalter AM, Bell JN, Cramer CL, Bailey JA, Varner JE, Lamb CJ (1985): Accumulation of hydroxyproline-rich glycoprotein mRNAs in response to fungal elicitor and infection. Proc Natl Acad Sci USA 82:6551–6555.

Sivan A, Chet I (1989): Cell wall composition of *Fusarium oxysporum*. Soil Biol Biochem 21:869–876.

Swegle M, Huang J-K, Lee G, Muthukrishnan S (1989): Identification of an endochitinase cDNA clone from barley aleurone cells. Plant Mol Biol 12:403–412.

Van den Bulcke M, Bauw G, Castresana C, Van Montagu M, Vandekerckhove J (1989): Characterization of vacuolar and extracellular β-1,3-glucanases of tobacco: Evidence for a strictly compartmentalized plant defense system. Proc Natl Acad Sci USA 2673–2677.

Van Loon LC (1985): Pathogenesis-related proteins. Plant Mol Biol 4:111–116.

Vance CP, Kirk TK, Sherwood RT (1980): Lignification as a mechanism of disease resistance. Annu Rev Phytopathol 18:259–288.

Vogeli-Lange R, Hansen-Gehri A, Boller T, Meins Jr F (1988): Induction of the defense-related glucanohydrolases, β-1,3-glucanase and chitinase, by tobacco mosaic virus infection of tobacco leaves. Plant Sci 54:171–176.

Wessels JGH, Sietsma JH (1981): Fungal cell walls: A survey. In Tanner W, Loewus FA (eds): Encyclopedia of Plant Physiology, New Series, Plant Carbohydrates II, Vol 13B. New York: Springer-Verlag, pp 352–394.

Young DH, Pegg GF (1982): The action of tomato and *Verticillium albo-atrum* glycosidases on the hypal wall of *V. albo-atrum*. Physiol Plant Pathol 21:411–423.

9

PROTEINASE INHIBITORS IN THE POTATO RESPONSE TO WOUNDING

JOSE J. SÁNCHEZ-SERRANO
SIMONE AMATI
CHRISTIAN DAMMANN
MARCUS EBNETH
KARIN HERBERS
THOMAS HILDMANN
RUTH LORBERTH
SALOMÉ PRAT
LOTHAR WILLMITZER

Institut für Genbiologische Forschung Berlin GmbH, 1000 Berlin 33, Germany

1. INTRODUCTION

Plant diseases annually cause important economical losses in all major crops worldwide. This result is a major concern especially in underdeveloped countries where rudimentary agricultural practices, often associated with warmer climates, promote pathogen proliferation. The ensuing disease outbreaks result in elevated prices of farm products and occasionally in food shortages in populations already on the edge of famine.

In their natural environment plants are normally surrounded by myriads of potential pathogens without disease being actually triggered. This phenomena occurs as a result of the presence of protective layers and other passive defense mecha-

Biotechnology in Plant Disease Control, pages 157–173
© 1993 Wiley-Liss, Inc.

nisms developed in plants throughout evolution that help to hinder the penetration of any infectious agent. However, pathogens eventually manage to gain access into the plant through natural openings (stomata) or through wounds that are caused by withering or insect attack. Most pathogens require the presence of a wound site to infect plants as is the case of some molds, and all bacteria and viruses. Infections of the latter are usually transmitted by insect and nematode vectors while feeding on the plants (Agrios, 1988).

Breeding programs are generally aimed at high crop yield, which appears to negatively correlate to disease resistance. Most of these high yield commercial varieties obtained by conventional breeding are only suited for growth under very favorable conditions, but the low yield of local varieties exhibiting resistance to endemic pathogens compromise their use. Traditional breeding programs combining high yield with resistance are extraordinarily time consuming, and alternative methods for introducing resistance traits in high yield varieties would be warmly welcomed. Molecular breeding, the introduction of single genetic traits by molecular biological techniques, is proving to be a sensible alternative to circumvent these problems.

While no plant disease resistance gene has been isolated so far, efforts on plant improvement through newly available techniques have primarily concentrated on the molecular characterization of some features of the plant–pathogen interaction, which have already been described for quite a long time. For example, infection with a virulent virus strain appears in some cases to be prevented by previous infection of a milder one. This protection is associated with the production of the viral coat protein, and is to a certain extent unspecific, preventing subsequent infections not only of the same virus but also of distantly related ones (Beachy et al., 1990).

Other protective strategies include the over-expression of nonplant proteins toxic to insects and their larvae. While reducing the damage caused by insect attack itself, an additional advantage is the reduction of the diseases they often transmit. Preliminary tests reported the success of this approach (Delannay et al., 1989) and made us hopeful that the impact of disease on crop yield will be greatly reduced in the near future.

Before any of these products can be of commercial application, a stricter control on the expression of the foreign protein is clearly needed. Its actual presence in plants only upon pathogen attack would be highly desirable, and it would therefore be very useful to set the transcription of the introduced gene under the control of pathogen-induced plant promoters. However, unrelated plant pathogens are likely to induce different sets of genes. Thus a major caveat would be the requirement of different transgenic plants for each and every possible disease agent. In this regard, plant wound-inducible promoters are a sensible alternative to drive expression of resistance genes since wounding is a prerequisite for a number of unrelated diseases. These promoters would be only active in instances of pathogen attack, thus preventing the unnecessary accummulation of toxins, or putative side effects, such as a penalty in yield, due to the high expression of the foreign protein or interferences with the plant metabolism. Moreover, the product of the transgene would be absent from those plants that had not suffered any attempted attack, and this result would probably help to comply with future regulations on the marketing of transgenic products for human nutrition.

2. THE PROTEINASE INHIBITOR II GENE FAMILY

Plants react to wounding by activating a set of genes, most of them playing a role in wound healing and prevention of any subsequent pathogen invasion. While the expression of some of them is restricted to the close vicinity of the wound site, others are also activated in the nondamaged parts of the plant. This systemic accumulation of defense products would then serve a preventive function checking the spread of the damaging agent to the healthy part of the plant (Bowles, 1990). The tomato and potato proteinase inhibitor II (pin2) families are the best studied examples of genes systemically induced upon wounding.

2.1. Gene Family

Pin2 is a 12-kD double-headed serine proteinase inhibitor polypeptide with specificity against trypsin and chymotrypsin activities. Several pin2 complementary deoxyribonucleic acids (cDNAs) and genomic clones have been isolated from both tomato and potato (Graham et al., 1985; Sánchez-Serrano et al., 1986), where it belongs to gene families consisting of three to five members per haploid genome. Potato and tomato clones share extensive sequence identity (>80%). Homologs to pin2 have been also found in eggplant and alfalfa, but tobacco and *Arabidopsis,* for instance, are devoid of any gene with significant similarity to pin2. The deduced amino acid sequence of a carboxypeptidase inhibitor present in potato tubers is colinear to part of the potato pin2 cDNA, and it has been suggested that this carboxypeptidase inhibitor may arise from posttranslational processing of the pin2 protein. Comparison of the cDNA sequences with the corresponding genomic clones indicate the presence of a single intron interrupting the pin2 open reading frame within a sequence with characteristics of a signal peptide, necessary for importing the pin2 protein through the endoplasmic reticulum membrane to its final destination at the cell vacuole. Intron splicing does not appear to play a role in the regulation of pin2 expression, since a chimeric gene with the intron sequences removed was correctly regulated upon wounding in transgenic plants (Keil et al., 1990).

2.2. Developmental and Environmental Factors Influence pin2 Gene Expression Pattern

Greenhouse grown potato plants have a high content of pin2 in tubers (Fig. 9.1). Its messenger ribonucleic acid (mRNA) appears in stolons of plants induced to tuberize alongside other tuber-specific genes, such as patatin, and it is present throughout tuber development. After dormancy, no pin2 mRNA is detected in sprouting tubers. This temporal regulation suggests a role for pin2 in preventing premature degradation of tuber storage proteins but no endogenous proteinase has been so far found that would be inhibited by pin2.

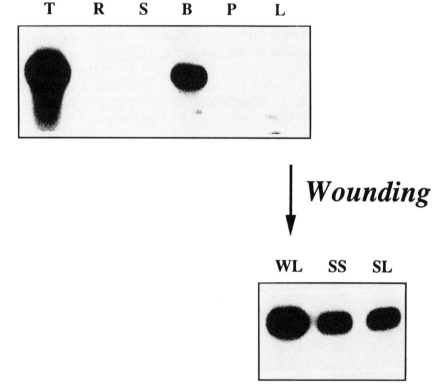

Fig. 9.1. During potato development pin2 messenger ribonucleic acid (mRNA) accumulates in tubers T and young floral buds B, being absent in the different organs of fully developed flowers, for instance in petals P. In healthy, nonstressed plants no pin2 mRNA is detectable in roots R, leaves L, or shoots S. However, pin2 transcription is activated upon mechanical wounding of the plant foliage, and its mRNA accumulates not only in the directly damaged leaves WL but also in nondamaged, systemically induced leaves SL and shoots SS.

In addition, pin2 is also constitutively expressed in potato and tomato flowers. In potato, pin2 mRNA accumulates in young flower buds but it is below the limits of detection in fully developed flowers (Fig. 9.1). Analyses of plants transgenic for a pin2 promoter-β-glucuronidase gene fusion indicated that pin2 promoter activity was primarily restricted to developing ovules. In contrast to this stage-specific activity, pin2 mRNA is present throughout development of the tomato flower (Peña-Cortés et al., 1991). This discrepancy in expression pattern may reflect the differential activity of members of the tomato pin2 gene family or, alternatively, differences in persistence of the putative signals responsible for pin2 gene activation in flowers, which might be related to differences in the metabolism of potato and tomato.

In addition to the above mentioned constitutive expression, pin2 mRNA accumulates to a large extent in the foliage of plants subject to mechanical damage. This wound-induced accumulation is not restricted to the tissue surrounding the wound site, but rather a systemical accumulation is also observed in the distal, nondamaged tissue. The accumulation of pin2 upon mechanical injury, together with its constitutive presence in tissues important in plant reproduction, lends support to a putative role of pin2 in the plant defense strategy.

The pattern of pin2 expression after wounding suggests the existence of a wound signal, synthesized or released at the wound site, and spreading throughout the plant to trigger systemic pin2 gene activation. It has been shown that most organs of the potato plant are able to produce a signal upon wounding. Potato roots offer a particularly interesting example of the complexity of pin2 regulation by environmental factors. Wounding of roots by mechanical damage (Peña-Cortés et al., 1988) or nematode attack (Hammond-Kosack et al., 1990) triggers pin2 accumulation in the aerial part of the plant (indicative of the presence of the wound signal) but not in roots themselves. Roots are therefore competent for the production of the wound signal but not for its perception.

In addition, pin2 expression in tubers appears to be affected by environmental stimuli, since wounding of the tuber under aerobic conditions results in disappearance of pin2 mRNA (Sánchez-Serrano et al., 1986). Wounding under anaerobiosis has, however, no effects on pin2 mRNA levels (Peña-Cortés et al., 1988).

Pin2 activation upon wounding thus appears to result from the interaction of one or more signals released or synthesized upon wounding with cell-specific receptors absent in some parts of the plant. Several substances have been put forward as putative wound signals, and it is likely that most of them play a role in wound-induced gene activation. Oligosaccharides have been repeatedly mentioned as putative elicitors of wound responses, alongside their modulating role in plant development (Ryan, 1991). These molecules would be released at the wound site upon contact with endogenous *endo*-polygalacturonases released from intracellular stores by mechanical disruption.

Indeed, oligosaccharides such as pectic fragments derived from the plant cell wall as well as chitosan, a β-1,4-glucosamine polymer normally found in fungal cell walls, greatly enhance the accumulation of pin2 mRNA upon application to detached potato or tomato leaves. Their limited solubility in the plant phloem (Baydoun and Fry, 1985), however, is not consistent with the rapid systemic pin2 induction observed many centimeters away from the wound site, and seems to preclude them as the actual factor responsible for pin2 systemic induction. In fact, the protein pattern observed upon oligosaccharide treatment of tomato leaves closely resembles the pattern observed upon local wounding rather than the pattern observed in systemically induced leaves (Dalkin and Bowles, 1989). Thus, these polymers appear to be related to the local events at the wound site but are most likely not responsible for gene activation at more distal sites. In this case, they might function by triggering the local synthesis of a more diffusible substance that would subsequently act as a systemic wound signal. However, no endogenous polygalacturonase activity has been detected so far, and thus the *in vivo* role of oligosaccharides in wound induction of pin2 remains to be determined.

The plant hormone abscisic acid (ABA) has been often associated with plant responses to different stress situations (Skriver and Mundy, 1990). Several lines of evidence indicate that ABA also plays a role in pin2 induction upon wounding. First, the expression of the pin2 gene family is not induced in wounded leaves of tomato and potato mutants deficient in ABA synthesis, and this lack of expression is lifted by supplying ABA through the petioles of detached leaves. Second, ABA treatment can on its own induce pin2 accumulation in the foliage of ABA-sprayed potato plants, in the absence of any damage (Peña-Cortés et al., 1989). Determination of endogenous ABA concentrations upon wounding revealed that this ABA-induced pin2 activation is physiologically meaningful. There is a six-fold rise in endogenous ABA concentration upon wounding in both tomato and potato leaves. These levels peak before maximal pin2 mRNA accumulation is reached.

While ABA is involved at some stage during wound-induced gene activation, constitutive pin2 expression in potato flowers and tubers is apparently not affected by the block in the ABA biosynthetic pathway. Thus the ABA-deficient droopy mutant has wild-type pin2 mRNA levels in tubers and flower buds, suggesting the presence of different or additional factors for pin2 gene activation in tubers and flowers (Peña-Cortés et al., 1991). Because pin2 expression is not induced in water deficit situations, in spite of an 8–10-fold increase in the endogenous ABA concentration, encourages the argument for the existence of other signals modulating the effect of ABA on pin2 gene activation. The linolenic acid derivative jasmonic acid (JA) has been shown to trigger molecular responses similar to those induced by ABA treatment in several plant species, and thus has been associated with plant stress reactions (Parthier, 1990). More precisely, JA involvement in plant responses to mechanical damage has been conclusively established for the wound-induced expression of vegetative storage proteins (VSP) in soybean leaves. In this case, JA treatment triggered VSP accumulation in the plant foliage in the absence of any damage. Moreover, inhibitors of JA synthesis blocked wound-induced VSP accumulation, but this block was lifted by JA directly applied to the foliage (Staswick et al., 1991).

Pin2 expression is also triggered in the foliage of potato plants treated with JA or its methyl ester (Farmer and Ryan, 1990). This JA-induced pin2 accumulation also occurs in the foliage of the ABA-deficient potato and tomato mutants (Ebneth M and Sánchez-Serrano JJ, unpublished). These data suggest that the step where JA is involved is located downstream to the ABA requirement in the pathway that links wounding to pin2 gene activation. This result would be consistent with the association often found in plant responses to JA and ABA treatments. Yet the possibility that the steps involving JA and ABA belong to different signal pathways that eventually converge in pin2 activation cannot be ruled out.

In a manner similar to other responses to plant hormones, pin2 transcription is likely to respond to the imbalance of the different plant growth regulators exerting additive or antagonistic effects. In this respect, it has been shown that high auxin concentrations repress pin2 expression (Kernan and Thornburg, 1989). Moreover, indole 3 acetic acid levels decline after wounding suggesting the involvement of auxins in the regulation of pin2 activity (Thornburg and Li, 1991).

Recently, a peptide of 18 amino acid residues has been shown to induce pin2 expression if applied through a cut stem of tomato plant, while cutting and further incubation in water alone do not trigger detectable pin2 accumulation (Pearce et al., 1991). Moreover, this peptide, which has been termed systemin, can be transported out of the wound site suggesting its involvement in pin2 systemic activation. This peptide is likely to be derived from a precursor protein suggesting posttranslational events as possible regulatory steps. How these different signals integrate with one another in a signal transduction pathway or if they at all belong to the same pathway is not well established. As mentioned previously, ABA appears to precede JA in the transduction pathway. Its similarity to animal prostaglandins, both in chemical structure and in biosynthetic pathway, suggests that JA might act as a short-range signal, as prostaglandins in animals. However, Pin2 mRNA accumulates upon exogenous JA application either by spraying or through cut petioles indicating that under these circumstances JA is able to migrate and trigger systemic pin2 expression (Ebneth M and Sánchez-Serrano JJ, unpublished). Thus its role as an intracellular signal has to be taken with caution.

2.3. Promoter Elements

As mentioned previously, pin2 belongs to a small multigene family and, therefore, its different modes of expression (constitutive in tubers and flowers, and inducible in leaves upon wounding) might reflect differential activity of members of the gene family. Alternatively, one or several members of the gene family may be active in part or all of these situations. This issue was addressed by introducing into potato a gene fusion consisting of a pin2 promoter hooked to a β-glucuronidase (GUS) reporter. This construction endowed the transgenic plants with constitutive GUS activity in tubers and floral buds, and wound-induced activity in leaves (Keil et al., 1989). A single promoter element therefore mirrors the expression pattern of the whole gene family indicating the presence within this promoter of all cis-acting sequences responsible for the complex regulation of pin2 by environmental and developmental factors. The activity of another highly similar pin2 promoter has also been shown to be wound inducible in transgenic tobacco plants (Thornburg et al., 1987). Its activity in potato so far has not been tested and thus it is not known yet if it is a general rule that pin2 promoters are active in both tubers and wounded leaves.

The cis-acting elements required for correct regulation of the pin2 promoter have been defined by determining the activity of promoters with 5', 3', and internal deletions, in transgenic tobacco and potato plants. These analyses have revealed that pin2 promoter apparently consists of at least two functionally distinct elements: an upstream quantitative enhancer, required for maximal levels of expression, whose activity is modulated by downstream regulatory elements. The basic promoter elements located close to the transcription start are apparently not involved in the developmental and environmental regulation of pin2 promoter activity. Thus, truncated promoters derived from either the Cauliflower Mosaic Virus (CaMV) 35S promoter (Keil et al., 1990) or the nopaline synthase promoter (Palm et al., 1990) can substitute for the endogenous pin2 sequences, and be activated by upstream regions of the pin2 promoter in a fashion identical to the authentic pin2 promoter itself. This pin2 promoter upstream region moreover has the feature of an enhancer, since it can activate truncated promoters when fused in either orientation. The quantitative enhancer function maps to the promoter region from -1300 to -700 with respect to the pin2 transcription start site, and is required for maximal promoter activity in potato tubers and wounded leaves of transgenic tobacco and potato plants. In the latter organs, this enhancing region could be narrowed down to sequences within -801 to -699. This region was shown to differ in this respect from the sequence requirements for maximal pin2 expression in flowers, which appears to depend also on other sequences located elsewhere within the quantitative enhancer (Lorberth et al., 1992). To confer maximal activity, this enhancer element requires the presence of the downstream regulatory region, and it thus cannot activate a silent truncated 35S promoter on its own, while activity is recovered by addition of the region between -624 and -405. This recovery occurs in tubers and wounded leaves in a concerted fashion, and no promoter deletion so far analyzed could separate tuber-specific from wound-specific cis-acting sequences. Clearly, a more refined analysis, for example, linker-scanning mutagenesis, is needed to discriminate whether expression in tubers and wounded leaves is mediated by closely located, but different, cis-regulatory sequences or a single cis-acting sequence responding to a signal common to tuberization and wounding.

Sequences downstream to -405 are also able to interact with the quantitative enhancer element and activate a truncated 35S promoter in flowers in the absence of the -624 to -405-region. These flower-specific cis-acting sequences are thus redundant with respect to those located within the -624 to -405 regulatory region, and have to be therefore different from those regulating pin2 expression in tubers and wounded leaves (Lorberth et al., 1992). This result, together with the different requirements concerning the quantitative enhancer region for maximal expression, suggests that pin2 promoter activity in flowers is regulated through a signal transduction pathway different from the one leading to pin2 activity in tubers and wounded leaves.

It is worth mentioning that the effect of JA and ABA on pin2 mRNA accumulation in leaves requires the presence of the -624 to -405-region. In contrast to wild-type promoters, leaves of plants transgenic for a promoter with this region deleted were not induced to accumulate the reporter GUS mRNA upon JA or ABA treatment, in spite of the fact that this mRNA was readily detected in the floral buds of these plants. Expression in flowers has to follow a signal pathway mediated by substances different from, or in addition to, JA and ABA.

As mentioned previously, pin2 expression in tubers and flowers is not affected by the ABA deficiency in the potato droopy mutants, suggesting the existence of an ABA-independent transduction pathway for constitutive pin2 activity in tubers and flowers, and an ABA-dependent pathway for the transduction of the wound signal in leaves. The effect of JA on pin2 promoter activity also may be related to tuber expression, as suggested by the similar chemical structure of JA with a recently isolated tuber-inducing substance, named tuberonic acid (Yoshihara et al., 1989). Indeed, JA itself induces tuberization at similar concentrations (Koda et al., 1991). A direct effect on pin2 expression in tubers is difficult to demonstrate, since no JA-deficient mutant is available to date. However, involvement of JA in wound induction is well supported by data on pin2 expression in JA-treated leaves and the blocking of wound-induced VSP expression in soybean by inhibitors of JA synthesis. Although JA may, therefore, well be the compound linking pin2 activity in tubers and wounded leaves, conclusive evidence is still missing.

3. OTHER POTATO GENES INDUCED IN RESPONSE TO WOUNDING

Mechanical wounding leads to dramatic changes in the plant's gene expression pattern with the subsequent appearance of newly synthesized protein products. These are normally absent in this tissue prior to the injury, but may be constitutively present in other nondamaged organs. In potato, for instance, the appearance of several tuber-specific proteins in leaves is induced upon wounding. Quite remarkably, all products characterized so far have inhibitory activity towards proteinases, with proteinase inhibitors I and II being the best studied examples among them. Also, the tuber-abundant 22-kD protein has been recently shown to inhibit serine-proteinases like trypsin and chymotrypsin, and to accumulate after mechanical damage in the plant foliage in a systemic manner (Suh et al., 1991). Likewise, a small metalloproteinase inhibitor, constitutively expressed in tubers, accumulates in leaves upon wounding (Martineau et al., 1991). Although wound induction of tuber-specific genes is a repeated motif in the potato defense strategy, the major tuber protein, patatin, is not wound inducible in the plant foliage where it, however, responds to alterations in nitrogen and carbon levels (Peña-Cortes et al., 1992) in a manner reminiscent of the soybean vegetative storage protein, which is itself inducible in leaves upon wounding.

Work on the wound induction of the pin2 gene family has established the role of ABA as an early signaling event in the potato response to mechanical wounding. To validate this point, a search for other genes that would fit into this model was undertaken. To this end, a cDNA library for ABA-treated, nondamaged potato leaves was constructed. Clones were isolated from this library by differential screening with radioactive cDNA probes from wounded versus nonwounded leaves. Four of them were characterized in more detail (Hildmann et al., 1992), and are described later (Fig. 9.2). The identification of the nature of their encoded products has lent more support to the pivotal role of proteinase inhibitors in potato response to wounding. The identity of some of them, however, clearly indicates

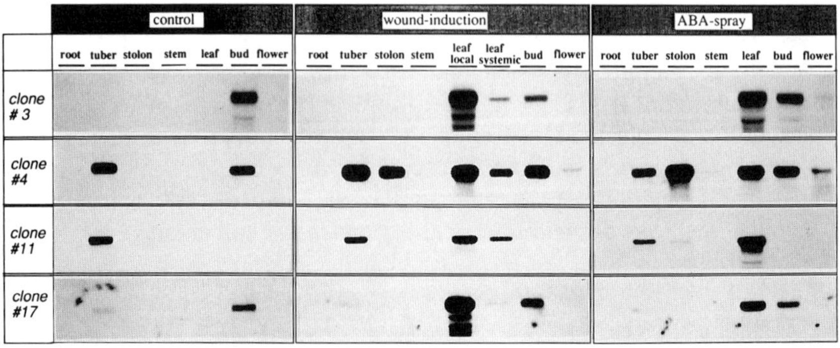

Fig. 9.2. Messenger RNAs encoding threonine dehydrase (#3), cathepsin D inhibitor (#4), a thiol proteinase inhibitor (#11), and a leucine aminopeptidase (#17) in the potato response to mechanical damage (wound induction). These genes are also active upon ABA treatment of the plant foliage (ABA spray). These mRNAs also constitutively accumulate in tubers (#4 and #11) and flower buds (#3, #4, and #17) of noninduced potato plants.

the involvement of other activities in plant defense strategy and may reflect profound alterations in plant metabolic pathways as a result of mechanical damage. As ABA is but one of the signal molecules produced upon wounding, it is likely that genes are induced upon wounding that, however, are not ABA responsive. These genes may have escaped this selection procedure, and other lower expressed ABA-responsive genes could have also remained undetected.

3.1. Cathepsin D Inhibitor

The sequence of one of these ABA-responsive, wound-inducible clones that were isolated exhibited over 90% identity with the recently described potato cathepsin D inhibitor (isolated from a different potato variety) (Strukelj et al., 1990). It belongs to a small multigene family in potato with a highly similar counterpart in the tomato genome.

Cathepsin D inhibitor is encoded in a 800 nucleotides (nt) mRNA, which exhibits an expression pattern identical to pin2, being both constitutively expressed to high levels in tubers and flower buds, and in the foliage upon wounding. A systemic accumulation of the cathepsin D inhibitor mRNA after wounding is also observed. In a similar manner to pin2, cathepsin D inhibitor accumulation is also triggered in the foliage upon ABA or JA treatment. In addition, tomato plants accumulate this inhibitor in the mechanically damaged leaves, or in JA treated leaves. The function of this aspartate proteinase inhibitor may complement the action of the serine proteinase inhibitors pin1 and pin2, covering a broader range of proteinases that might be present in the digestive tract of plant feeding insects.

3.2. Thiol Proteinase Inhibitor

Another clone hybridized against a 2.5-kb message present in wounded, ABA-, or JA-treated leaves, but not in nonwounded control ones. The cloned fragment represents only a small part of this message. Sequencing of this insert revealed some degree of similarity with the rice thiol proteinase inhibitor oryzacistatin (Abe et al., 1987), including the QVVAG amino acid motif apparently required for their inhibitory activity. This mRNA would have coding capacity for a 80-kD protein, thus matching the size of a thiol proteinase inhibitor found in potato tubers (Rodis and Hoff, 1984). Indeed, the mRNA constitutively accumulates in tubers but, in contrast to pin2, no accumulation can be detected in flower buds. This putative thiol proteinase inhibitor also accumulates in both damaged leaves and systemically induced ones.

Hybridization to potato genomic DNA shows that this thiol proteinase inhibitor belongs to a small gene family, which has a homologous counterpart in tomato. Indeed, a homologous message also accumulates in wounded tomato leaves.

As in the case of cathepsin D inhibitor, expression of the thiol proteinase inhibitor might help to completely inactivate the different proteinases present in the insect's gut.

3.3. Leucine Aminopeptidase

A third clone was characterized and shown to hybridize to a 2.0-kb message present in ABA- or JA-treated leaves, as well as in wounded leaves. The size of the cloned insert was 1.7 kb, thus missing the sequences coding for the NH_2 terminal region of the protein. Sequence comparisons revealed high similarity to the bovine lens and yeast leucine aminopeptidase. This homology is especially high around the active center of the enzyme, where the homology between yeast and bovine proteins is also highest (Stirling et al., 1989).

As mentioned earlier, mRNA homologous to this leucine aminopeptidase is detected in directly wounded leaves. In addition, a very weak systemic induction can be detected. Its message is constitutively present in flower buds but only very weakly in tubers. As for the other clones, homologous genes are also found in tomato, where they are activated in leaves upon wounding.

The role of this enzyme in the plant response to wounding is not known. It may help to reduce to amino acids proteins that have already been attacked by endopeptidases, and thus help to allocate energy resources to the synthesis of newly needed compounds. However, as has been postulated already in the case of the bovine lens protein, a structural role cannot be discarded. In that case, leucine aminopeptidase would be an ancillary activity of a protein overexpressed for a different structural, nonenzymatic role.

3.4. Threonine Dehydrase

The last clone characterized hybridizes to a 2.0-kb message present as the others in wounded potato leaves and nonwounded leaves treated with either ABA or JA. Sequence comparison detected 90% similarity to a recently described abundant tomato flower protein, namely, threonine dehydrase (Samach et al., 1991). As in tomato, its mRNA accumulates in potato flowers but, in contrast to the other clones described, it is not detectable in tubers.

Threonine dehydrase accumulates in directly damaged leaves but displays only very weak, if at all, systemic accumulation in the nonwounded leaves. As in the case of the leucine aminopeptidase, the role of threonine dehydrase in plant response to wounding is difficult to envision. A major effect of its over-expression might be the enhanced synthesis of side chain bearing amino acids like isoleucine. However, a degradative role, converting threonine in the easily metabolizable α-ketobutyrate, has a better appeal in regard to energy redistribution after wounding.

4. POSSIBLE APPLICATIONS

4.1. The Wound Response and the Accumulation of Proteinase Inhibitors

Plant response to mechanical damage results in the transcriptional activation of genes whose function is devoted to preventing further damage and restoring the conditions present prior to wounding, subsequently returning to noninduced transcription levels. In potato, the products of several of these genes induced in the plant foliage as a result of mechanical wounding have inhibitory activity towards proteinases, and may thus be involved in the plant's defense strategy against feeding pests by inactivating proteinases present in their digestive tract. In this case, inhibitors of serine-, thiol-, aspartate-, and metalloproteinases accumulate in the plant foliage upon wounding. This accumulation also occurs in the nondamaged part of the plant. Thus, the plant responds to insect feeding by accumulating products whose action greatly decreases the nutritional value of the ingested tissue. Their accumulation in systemically induced tissues precedes insect feeding. Although experiments with artificial diets have shown that ingestion of proteinase inhibitors might lead to death of insect larvae, wound-induced accumulation might not reach such toxic levels. A complementary mechanism involving overproduction of the endogenous insect proteinases upon inhibition by the plant proteinase inhibitors is also apparent (Broadway and Duffey, 1986). Thus, a decrease in fitness due to undernurturing may be expected, with subsequent reduced numbers in the progeny controlling pest population to levels that do not compromise the survival of the plant species. However, monocultivar agricultural practices together with environmental damage due to human activities, which have led to a decrease in the natural predators, have resulted in massive pest attacks that cause significant damage to the crop before these natural mechanisms may be put into action.

One can thus envision the use of proteinase inhibitors to check pest populations by over-expressing them in plants prior to pest attack. This procedure is unlikely to result in the appearance of resistant subpopulations because of the absolute requirement of proteinase function by the insect, and the evolutionary conservation of the enzymatic activity and the mechanisms of inhibition. Although a high degree of control in a given plant–pest interaction may require the study of the proteinases present in the pest gut, and the transformation of the plant with appropriate inhibitors, a strategy based on expression of several inhibitors with different specificities, as is the case in potato, might result in a valuable degree of protection against a wide range of interactions.

4.2. Delineation of Elements to Construct Chimeric Promoters

In several instances, a faster elimination of the pest may be required, and thus methods tending to establish a new equilibrium dynamics in the pest and crop populations might be too slow. The expression of toxic products highly aggressive towards the pest might be obtained in transgenic plants for essentially all major dicotyledonous crops in the near future. This method poses a great environmental threat since over-expression of the product for long periods may result in the fast appearance of resistant mutants. Since the pool of wild-type individuals would constantly decrease, the spread of the mutant phenotype might be anticipated to be very rapid. Also, access of the toxins to food webs including other nondamaging insects may substantially alter ecological balances and lead to greater outbreaks of pest attacks. On the other hand, increasingly tighter regulations in plant production for human feeding may make highly desirable, and thus economically profitable, the expression of these toxic products only under circumstances of actual pest attack.

Wound-inducible promoters are therefore a natural choice for the over-expression of these products in transgenic plants. Systemically induced promoters, as is the case for pin2, offer a greater degree of protection. Thus, definition of the sequences involved in directing wound-inducible promoter activity is required for the design of appropriate expression cassettes. In addition, studies on the organ-specific activity of this promoter are also required to reduce, if required, its constitutive expression in tubers or to deal, for example, with the absence of promoter activity in systemically induced roots. This organ specificity is a particular problem in plant protection against disease, since pin2 promoter based constructions would be ineffective against nematodes and associated diseases. Moreover, higher levels of expression of the toxin might be required for effective protection. Thus, construction of chimeric promoters involving cis-acting wound-regulatory sequences fused to strong enhancer elements, such as the enhancer element of the CaMV 35S promoter, could be necessary.

4.3. Manipulation of the Signal Transduction Pathway to Generate Hyperreactive Plants

Plants and pathogens have coevolved to reach a balanced coexistence that rarely results in disease outbreak. However, agricultural practices have introduced monocultivar farming, allowing larger numbers of the putative pathogens to grow in more limited areas. Also, breeding for high yield production has incurred penalties for plant resistance. In cases where plant–pathogen interaction results in disease, pathogens are more often successful because a delayed defense reaction occurs (Dixon and Lamb, 1990), rather than inactivation of the activated defensive mechanisms or colonization of plant varieties devoid of these deterring systems.

Plants offer an array of defensive substances that can be used for breeding programs aimed to improved resistance. However, higher resistance also might be achieved by studying, and subsequently manipulating, the mechanisms leading to gene activation upon pathogen attack. Thus, a faster response towards a pathogen may render resistant a previously susceptible plant variety. In this case, no expression of toxic products other than those normally occurring in this precise plant species will occur. This potentiation of the plant's own defense mechanisms can be achieved by expression of extremely low levels of the appropriate enzymes, for example, those involved in the bottleneck steps of synthesis of stress hormones, or hormonal receptors.

5. PROSPECTS

The application of the molecular biological techniques currently available is bound to have a major impact in plant breeding. Plant resistance to disease is one of the aspects where the best prospects are set for the generation of commercially available varieties in the near future, which are currently subject to field tests. The use of transgenic varieties will allow an environmentally cleaner approach to crop management in pest protection due to the reduction of pesticide release to the environment. Moreover, these plants will hopefully allow a better discrimination of the target pest in contrast to indiscriminate destruction with pesticide treatment. Other risks, however, will have to be faced that were not present before. The persistence of the toxins in the plants may pose a high selective pressure favoring the early appearance of resistant mutants, and their presence in products destined for human consumption may render the marketing of these plant varieties difficult. The use of wound-inducible promoters, from which the pin2 promoter is one of the best studied examples, may help to improve these situations. On one hand, accumulation of the protein under circumstances of actual attack will target the toxic products to the actual pest, and may help in preventing appearance of resistance by retaining a pool of the wild-type population, which would always survive at the early stages of pest attack. Healthy plants would, on the other hand, not accumulate the toxic products, and are most likely candidates to easily comply with future regulations of commercial use of transgenic food products.

ACKNOWLEDGMENT

We are especially grateful to Dr. Peter Morris for useful comments on the manuscript.

REFERENCES

Abe K, Emori Y, Kondo H, Suzuki K, Arai S (1987): Molecular cloning of a cysteine proteinase inhibitor of rice (Oryzacystatin). J Biol Chem 35:16793–16797.

172   SÁNCHEZ-SERRANO ET AL.

Agrios GN (1988): Plant Pathology. San Diego, CA: Academic Press.

Baydoun EA, Fry SC (1985): The immobility of pectic substances in injured tomato leaves and its bearing on the identity of the wound hormone. Planta 165:269–276.

Beachy RN, Loesch-Fries S, Tumer NE (1990): Coat protein-mediated resistance against virus infection. Annu Rev Phytopathol 28:451–474.

Bowles DJ (1990): Defense-related proteins in higher plants. Annu Rev Biochem 59:873–907.

Broadway RM, Duffey SS (1986): Plant proteinase inhibitors: mechanism of action and effect on the growth and digestive physiology of larval *Heliothis zea* and *Spodoptera exiqua*. J Insect Physiol 32:827–833.

Dalkin K, Bowles DJ (1989): Local and systemic changes in gene expression induced in tomato plants by wounding and by elicitor treatment. Planta 179:367–375.

Delannay X, LaVallee BJ, Proksch RK, Fuchs RL, Sims SR, Greenplate JT, Marrone PG, Dodson RB, Augustine JJ, Layton JG, and Fischhoff D (1989): Field performance of transgenic tomato plants expressing the *Bacillus thuringensis* var. Kurstaki insect control protein. Biotechnology 7:1265–1269.

Dixon RA, Lamb CJ (1990): Molecular communications in interactions between plants and microbial pathogens. Annu Rev Plant Physiol Plant Mol Biol 41:339–367.

Farmer EE, Ryan CA (1990): Interplant communication: airborne methyl jasmonate induces synthesis of proteinase inhibitors in plant leaves. Proc Natl Acad Sci USA 87:7713–7716.

Graham JS, Pearce G, Merryweather J, Titani K, Ericsson LH, Ryan CA (1985): Wound-induced proteinase inhibitors from tomato leaves. II: The cDNA-deduced primary structure of pre-inhibitor II. J Biol Chem 260:6561–6564.

Hammond-Kosack KE, Atkinson HJ, Bowles DJ (1990): Changes in abundance of translatable mRNA species in potato roots and leaves following root invasion by cyst-nematode *G. rostochiensis* pathotypes. Physiol Mol Plant Pathol 37:339–354.

Hildmann T, Ebneth M, Peña-Cortes H, Sanchez-Serrano JJ, Willmitzer L, Prat S (1992): General roles of abscisic and jastionic acids in gene activation as a result of mechanical wounding. The Plant Cell (in press).

Keil M, Sánchez-Serrano JJ, Schell J, Willmitzer L (1990): Localization of elements important for the wound-inducible expression of a chimeric potato proteinase inhibitor II-CAT gene in transgenic tobacco plants. Plant Cell 2:61–70.

Keil M, Sánchez-Serrano JJ, Willmitzer L (1989): Both wound-inducible and tuber-specific expression are mediated by the promoter of a single member of the potato proteinase inhibitor II gene family. EMBO J 8:1323–1330.

Kernan A, Thornburg RW (1989): Auxin levels regulate the expression of a wound-inducible proteinase inhibitor II-chloramphenicol acetyl transferase gene fusion *in vitro* and *in vivo*. Plant Physiol 91:73–78.

Koda Y, Kikuta Y, Tazaki H, Tsujino Y, Sakamura S, Yoshihara T (1991): Potato tuber-inducing activities of jasmonic acid and related compounds. Phytochemistry 30:1435–1438.

Lorberth R, Dammann C, Ebneth M, Amati S, Sanchez-Serrano JJ (1992): Promoter elements involved in environmental and developmental control of potato proteinase inhibitor II expression. Plant Journal 2:477–486.

Martineau B, McBride KE, Houck CM (1991): Regulation of metallocarboxypeptidase inhibitor gene expression in tomato. Mol Gen Genet 228:281–286.

Palm CJ, Costa MA, An G, Ryan CA (1990): Wound-inducible nuclear protein binds DNA fragments that regulate a proteinase inhibitor II gene from potato. Proc Natl Acad Sci USA 87:603–607.

Parthier B (1990): Jasmonates: hormonal regulators or stress factors in leaf senescence? J Plant Growth Regul 9:57–63.

Pearce G, Strydom D, Johnson S, Ryan CA (1991): A polypeptide from tomato leaves induces wound-inducible proteinase inhibitor proteins. Science 253:895–898.

Peña-Cortés H, Liu X, Sánchez-Serrano JJ, Schmid R, Willmitzer L (1992): Factors affecting gene expression of patatin and proteinase inhibitor II gene families in detached potato leaves: Implications for their coexpression in developing tubers. Planta, 186:495–502.

Peña-Cortés H, Sánchez-Serrano JJ, Mertens R, Willmitzer L, Prat S (1989): Abscisic acid is involved in the wound-induced expression of the proteinase inhibitor II gene in potato and tomato. Proc Natl Acad Sci USA 86:9851–9855.

Peña-Cortés H, Sánchez-Serrano JJ, Rocha-Sosa M, Willmitzer L (1988): Systemic induction of proteinase inhibitor II gene expression in potato plants by wounding. Planta 174:84–89.

Peña-Cortés H, Willmitzer L, Sánchez-Serrano JJ (1991): Abscisic acid mediates wound induction but not developmental-specific expression of the proteinase inhibitor II gene family. Plant Cell 3:963–972.

Rodis P, Hoff JE (1984): Naturally occurring protein crystals in the potato. Plant Physiol 74:907–911.

Ryan CA, Farmer EE (1991): Oligosaccharide signals in plants: a current assessment. Annu Rev Plant Physiol Mol Biol 42:651–674.

Samach A, Hareven D, Gutfinger T, Ken-Dror S, Lifschitz E (1991): Biosynthetic threonine deaminase gene of tomato: isolation, structure, and upregulation in floral organs. Proc Natl Acad Sci USA 88:2678–2682.

Sánchez-Serrano JJ, Schmidt R, Schell J, Willmitzer L (1986): Nucleotide sequence of proteinase inhibitor II encoding cDNA of potato (*Solanum tuberosum*) and its mode of expression. Mol Gen Genet 203:15–20.

Skriver K, Mundy J (1990): Gene expression in response to abscisic acid and osmotic stress. Plant Cell 2:503–512.

Staswick PE, Huang J-F, Rhee Y (1991): Nitrogen and methyl jasmonate induction of soybean vegetative storage protein genes. Plant Physiol 96:130–136.

Stirling CJ, Collomns SD, Collins JF, Szatmari G, Sherratt DJ (1989): *xerB*, an *Escherichia coli* gene required for plasmid ColE1 site-specific recombination, is identical to *pepA*, encoding aminopeptidase A, a protein with substantial similarity to bovine lens leucine aminopeptidase. EMBO J 8, 1623–1627.

Strukelj B, Pungercar J, Ritonja A, Krizaj I, Gubensek F, Kregar I, Turk V (1990): Nucleotide and deduced amino acid sequence of an aspartic proteinase inhibitor homologue from potato tubers (*Solanum tuberosum* L.). Nucleic Acids Res 18:4605.

Suh S-G, Stiekema WJ, Hannapel DJ (1991): Proteinase-inhibitor activity and wound-inducible gene expression of the 22-kDa potato-tuber proteins. Planta 184:423–430.

Thornburg RW, An G, Cleveland TE, Johnson R, Ryan CA (1987): Wound-inducible expression of a potato inhibitor II-chloramphenicol acetyltransferase gene fusion in transgenic tobacco plants. Proc Natl Acad Sci USA 84:744–748.

Thornburg RW, Li X (1991): Wounding *Nicotiana tabacum* leaves causes a decline in endogenous indole-3-acetic acid. Plant Physiol 96, 802–805.

Yoshihara T, Omer E-SA, Koshino H, Sakamura S, Kikuta Y, Koda Y (1989): Structure of a tuber-inducing stimulus from potato leaves (*Solanum tuberosum* L.). Agric Biol Chem 53:2835–2837.

10

USING GENES ENCODING NOVEL PEPTIDES AND PROTEINS TO ENHANCE DISEASE RESISTANCE IN PLANTS

LUIS DESTÉFANO-BELTRÁN

Laboratorium voor Genetica, Rijksuniversiteit Gent, B-9000 Gent, Belgium

PABLITO G. NAGPALA

Plant Molecular Biology, The Upjohn Co., Kalamazoo, Michigan

SELIM M. CETINER

Cukurova University, Adana, Turkey

TIMOTHY DENNY

Department of Plant Pathology, University of Georgia, Athens, Georgia

JESSE M. JAYNES

Department of Biochemistry, Louisiana State University and Agricultural and Mechanical College, Baton Rouge, Louisiana 70803

1. INTRODUCTION

It has been said that plant disease is the exception rather than the rule (Lamb et al., 1989). In effect, only the interaction between a virulent race and a susceptible cultivar leads to disease. Despite this delicate interplay between host and patho-

Biotechnology in Plant Disease Control, pages 175–189
© 1993 Wiley-Liss, Inc.

gen, plant disease is one of the leading causes of crop loss in the world. Indeed, it has been estimated that as much as one-third of the total crop losses in the world can be directly attributed to plant disease (Agrios, 1978). For example, losses in potato, associated with bacterial disease, can be as high as 25% of the total worldwide production, with developing countries carrying most of the losses (Sawyer, 1984).

Plants under microbial attack elaborate several inducible potential defense responses of both a structural and biochemical nature. Among them are the synthesis of hydrolytic enzymes, such as chitinases and β-glucanases, and proteinase inhibitors; the modification of the plant cell wall by lignification and accumulation of callose, a β-1,3-glucan, and hydroxyproline-rich glycoproteins (HGRPs); and the synthesis of phytoalexins, low molecular weight compounds with antimicrobial activity (Collinge and Slusarenko, 1987).

Recent advances in genetic engineering have made possible the development of plants with new predictable phenotypes. Thus, the expression of the Tobacco Mosaic Virus coat protein (TMV CP) in transgenic tobacco plants has been shown to cause delay in the development of symptoms upon infection by the virus in concordance with the idea of cross-protection (Abel et al., 1986). Also, expression of an insecticidal protein, the B.t. protein from *Bacillus thuringiensis,* has rendered plants resistant to insect attack (Vaeck et al., 1987). Therefore, appropriate small modifications in the biochemistry of the plant defense arsenal may significantly augment its resistance to the action of a pathogen.

Jaynes et al. (1987), and more recently, Casteels et al. (1989), proposed the idea of using the genes found in insects, encoding for proteins with antimicrobial activity, to enhance bacterial disease resistance in plants. The results reviewed in this chapter indicate that this idea will have practical application in the development of new plant cultivars with increased microbial disease resistance. These results also suggest an entirely new approach as newly designed lytic peptides have become available with broader and in some cases new lytic activities.

2. BACKGROUND

Surface-active peptides, that is, peptides that bind to and affect amphipathic surfaces, such as membranes and receptors, have been extensively studied in recent years (DeGrado et al., 1981; Kaiser and Kézdy, 1983). A common feature of many of these peptides is their characteristic helical amphipathic secondary structure, which is often induced by the respective target. An amphipathic helix is defined as an alpha-helix with opposing polar and nonpolar faces oriented along the long axis of the helix (Segrest et al., 1990).

The affinity of these peptides to the target surface is intensified in parallel to the increase in complementarity between the two counterparts (Kaiser and Kézdy,

1987). Some of these peptides, like hormones, which put forth their biological activity through specific membrane-bound receptors demand meticulous conformation for manifestation of optimal function. The action of others, however, is far less dependent on strict structural requirements. Among the latter family of compounds are cationic lytic peptides isolated from mammalian phagocytes, for example, defensins (Selsted et al., 1985); insects, for example, melittins (Habermann, 1972), cecropins (Steiner et al., 1981), sarcotoxins (Okada and Natori, 1985a); and amphibians, for example, magainins (Zasloff, 1987). The target of these surface-active peptides appears to be the cellular lipid bilayer membrane. The peptides have been reported to act almost exclusively by virtue of their unique structural features: a cationic domain flanked by a hydrophobic surface (Kini and Evans, 1989), which allow them to associate with cell membranes altering their membrane potential and permeability, which in turn leads to rapid osmotic lysis. Mediation of these events through pore formation in the plasma membrane, as deduced from the generation of large time-variant, voltage-dependent ion channels in planar lipid membranes, has been proposed (Okada and Natori, 1985b; Christensen et al., 1988).

Most of the lytic peptides appear to be major components of the antimicrobial defense systems of a number of animal species (Boman, 1991). The presence of these peptides in other higher organisms is under intensive investigation, and it has been proposed that they will be found in all mammals, including humans (Berkowitz et al., 1990). Defensins, a family of small molecular weight antimicrobial peptides found in mammals, are a separate kind of lytic peptide because their action seems to be intracellularly localized following phagocytosis. All defensins contain 29–34 amino acid residues, have three intramolecular disulfide bonds, and are quite basic. Defensins seem to act on a wide variety of bacteria but usually more effectively on gram-positive than on gram-negative bacteria (Selsted et al., 1985).

On the other hand, magainins, melittins, and cecropins constitute *bona fide* lytic peptides, however, distinct from each other (manuscript in preparation, Jaynes JM) based on the arrangement of amphipathy and high positive charge within the molecule. Magainins are only 23 residues, devoid of cysteine, and form an amphipathic alpha-helix the full length of the molecule. They act on both gram-positive and gram-negative bacteria, fungi, and protozoa (Bevins and Zasloff, 1990). Melittins are 26 amino acid hemolytic peptides present in bee venoms (*Apis* spp). The amino-terminal sequence is hydrophobic, while the carboxy-terminal is hydrophilic. X-ray diffraction studies have shown that melittin forms an amphipathic alpha-helix with the hydrophobic residues located on one surface (Kini and Evans, 1989). Cecropins are 31–39 amino acid lytic peptides present in the hemolymph of several lepidoptera and diptera species and in at least one mammal (pig intestine, Lee et al., 1989), are devoid of cysteine, and have a strongly basic amino-terminal region and a long hydrophobic stretch in the carboxy-terminal one-half. The structure of cecropins, as deduced from nuclear magnetic resonance (NMR) studies, is two amphipathic alpha-helices joined by a hinge region containing Gly-Pro (Boman, 1991).

The physical attributes just discussed appear to be absolute requirements for lytic activity. An additional criterion of length based on studies with synthetic analogs (see below) must also be met: the peptides must be at least 20 amino acids long. This requirement corresponds well to the minimum length necessary for an alpha-helix to span the cell membrane (Jaynes JM, unpublished).

3. ENGINEERING FOR DISEASE RESISTANCE USING ANTIBACTERIAL GENES

Preliminary *in vitro* experiments showed cecropin B (35 amino acids long) to be an ideal candidate for genetic engineering for bacterial disease resistance, because it demonstrated to be bactericidal against a wide range of economically important phytopathogenic bacteria including *Pseudomonas syringae* pv. *tabaci, pseudomonas solanacearum, Erwinia carotovora* subsp. *carotovora, Erwinia chrysanthemi, Xanthomonas campestris* pv. *campestris, Clavibacter michiganense* subsp. *michiganense* (Destéfano-Beltrán et al., 1990; Destéfano-Beltrán, 1991; Destéfano-Beltrán et al., 1991). Most of the LD_{50} values are in the submicromolar range. However, *P. solanacearum* consistently proved to be most resistant to the action of the peptide with LD_{50} values almost 60-fold higher. An intriguing result was obtained when very low concentrations of the peptide were assayed: A proliferative rather than a lytic effect was observed in all bacteria tested. This stimulatory effect also has been demonstrated in other eukaryotic systems (Jaynes JM, unpublished).

The mature form of cecropin B is processed from a 62-amino acid residue precursor that has been elegantly demonstrated to take place in at least three steps (Boman et al. 1990). First, a signal–leader peptidase cleaves between position -5 and -4. Second, a specific dipeptidyl aminopeptidase removes, in a stepwise manner, the dipeptides Ala-Pro (residues -4 and -3) and Glu-Pro (residues -2 and -1). Finally, although this step is not yet fully characterized, the amidated group in the carboxy-terminal end of the mature form is derived from the glycine residue that ends the coding sequence.

Consequently, it became necessary to introduce modifications in the mature sequence of the peptide in order to obtain proper expression of its gene in the plant: introduction of a methionine residue at position $+1$ and a glycine at the carboxy-terminal end. Later, it became evident that we must introduce a proline residue at position $+2$ and replace the methionine at position $+11$ by a valine residue to result in a new peptide 38-amino acids long, SB-37. All these modifications did not alter the bactericidal activity, contrary to results obtained in other laboratories (Andreu et al., 1985). This derivative showed the same levels of activity against the different phytopathogenic bacteria as the natural cecropin B. Furthermore, it also proved to have growth inhibitory activity to a series of phytopathogenic fungi of economical importance including *Rhizoctonia solani* and *Phytophtora cinnamoni* (Destéfano-Beltrán et al., 1990; Cetiner, 1990). For all practical purposes cecropin B and SB-37 display the same lytic activities against all phytopathogens tested.

The SB-37 gene used to produce transgenic plants was assembled from six overlapping phosphorylated oligonucleotides and cloned into the *BglII/EcoRI* site of pMON316. This cointegrate vector (kindly given by S. Rogers, Monsanto, Co.) carries a cauliflower mosaic virus (CaMV) 35S promoter followed by a polylinker and a nopaline synthase (NOS) 3′ polyadenylation signal. Later the assembled gene was transferred to pUC19 for dideoxy double-stranded sequencing. The first crop plant chosen for transformation was tobacco (*Nicotiana tabacum* cultivar xanthi). The construct, pLDB7, was subsequently introduced into *Agrobacterium tumefaciens* GV3111SE (pTiB6S3SE) (Fraley et al., 1985). The binary vector pBI121 (Jefferson et al., 1987), mobilized into *A. tumefaciens* LBA4404, was used to produce control transformants. Leaf disks of *N. tabacum* were transformed as previously described by Horsch et al. (1985). Transformants were selected by their resistance to kanamycin (300 μg/ml) and transformed plants were regenerated from shootlets by transfer to a root-inducing media containing lower levels of kanamycin (100 μg/ml). Rooted plants were transferred to the greenhouse for further analysis and production of seeds.

Southern genomic blots of the primary transformants confirmed the presence of at least one nonrearranged copy of the SB-37 gene in five independent lines from a total of eight analyzed. The presence and levels of SB-37 production in the original transformants were measured by western blot analysis using polyclonal antibodies raised in rabbits against pure SB-37. The levels of expression in leaves from all five different individual pLDB7 transformants were pretty close to the limit of detection, about 10 ng/100 μg of total soluble protein or about 0.01%, under the established conditions. It was rationalized that factors in the design of the coding sequence might have affected the expression of the gene at the protein level (Destéfano-Beltrán, 1991).

A newly designed peptide became available in our laboratory during the course of these experiments. The peptide, Shiva-1, was designed to have significant differences in sequence with the natural cecropin B (46% homology), while conserving the overall charge distribution, amphipathy, and hydrophobic properties of the natural peptide. Bactericidal *in vitro* activities of Shiva-1 were, in all cases but one, severalfold higher than SB-37 values (Destéfano-Beltrán, 1990; Nagpala, 1990; Jaynes et al., 1991). Moreover, the peptide showed comparable activities against different phytopathogenic fungi. Based on these results it was decided to clone and express a synthetic gene encoding for Shiva-1 in transgenic tobacco plants.

The Shiva-1 gene, assembled as for SB-37, was eventually placed under the control of the proteinase inhibitor II (PiII) promoter (graciously provided by L. Willmitzer) derived from potato. In nonwounded potato plants PiII accumulates in the tubers with nondetectable levels of protein in leaves, stem, or roots. When the leaves are wounded, expression of the gene is induced not only in the wounded leaves, but also in nonwounded upper and lower leaves and in the upper part of the stem (Peña-Cortés et al., 1988). Detailed histological analysis, using the β-glucuronidase gene under the control of the PiII promoter, has revealed that induced expression is strongest in cells closest to the vascular tissue. This result suggests that the signal mediating the wounding response is transported by this tissue (Keil et al., 1989). When the PiII gene is transferred to tobacco, it is regulated in the same way as it is in potato, indicating that tobacco plants contain similar trans-activating factors that recognize the cis-elements present in the promoter (Sánchez-Serrano et al., 1987). The PiII/Shiva-1 construct, subcloned in pBI121, was mobilized into *A. tumefaciens* LBA4404 and used to produce tobacco transgenic plants (Jaynes et al., 1991).

Fertile transgenic tobacco plants were obtained and tested for the presence of intact copies by standard Southern blot analysis (data not shown). Transgenic lines exhibiting nonrearranged single-copy gene insertions were selected for further study. Western blot analysis of total protein extracts isolated from wounded plants revealed the presence of a peptide with a molecular weight similar to that of Shiva-1 when the blots were probed with antibody raised against chemically synthesized Shiva-1 peptide. Expression level in the best lines was about 0.1% and did not have any obvious detrimental effect on the plants, since they proved to have an indistinguishable phenotype from nontransformed controls. No Shiva-1 immunoreactive band was found in transgenic control lines (wounded or nonwounded) and in nonwounded Shiva-1 containing plants. Northern analysis in wounded versus nonwounded plants demonstrated similar tight control for the Shiva-1 messenger ribonucleic acid (mRNA) transcription. Two lines, Sh-3 and Sh-4 were selected for further study.

Preliminary bacterial challenge results, utilizing a highly pathogenic strain of *P. solanacearum* (a vascular pathogen that causes severe wilting and eventually death) indicated that plants expressing the Shiva-1 peptide exhibited a delayed appearance of symptoms, which were less severe than those showed by the control plants. Furthermore, there was a dramatic difference in the mortality of Shiva-1 plants when compared to control plants 3 weeks after infection. No enhanced resistance was observed among the plants producing SB-37, presumably because of its low activity against this pathogen. Interestingly, however, two lines consistently showed symptoms earlier than control plants and wilted first. It is tempting to speculate that in the challenge assays, a high ratio of inoculum to peptide took place due to the very low levels of expression. As a result, a cell-proliferative effect of the SB-37 peptide on the bacteria occurred in the transgenic plants causing an earlier and more "pathogenic" action of the bacteria (Fig. 10.1).

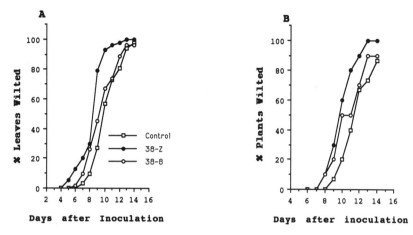

Fig. 10.1. Disease susceptibility assay of tobacco plants containing the SB-37 gene. F1 tobacco seed-lings were root-inoculated with 5 ml of a suspension of a virulent strain of *P. solanacearum* UW77 (10^6 cfu/ml). The percentage of leaves for each plant on a daily basis, and the mean percentage of leaves wilted calculated. When all leaves were wilted the plant was considered completely wilted (for more details see Destéfano-Beltrán, 1991).

In a more rigorous set of single-blind bacterial challenge experiments it was decided to focus on the Shiva-1 transgenic lines. Two inoculation methods were used to evaluate the two selected lines, Sh-3 and Sh-4, for their susceptibility to *P. solanacearum*. In a wounded root assay, assessment of the percentage of leaves wilted revealed that both lines showed symptoms before the control. However, when examined over time, line Sh-3 was significantly slower to wilt than the control, whereas line Sh-4 was only marginally different. A similar trend was observed in the percentage of plants wilted after inoculation. The same experiments, performed with 100-fold more inoculum, gave similar results; line Sh-3 wilted more slowly while line Sh-4 wilted and died almost as fast as the control (data not shown).

When tested with a stem-inoculation assay, lines Sh-3 and Sh-4 were mark-edly less susceptible to the bacterial pathogen. Both lines showed significantly fewer wilted leaves in individual plants, and overall fewer plants wilted by the end of the experiment. Furthermore, line Sh-4 showed no wilted plants at all com-pared to almost 85% wilted plants in the control. Here too, experiments with 100 times more inoculum gave similar results. Prior wounding of the plants had no effect on the overall outcome of these experiments (Fig. 10.2).

The enhanced resistance of Shiva-1 transgenic plants tested by stem inocula-tion is consistent with previous studies of the wound-inducible expression of PiII in potato and in transgenic tobacco plants. Keil et al. (1989) reported that after

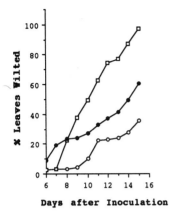 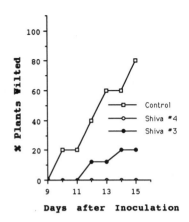

Fig. 10.2. Disease susceptibility assay of tobacco plants containing the Shiva-1 gene. F1 tobacco seedlings were stem inoculated with a 20-µl inoculum of a virulent strain of *P. solanacearum* UW77 (10^6 cfu/ml). The percentage of leaves that were completely wilted was recorded for each plant on a daily basis, and the mean percentage of leaves wilted calculated as before. When all leaves were wilted the plant was considered completely wilted (for more details see Jaynes et al., 1991).

wounding a single leaf there is a systemic activation of the PiII gene in the upper part of the stem, but not in the lower part of the stem and roots. The authors proposed that the signal responsible for the systemic induction of PiII must move both acropetally and basipetally. Assuming the expression of Shiva-1 is regulated in the same way at the cellular level, there will be a higher expression of the peptide in the cells surrounding the vascular tissue of the stem, which could explain the overall better response of the plants to the infection when plants were stem inoculated rather than root inoculated.

From the evidence presented on the use of antibacterial genes, it appears that Shiva-1 offers the best potential in the production of bacterial resistance crops. However, potato transgenic plants containing the SB-37 gene have shown promising results in preliminary experiments when infected with *E. carotovora* var. *atroseptica* (the blackleg/storage soft rot pathogen) (Belknap W, personal communication). These results are not totally unexpected, since a related pathogen, *E. carotovora* subsp *carotovora,* is one of the most sensitive bacteria to SB-37 with LD_{50} in the micromolar range (Destéfano-Beltrán et al., 1990, Destéfano-Beltrán, 1991).

Currently, two different signal sequences have been spliced into the SB-37 and Shiva-1 genes. This modification intends to improve their delivery into the plant tissues. The constructs are already *in planta* and soon it will be possible to evaluate their possible advantage over the first constructs.

4. SYNTHETIC ANTIMICROBIAL PEPTIDES

Lately, the design of peptides with predetermined structures and properties has met with some success (Chen et al. 1988; Jaynes et al., 1988; Jaynes et al., 1989; Boman et al., 1989; Fink et al., 1989; Jaynes et al., 1991). Advances in the synthesis and purification of peptides have increased the ease with which newer sequences can be made. As a natural extension of our research it was therefore interesting to design new lytic peptides whose genes eventually could be used to enhance bacterial disease resistance. The results are encouraging, since some of them also show a high in vitro activity against fungi and nematodes.

Based on the three distinct classes of peptides described (see above) three series of *de novo* designed analogs have been characterized (Destéfano-Beltrán, 1991; Jaynes et al., manuscript in preparation). All series have almost the same amino acid composition but are arranged both in charge density as well as amphipathy in such a way that they structurally mimic their natural counterparts. Hecate-1, Anubis-2, and Shiva-11 series are based on melittin, magainin-2, and cecropin B, respectively.

All peptides were synthesized following the Fmoc chemistry strategy (Fields and Noble, 1990), found to be homogeneous by analytical high-performance liquid chromatography (HPLC) and with expected molecular weights by mass spectrometry (MS). The peptides were then assayed quantitatively for antibacterial activity against several bacteria, plant, and human pathogens. The results in Table 10.1 show that under the present assay conditions cecropin B is the best lytic peptide. However, against other bacteria (i.e., *Staphylococcus aureus*) the synthetic analogs proved to be better antibiotics (data to be published elsewhere). Interestingly, the three synthetic peptides also showed a proliferative effect when tested at low concentrations, although the "proliferative" concentration varied from bacteria to bacteria. The analog Anubis-2 showed better lytic activities than its natural counterpart, magainin-2.

The peptides were also tested for their ability to lyse human red blood cells. As expected, mellitin had the highest hemolytic activity. Its synthetic analog, Hecate-1 showed partial hemolytic activity but at concentrations 50-fold higher than the used for mellitin. At the same concentrations, the rest of the peptides showed no significant activity (Destéfano-Beltrán, 1991).

The solution conformation of the different peptides were measured by circular dichroism as a function of solvent composition, 0 and 20% (v/v) hexafluoro-2-propanol (HFP). In aqueous solutions the peptides were largely random coils, with no observable alpha-helix. The only exception was Shiva-11, which surprisingly exhibits an alpha-helical conformation in aqueous solution. As predicted, helicity increased when the spectra were recorded in the presence of 20% HFP, a solvent commonly used to induce helicity in single-chain alpha-helical peptides (Wade et al., 1990), indicating that a conformation transition takes place as the

TABLE 10.1. Summary of Activities Against *E. carotovora* subsp. *carotovora*, Ecc-13, and *E. coli*, Strain D31

Peptide (µ*M*)	Ecc-13[a]	D31[a]
Mellitin		
1.0	4.6 (4.6)	5.9 (5.9)
0.5	4.6 (4.6)	0.6 (5.9)
0.1	1.39 (4.6)	
Hecate-1		
1.0	4.6 (4.6)	5.9 (5.9)
0.5	0.03 (4.6)	14.9%
0.1	61.4%	
Cecropin B		
1.0	4.6 (4.6)	5.9 (5.9)
0.5	4.6 (4.6)	1.53 (5.9)
0.1	0.06 (4.6)	
Shiva-11		
1.0	4.6 (4.6)	5.9 (5.9)
0.5	0.08 (4.6)	5.7%
0.1	52%	
Magainin-2		
1.0	0.23 (4.6)	8.3% (5.9)
0.5	8 %	8.5% (5.9)
0.1	19.3%	
Anubis-2		
1.0	1.78 (4.6)	5.9 (5.9)
0.5	0.44 (4.6)	1.2%
0.1	39.4%	

[a]The number in parentheses indicates the log of the initial number of cells. Percentage indicates an increase over the initial number of cells.

hydrophobicity of the solution is increased (Fig. 10.3, Destéfano-Beltrán, 1991).

The peptides described are currently being evaluated and in at least one case, a gene has been constructed and introduced into plants.

5. EXPRESSION OF CHICKEN EGG-WHITE LYSOZYME IN TOBACCO TRANSGENIC PLANTS

During our preliminary experiments with SB-37 and cecropin B it became interesting to try to reconstruct *in vitro* the coordinate action of more than one lytic protein reported in the humoral immune system of the cecropia moth, *Hyalophora cecropia* (Boman and Hultmark, 1987). When the peptides were incubated in the

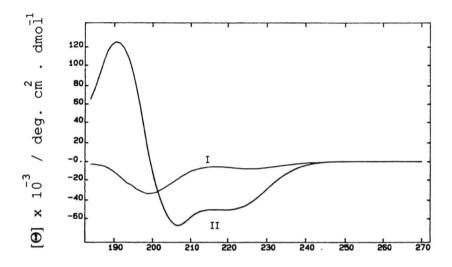

Fig. 10.3. Mean residue ellipticity of Anubis-2 in (I) phosphate buffer and in (II) 20% HFP (v/v) in the same buffer.

presence of chicken egg-white lysozyme (CEWL) a significant decrease in the LD_{50} values was observed (Destéfano-Beltrán et al., 1990; Destéfano-Beltrán, 1991). Most likely a loss of integrity of the peptidoglycan bacterial cell wall, combined with the lytic activity of the peptides, created a synergistic interaction similar to the concerted action of cecropins, attacins, and lysozyme in the cecropia moth (Boman and Hultmark, 1987). This result supported the idea of using the combined action of lytic peptides and CEWL within the same line of transgenic plants to allow for a potent synergy to develop and could permit an even higher level of plant resistance.

The cDNA gene of CEWL, including its signal sequence, contained in plys1023 (graciously provided by G. Schütz) was properly manipulated and placed under the control of the CaMV35S and Double-35S (provided by J. MacPherson) promoters to yield constructs pLDB9 and pLDB12 based on pMON316 and pBI121, respectively (Destéfano-Beltrán, 1991). Following standard *A. tumefaciens* leaf disk infections and regeneration protocols (Horsch et al., 1985), several transgenic tobacco lines were obtained. Southern analysis confirmed the presence of intact copies of the T-constructs and five lines containing single inserts were selected for further study.

Transgenic plants expressed the CEWL gene at the protein level as determined by binding of a highly specific, HyHEL5, anti-CEWL, monoclonal antibody (a gift of S. Smith-Hill) to protein gel blots. Expression levels varied from line to line reaching at their best, 0.3–0.4% of total leaf protein (Fig. 10.4). The expected higher expression, on average, of construct pLDB12 (Double 35S/CEWL) could not be evaluated fully, since only two single-copy transformants were analyzed. Several more transformants harboring single-copy insertions would have been necessary to establish the higher strength of this promoter (Kay et al., 1987).

The expression of CEWL, at the protein level, does not imply the correct recognition and processing of its signal sequence. However, an apparently correct processing of CEWL has been reported in yeast (Oberto and Davison, 1985); Oberto J, personal communication). Correct processing and secretion of plant and animal proteins in transgenic plants have been reported with mixed results where heterologous plant signal sequences are correctly recognized but, in at least two cases, animal signal sequences are only partially processed (Denecke et al., 1990; Hoekema et al., 1990). Unequivocal proof of correct processing of CEWL in tobacco plants will suppose both its purification from crude leaf extracts and the determination of its amino terminal sequence by microsequencing.

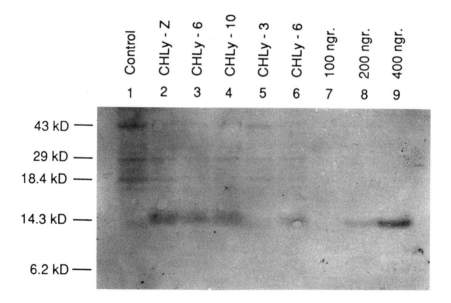

Fig. 10.4. Protein gel blot of total tobacco leaf proteins probed with anti-CEWL monoclonal antibody. A 100 μg sample of total leaf proteins were loaded per lane. Control is a nontransformed tobacco plant. Lanes 2–4 are lines containing CaMV35S/CEWL and lanes 5–6, lines containing double 35S/CEWL. Lanes 7–9 contain 100, 200, and 400 ng of purified CEWL.

The expression of CEWL in tobacco plants allows us now to test the possibility that plants containing both CEWL and Shiva-1 (or SB-37) would show a higher resistance to *P. solanacearum* than those expressing only the peptide (Jaynes et al., 1991). Currently, homozygous lines for both genes are being obtained and eventually crossing of these two should generate plants expressing both genes, which will be evaluated as described before.

The use of other lysozyme genes might prove resourceful. The lysozyme of *Parthenocissus quinquifolia* has been reported to have a high lysozyme and chitinase activity (Bernasconi et al., 1987). Indeed, the activity of 1 μg of the purified enzyme (30.3 kD) was described to correspond to 17 μg of purified CEWL (14.4 kD). Also, this enzyme shows a higher chitinase activity than CEWL. According to this report, it would be very interesting to see if this protein will behave in a similar way as CEWL when assayed in the presence of SB-37 or Shiva-1. Eventually, the introduction of its gene in tobacco plants could follow. An enzyme with high lysozyme and chitinase activity expressed in a constitutive manner together with a lytic peptide, Shiva-1, in plants might be instrumental in obtaining protection against bacteria and fungi attack.

ACKNOWLEDGMENTS

We wish to acknowledge the expert technical assistance in peptide synthesis and purification provided by Martha Juban. The challenging experiments were carried out by T. Denny from the Department of Plant Pathology at the University of Georgia, Athens, GA. This research was supported, in part, by funds from the Louisiana State University Agricultural Experimental Station.

REFERENCES

Abel PP, Nelson RS, De B, Hoffman N, Rogers SG, Fraley RT, Beachy RN (1986): Delay of disease development in transgenic plants that express the tobacco mosaic virus coat protein gene. Science 232:738–743.

Agrios GN (1978): Plant Pathology, 2nd ed. New York: Academic Press.

Andreu D, Merrifield RB, Steiner H, Boman HG (1985): N-terminal analogues of cecropin A: synthesis, antibacterial activity, and conformational properties. Biochemistry 24:1683–1688.

Berkowitz BA, Bevins CL, Zasloff M (1990): Magainins: A new family of membrane active host defense peptides. Biochem Pharmacol 39:625–629.

Bernasconi P, Locher R, Pilet PE, Jolles J, Jolles P (1987): Purification and N-terminal amino acid sequence of a basic lysozyme from *Parthenocissus quinquifolia*. Biochim Biophys Acta 916:254–260.

Bevins CL, Zasloff M (1990): Peptides from frog skin. Annu Rev Biochem 59:395–414.

Boman HG (1991): Antibacterial peptides: key components needed in immunity. Cell 65:205–207.

Boman HG, Boman A, Andreu D, Li Z, Merrifield RB, Schelenstedt G, Zimmermann R (1990): Chemical synthesis and enzymic processing of precursor forms of CecropinS A and B. J Biol Chem 264:5852–5860.

Boman HG, Hultmark D (1987): Cell-free immunity in insects. Annu Rev Microbiol 41:103–126.

Boman HG, Wade D, Boman IA, Wåhlin B, Merrifield RB (1989): Antibacterial and antimalarial properties of peptides that are cecropin-melittin hybrids. FEBS Lett 259:103–106.

Casteels P, Ampe C, Jacobs F, Vaeck M, Tempst P (1989): Apidaecins: antibacterials peptides from honeybees. EMBO J 8:2387–2391.

Cetiner S (1990): The introduction into tobacco plants with genes which encode some of the natural components of the humora; immune response of *Hyalopliora cecropia*. Ph.D. Dissertation, Louisiana State University.

Chen CH, Brown JH, Morell J, Huang CM (1988): Synthetic magainin analogues with improved antimicrobial activity. FEBS Lett 236:462–466.

Christensen B, Fink J, Merrifield RB, Mauzerall D (1988): Channel-forming properties of cecropins and related model compounds incorporated into planar lipid membranes. Proc Natl Acad Sci USA 85:5072–5076.

Collinge DB, Slusarenko AJ (1987): Plant gene expression in response to pathogens. Plant Mol Biol 9:389–410.

DeGrado WF, Kézdy FJ, Kaiser ET (1981): Design, synthesis and characterization of a cytotoxic peptide with melittin-like activity. J Am Chem Soc 103:679–681.

Denecke J, Botterman J, Deblaere R (1990): Protein secretion in plant cells can occur via a default pathway. Plant Cell 2:51–59.

Destéfano-Beltrán L (1991): Characterization of regenerated transformed plants derived from lytic peptide gene inserting. Ph.D. Dissertation, Louisiana State University.

Destéfano-Beltrán L, Nagpala P, Cetiner S, Dodds J, Jaynes J (1990): Enhancing bacterial and fungal disease resistance in plants: application to potato. In Vayda M, Park W (eds): The Molecular and Cellular Biology of the Potato. Wallingford: C.A.B. International, pp 205–221.

Destéfano-Beltrán L, Nagpala P, Kim J, Dodds J, Jaynes J (1991): Genetic transformation of potato to enhance nutritional value and confer disease resistance. In Dennis E, Llewellyn L (eds): Molecular Approaches to Crop Improvement. New York: Springer-Verlag, pp 17–32.

Fields GB, Noble RL (1990): Solid phase peptide synthesis utilizing 9-fluorenylmethoxycarbonyl amino acids. Int J Peptide Protein Res 35:161–214.

Fink J, Boman A, Boman HG, Merrifield RB (1989): Design, synthesis and antibacterial activity of cecropinlike model peptides. Int J Peptide Protein Res 33:412–421.

Fraley RT, Rogers SG, Horsch RB, Eichholtz DA, Flick JS, Fink CL, Hoffmann NL, Sanders PR (1985): The SEV System: A new disarmed Ti plasmic vector system for plant transformation. Bio/Technol 3:629–635.

Habermann E (1972): Bee and wasp venoms. Science 177:314–322.

Hoekema A, Dekker BM, Schrammeijer B, Verwoerd TC, van de Elzen JM, Sijmons PC (1990): Production of correctly processed Human Serum Albumin in transgenic plants. Abstract R414. J Cell Biochem Sup14E:333.

Horsch R, Fry J, Hoffmann NL, Wallroth M, Eichholtz D, Rogers SG, Fraley RT (1985): A simple and general method for transferring genes into plants. Science 227:1229–1231.

Jaynes JM, Burton CA, Barr SB, Jeffers GW, White KL, Enright FM, Klei TR, Laine RA, Julian GR (1988): In vitro cytocidal effect of novel lytic peptides on *Plasmodium falciparum* and *Trypanosoma cruzi*. FASEB J 2:2878–2883.

Jaynes JM, Julian GR, Jeffers GW, White KL, Enright FM (1989): *In vitro* cytocidal effect of lytic peptides on several transformed mammalian cell lines. Peptide Res 2:157–160.

Jaynes JM, Nagpala P, Destéfano-Beltrán L, Denny T, Clark C, Cetiner S, Kim J (1992): Expression of a highly sequence divergent Cecropin B analog in transgenic tobacco plants confers enhanced resistance to *Pseudomonas solanacearun*. Plant Science (submitted).

Jaynes JM, Xanthopoulos K, Destéfano-Beltrán L, Dodds J (1987): Increasing bacterial disease resistance in plants utilizing antibacterial genes from insects. Bioessays 6:263–270.

Jefferson RJ, Kavanagh TA, Bevan MW (1987): GUS fusions: β-glucuronidase as a sensitive and versatile gene fusion marker in higher plants. EMBO J 6:3901–3907.

Kaiser ET, Kézdy FJ (1983): Secondary structures of proteins and peptides in amphiphillic environments (a review). Proc Natl Acad Sci USA 80:1137–1143.

Kaiser ET, Kézdy FJ (1987): Peptides with affinity for membranes. Annu Rev Biophys Chem 16:561–581.

Kay R, Chan A, Daly M, McPherson J (1987): Duplication of CamV35S promoter sequences creates a strong enhancer for plant genes. Science 236:1299–1302.

Kleil M, Sanchez-Serrano JJ, Willmitzer L (1989): Both wound-inducible and tuber-specific expression are mediated by the promoter of a single member of the potato proteinase inhibitor II gene family, EMBO J 8:1323–1330.

Kini RM, Evans HJ (1989): A common cytolytic region in myotoxins, hemolysins, cardiotoxins and antibacterial peptides. Int J Peptide Protein Res 34:277–286.

Lamb CJ, Lawton MA, Dron M, Dixon R (1989): Signals and transduction mechanisms for activation of plant defenses against microbial attack. Cell 56:215–224.

Lee JY, Boman A, Chuanxin S, Andersson M, Jörnvall H, Mutt V, Boman HG (1989): Antibacterial peptides from pig intestine: Isolation of a mammalian cecropin. Proc Natl Acad Sci USA 86:9159–9162.

Nagpala PG (1990): The introduction of a gene encoding a novel peptide, into plants to increase plant bacterial resistance. Ph.D. Dissertation, Louisiana State University.

Oberto J, Davison J (1985): Expression of chicken egg white lysozyme by Saccharomyces cerevisiae. Gene 40:57–65.

Okada M, Natori S (1985a): Primary structure of sarcotoxin I, and antibacterial protein induced in hemolymph of Sarcophaga peregrina (flesh fly) larvae. J Biol Chem 260:7174–7177.

Okada M, Natori S (1985b): Ionophore activity of sarcotoxin I, a bactericidal protein of Sarcophaga peregrina. Biochem J 229:453–458.

Peña-Cortés H, Sanchez-Serrano J, Rocha-Sosa M, Willmitzer L (1988): Systemic induction of proteinase-inhibitor-II gene expression in potato plants by wounding. Planta 174:84–89.

Sanchez-Serrano JJ, Keil M, O'Connor A, Schell J, Willmitzer L (1987): Wound-induced expression of a potato proteinase inhibitor II gene in transgenic tobacco plants. EMBO J 3:303–306.

Sawyer RL (1984): Potatoes for the developing world. International Potato Center, Lima, Perú.

Segrest JP, De Loof H, Dohlma JG, Brouillette CG, Anantharamaiah (1990): Amphipathic helix motif: classes and properties. Proteins 8:103–117.

Selsted M, Brown DM, Delange RJ, Harwig SL, Lehrer RI (1985): Primary structures of six antimicrobial peptides of rabbit peritoneal neutrophils. J Biol Chem 260:4579–4584.

Steiner H, Hultmark D, Engström A, Bennich H, Boman HG (1981): Sequence and specificity of two antibacterial proteins involved in insect immunity. Nature (London) 292:246–248.

Vaeck M, Reynaerts A, Höfte H, Jansens S, De Beuckeer M, Dean C, Zabeau M, Van Montagu M, Leemans J (1987): Transgenic plants protected from insect attack. Nature (London) 328:33–37.

Wade D, Boman A, Wåhlin B, Drain CM, Andreu D, Boman H, Merrifield RB (1990): All D-amino acid-containing channel-forming antibiotic peptides. Proc Natl Acad Sci USA 87:4761–4765.

Zasloff M (1987): Magainins, a class of antimicrobial peptides from Xenopus laevis skin: isolation, characterization of two active forms, and partial cDNA sequence of a precursor. Proc Natl Acad Sci USA 84:5449–5453.

≡ 11

ROLE OF PHENOLICS IN PLANT DISEASE RESISTANCE

J. P. MÉTRAUX

Institut de Biologie Végétale et de Phytochimie, Université de Fribourg Suisse, CH-1700 Fribourg, Switzerland

I. RASKIN

AgBiotech Center, Rutgers University, Cook College, New Brunswick, New Jersey 08903

1. INTRODUCTION

Phenolics are usually defined as substances that possess an aromatic ring bearing a hydroxyl group or its functional derivative. Phenolic compounds are amazingly diverse and extremely widespread in all plants (Harborne, 1980). In the past, plant phenolics have often been relegated to what is commonly referred to as secondary metabolites. The term ''secondary'' would indicate that such compounds are only of minor importance for the plant. However, over recent years, this opinion gradually has been replaced by a view that secondary metabolites are beneficial and even essential for the survival of the plant (Harborne, 1990). Phenolics enter in the composition of lignins, an important structural component in plant cell walls. Furthermore, phenolics have been associated extensively with the chemical defense of plants against microbes, insects, and other herbivores. A number of phenolics function as allelopathic compounds influencing germination and growth of neighboring plants (Einhellig, 1986). Phenolic molecules produced by plant roots are essential for germination, haustorium formation, and host attachment in parasitic *Striga* species (Lynn and Chang, 1990). Experimental evidence increasingly suggests that phenolics function as signals in plant development and

Biotechnology in Plant Disease Control, pages 191–209
© *1993 Wiley-Liss, Inc.*

in plant–microbe interactions (Lynn and Chang, 1990). In *Agrobacterium tumefaciens*, virulence gene expression was shown to be activated specifically by the plant-produced phenolic compounds acetosyringone and α-hydroxyacetosyringone (Stachel et al., 1985). These compounds are found in the exudates of wounded plant cells and probably allow *Agrobacterium* to recognize susceptible cells in nature. Species-specific flavonoids exuded from legume roots and seeds are essential for the induction of the *nod* genes of *Bradyrhizobium* and *Rhizobium* species (Long, 1989). The activation of *nod* genes is required for the early nodulation process and the establishment of symbiotic relationships between the host plant and the bacterium.

The chemistry and the biochemistry of plant phenolics has been a subject of a number of recent reviews (Barz et al., 1985; Friend, 1981, 1985; Harborne, 1980, 1990; Matern and Kneusel, 1988). This chapter will focus on the role of plant phenolics in the defense of plants against pathogens. First, we will discuss the role played by phenolic cell wall materials produced in response to infection and the importance of phenolics as preformed or induced antimicrobials. In addition, we will review recent findings indicating that salicylic acid may function as an endogenous signal compound for induced resistance in plants.

2. PHENOLICS AS COMPONENTS OF THE PLANT CELL WALL

Cell walls are a composite polymeric matrix of noncellulosic polysaccharides interwoven with crystalline cellulose, and in dicotyledonous species with a glycoprotein network (Cassab and Varner, 1988). In addition, cell walls contain lignin, a complex phenolic polymer. Lignin is a rigid water repellent network of substituted cinnamyl alcohols (Lewis and Yamamoto, 1990). Major lignin types include guaiacyl lignin (derived from coniferyl alcohol) in ferns and gymnosperms, guaiacylsyringyl lignins (derived from coniferyl and synapyl alcohol) in angiosperms and guaiacyl-syringyl-*p*-hydroxyphenyl lignins (polymer derived from coniferyl, synapyl, and *p*-coumaryl alcohols) in grasses. These components vary with species, tissue, and developmental stage. The deposition of lignin in the cell wall results in physical strengthening somewhat similar to the role of concrete in ferro-concrete. Lignin or lignin-like compounds appear to be of special importance for resistance against various fungal pathogens because of their inability to degrade this complex polymer with the exception of white rot fungi and litter decomposing basidiomycetes (Kirk, 1971). Qualitative and quantitative changes, as well as changes in the speed of lignin formation, were observed in incompatible plant–pathogen interactions (Bird, 1988; Friend, 1985; Legrand, 1983; Matern and Kneusel, 1988; Ride, 1983; Vance et al., 1980).

The composition of lignin produced in response to infection can be markedly different from that found in healthy plants. For example, when roots of Japanese radish are infected with *Peronospora parasitica,* there is a shift from guaiacyl-syringyl lignin to lignin enriched in *p*-hydroxyphenyl and guaiacyl propane units (Asada and Matsumoto, 1972). Changes in the composition of lignin induced after infection were also observed in wheat (Ride, 1975), muskmelon (Touze and Rossignol, 1977), and potato (Hammerschmidt, 1984) indicating that different pathways for lignin biosynthesis are activated after infection. Such compositional changes are likely to affect the regiochemistry of the cell wall, but how this affects penetration is not understood in detail.

Another question is whether enough lignin is produced sufficiently ahead of the spreading pathogen to act as an efficient barrier. In wheat lines carrying a resistance gene against leaf or stem rust, the deposition of lignin and lignin-like materials was localized at the sites of infection of the rust fungus (Beardmore et al., 1983; Southerton and Deverall, 1990; Tiburzy and Reisener, 1990). A similar response was observed at attempted sites of penetration by nonpathogens (Moersbacher et al., 1990; Ride and Pearce, 1979). In potato, infection by nonpathogens resulted in a faster accumulation of lignin (Hammerschmidt, 1984). Lignification was also shown to increase in systemically protected leaves and petioles of cucumber (Hammerschmidt and Kuc, 1982; Kovats et al., 1991), which demonstrated a greater ability to incorporate labeled precursors into lignin in response to wounding or to challenge infection (Dean and Kuc, 1987). These authors confirmed that the labeled wall material was lignin by nitrobenzene oxidation, which yielded both vanillic acid and *p*-hydroxybenzaldehyde. Labeling experiments showed that lignification occurs at an early stage of the infection process at the sites of attempted invasion in resistant interactions (Ride and Pearce, 1979; Beardmore et al., 1983; Tiburzy and Reisener, 1990).

Another approach to assess the importance of lignification is by measurements of enzymes involved in the synthesis and polymerization of lignin precursors, such as phenylalanine ammonia lyase (PAL), 4-coumarate-CoA ligase (4CL), cinnamyl alcohol dehydrogenase (CAD), *ortho*-diphenol-methyltransferase (OMT), and peroxidase (PO). Comparisons between resistant and susceptible wheat lines indicated that all the enzymes mentioned above were associated with the incompatible reaction in the resistant cultivar (Moersbacher et al., 1988, 1989). The activities of PAL and OMT were also found to increase in wheat infected by nonpathogens (Maule and Ride, 1976, 1983; Thorpe and Hall, 1984). In tobacco leaves reacting hypersensitively to tobacco mosaic virus (TMV), the increases in the activities of PAL, cinnamic acid hydroxylase, 4CL, and OMT correlated with increased lignification around the site of the necrotic hypersensitive reaction (HR), where most of the virus remained localized (Legrand et al., 1976, 1978).

Inhibitors of PAL such as α-aminooxyacetic acid (AOA) or α-aminooxy-β-phenylpropionic acid (AOPP) supplied to tobacco increased the size of the necrotic lesions in several resistant plant–virus combinations (Massala et al., 1987). Alpha-aminooxyacetic acid is not a specific inhibitor of PAL, and AOA, as well as the related AOPP, could affect other biochemical pathways branching off from cinnamic acid. Therefore, results obtained with such treatments remain only indicative for a role of lignification in resistance. The importance of lignification was further evaluated using specific suicide inhibitors of CAD, N-(O-aminophenyl)sulfinamoyl-tertiobutyl acetate (NH_2-PAS) and N-(O-hydroxyphenyl)sulfinamoyl-tertiobutyl acetate (OH-PAS) (Baltas et al., 1988; Grand et al., 1985). Treatment of wheat resistant to stem rust with such inhibitors decreases the frequency of lignified host cells with accompanying increases in cell penetration by pathogens (Moersbacher et al., 1990).

Despite all the studies suggesting a role for induced lignification in general resistance against pathogens, it still remains unclear how pathogen invasion is prevented. We are left with several hypotheses proposed by Ride (1978) on how lignification might prevent pathogen invasion: mechanical reinforcement of the cell wall; protection of other wall polymers from the action of hydrolytic enzymes from the pathogen; buildup of a hydrophobic barrier preventing leakage of solutes from plant cells to the pathogen or toxins from the pathogen; and blockage of fungal growth by lignification of hyphal tips. In addition, direct antimicrobial effects of unpolymerized lignin precursors, such as coniferyl alcohol, were also observed (Hammerschmidt and Kuc, 1982) and could generate an unfavorable environment for the pathogen even before a lignin barrier is established.

3. THE ROLE OF PHENOLICS AS PREFORMED ANTIMICROBIAL DEFENSE CHEMICALS

When discussing the role of antibiotic compounds in plant defense a distinction is usually made between those compounds that are preformed in the plant before infection and those that accumulate in response to infection. Preformed antimicrobials that are responsible for resistance against pathogens occur in those cells or tissues that are in contact with the invading microorganism, in concentrations high enough to affect the pathogen (Wood, 1967).

Preformed fungistatic or fungitoxic compounds are present in the cuticular waxes of a number of plants where they might influence the first and crucial events in the propagation of fungal spores (Blakeman and Atkinson, 1981). Uncharacterized phenolic inhibitors of fungal germination were found in surface wax of apples (Martin et al., 1957). Other spore germination inhibitors include luteone and wighteone in lupine leaves (Harborne et al., 1976; Ingham et al., 1977), catechol and protocatechuic acid in onion scales (Walker et al., 1955), and gallic acid in *Acer* leaves (Dix, 1979).

Most preformed antimicrobials are effective only after pathogen penetration into the tissue. Proanthocyanidins are enzyme-inhibiting tannins present in high concentrations in leaves and immature fruits of many wild and cultivated plants. They are made of subunits of anthocyanidins and form complexes with structural proteins or enzymes. These compounds have no direct antimicrobial effect but restrict invading microorganisms by denaturing their cell wall-degrading enzymes (Schlosser, 1988). Young grape berries are resistant to *Botrytis cinerea,* which causes the gray mould disease. A positive correlation was observed between the content of active proanthocyanidins in fruit tissue and its resistance to the fungus (Hill et al., 1981). Towards maturation grape berries become susceptible to *B. cinerea.* One explanation could be that proanthocyanidins have a higher degree of polymerization, which leads to less efficient binding to fungal hydrolases (Schlosser, 1988). Similar observations were made in strawberry fruits where *B. cinerea* invades the flower at an early stage but remains confined to the receptacle, which has a high concentration in proanthocyanidins. At maturation the protein denaturing potential of the proanthocyanidins decreases and the entire fruit becomes rapidly invaded (Jersch et al., 1989). Breeders have directed substantial effort to developing cultivars with greater resistance to *B. cinerea.* A typical characteristic of the resistant cultivars is the presence in mature fruits of a ring of white immature tissue around the stem. In this ring proanthocyanidins are polymerized to a lesser extent and inhibit the gray mould fungus (Jersch et al., 1989). Chlorogenic acid present in various potato tissues was often associated with disease resistance (reviewed by Friend, 1981). However, it is doubtful that the endogenus level is high enough to be fungitoxic (Friend, 1981, 1985).

Oxidation of phenolics often increases their fungitoxicity. An example is phloridzin, a phenolic glucoside in apple, or its aglycone phloretin, neither of which have strong activity against *Venturia inaequalis*. Upon oxidation by a polyphenoloxidase present in apple tissue, these compounds yielded *o*-quinones, which were potent fungicides (Overeem, 1976). In pear, oxidation of arbutin leads to an antibacterial compound hydroquinone, which was associated with resistance to *Erwinia amylovora* (Hildebrand et al., 1969). It is likely that in the incompatible host–pathogen interaction the production of such compounds follows the hypersensitive response (HR) taking place at the site of infection. The resulting loss in membrane function with accompanying decompartmentalization could then lead to the formation of these toxic compounds.

Hydroxamic acids are closely related to phenolics and are usually found in the shoots and roots of *Gramineae* with the exception of *Avena, Hordeum,* and *Oryza* (Niemeyer, 1988). The compounds occur as glycosides, which are hydrolyzed by a β-glucosidase upon loss in compartmentalization after infection or mechanical injury. Correlations have been observed between hydroxamic acid levels and resistance against fungal diseases in wheat, corn, and rye (review by Niemeyer, 1988). Cereals are also plagued by aphids, which are important vectors for the transmission of viral diseases. Hydroxamic acid contents in corn and wheat leaves were inversely related to aphid infestations of these plants. When incorporated into diets of cereal aphids, hydroxamic acid had both antibiotic and antifeedant effects (Argandona et al., 1983). Hydroxamic acids have also been reported to trigger the reproduction of grass-feeding mammals and to have allelopathic effects on cereals (see review by Niemeyer, 1988). The presence of hydroxamic acids was proposed as an agronomically useful trait to be considered in breeding for aphid resistance in wheat (Copaja et al., 1991).

4. THE ROLE OF PHENOLICS AS INDUCED ANTIMICROBIAL DEFENSE CHEMICALS

Antimicrobial compounds that are produced or accumulate after exposure of plants to microorganisms are called phytoalexins (Paxton, 1981) and include over 300 natural products distributed among phenolics, terpenoids, polyacetylenes, fatty acid derivatives, and other classes. Among this remarkably heterogeneous group of chemicals there is no conspicuous relationship between structure and toxicity. Phytoalexins have been ably reviewed in a number of excellent articles (Bailey and Mansfield, 1982; Barz et al., 1990; Darvill and Albersheim, 1984; Dixon, 1986; Dixon and Harrison, 1990; Ebel, 1986; Hahlbrock and Scheel, 1989). Here we focus on the phenolic phytoalexins, such as the isoflavonoids or stilbenes, which are predominantly but not exclusively synthesized in leguminous plants (van Etten and Pueppke, 1976).

Two lines of investigations support the notion that phytoalexins are factors involved in the resistance of plants to pathogens. One line concerns the site of phytoalexin production in relation to the location of the pathogen. The other deals with the detoxification of phytoalexins by pathogens. Microscopic observations monitoring the movement of a pathogen were combined with biochemical determinations of the phytoalexin level in soybean hypocotyls resistant or susceptible to race 1 of *Phytophthora megasperma* (Yoshikawa et al., 1978). In resistant interactions, the levels of glyceollin around invading hyphae 8 h after infection are high enough to inhibit fungal growth by 90% *in vitro*. In susceptible interactions, glyceollin is barely detectable at that stage. By the time levels of glyceollin increased, the fungus has grown far beyond the inoculation site (Yoshikawa et al., 1978). Another study used small unwounded soybean seedlings, the roots of which were inoculated with virulent or avirulent races of *P. megasperma* (Hahn et al., 1985). Glyceollin was detected by radioimmunoassay in cryotome sections of the roots. At the same time, the fungal progression was monitored using immunoflu-orescent stains for *P. megasperma* hyphae. In the resistant interaction glyceollin rapidly accumulated in the epidermis and in the cortex around the site of infec-tion exceeding the ED_{90} concentration 8 h after inoculation. Hyphae were not detected in tissues containing measurable levels of glyceollin. In this compatible interaction the fungus proliferated rapidly into the root stele with no or little glyceollin detected in the cortex. These observations support the notion that glyceollin accumulation is an important factor in the defense of the plant. How-ever, they do not exclude the existence of other resistance mechanisms that might be triggered by the incompatible fungal race.

The detoxification of kievitone, phaseollin, phaseollidine, and phaseollinisoflavan in bean by *Fusarium solani f.sp. phaseoli* (Smith et al., 1980; Zang and Smith, 1983), of maackiain and medicarpin in chickpea by *Ascochyta rabiei* (Hohl et al., 1989), and pisatin in pea by *Nectria haematococca* (van Etten et al., 1989) suggest that detoxification of phytoalexins is an important trait conferring patho-genicity (review by van Etten et al., 1989). In pea the detoxifying activity is due to an inducible cytochrome P-450 monooxygenase referred to as pisatin demethylase (pda). Analyses of crosses between pathogenic and nonpathogenic strains of *Nectria* indicated that pisatin demethylation is necessary for pathogenicity (van Etten et al., 1989). Expression of pda in pda⁻ strains of *Nectria* or in the corn pathogen *Cochliobolus heterostrophus* resulted in the ability to demethylate pisatin and to infect pea (Ciuffetti et al., 1988; Schafer et al., 1989). This work provided addi-tional experimental evidence indicating that phytoalexins are involved in disease resistance and that their detoxification may be required for pathogenicity.

The pathway and enzymes of general phenylpropanoid metabolism that pro-vide common precursors for isoflavonoids, coumarins, stilbenes, lignins, and var-ious other phenolics are well described in many plants (reviewed by Hahlbrock and Scheel, 1989). After infection or treatments with elicitors the synthesis of isoflavonoid phytoalexins is induced. This increase is associated with an increase in the amounts of the related biosynthetic enzymes and their messenger ribonu-cleic acids (mRNAs) (reviewed by Dixon and Harrison, 1990). The availability of antibodies and complimentary deoxyribonucleic acid (cDNA) for several of these enzymes has yielded new information on the spatial distribution of phyto-alexin biosynthesis. In bean inoculated with *Colletotrichum lindemuthianum*, tran-scripts of PAL were found in cells adjacent to the site of infection and to lesser extent in healthy tissue several millimeters away from the point of inoculation (Bell et al., 1986), suggesting the presence of some mechanism of intercellular signaling. In the leaves of young potato plants PAL mRNA was shown to accu-mulate in healthy cells around the site of an incompatible interaction with *Phytoph-thora infestans* (Cuypers and Hahlbrock, 1988). Similar observations were made in parsley leaves infected with an incompatible pathogen *P. megasperma f.sp. glycinea* (Schmelzer et al., 1989; Somssisch et al., 1988). Tobacco was transformed with a bean chalcone synthase promoter fused with the reporter gene β-glucuronidase (GUS). Strong expression of the chalcone synthase promoter was observed at the site of the hypersensitive reaction against *Pseudomonas syringae*. Lower levels of expression were detected throughout the inoculated leaf (Stermer et al., 1990). Thus a promoter from the gene encoding the bean enzyme could respond to defense signals from tobacco indicating similarity in the signal transduction processes in various plants.

Little is known about the mode of action of phytoalexins, despite abundant work demonstrating their toxic activity against fungal, plant, and even mammalian cells. The structural requirements for activity are unknown; however, a certain degree of lipophilicity is required. Some evidence indicates that phytoalexins could act by uncoupling of oxidative phosphorylation (Laks and Pruner, 1989). In addition, pterocarpan phytoalexins might also cause proton leakage at the tonoplast membrane (Giannini et al., 1991).

5. PHENOLICS AS SIGNALS FOR INDUCED RESISTANCE

Induced or acquired resistance is a response of the plant to pathogen infection. Initial inoculation with a necrotizing pathogen results in resistance towards subsequent infections, even in tissue untouched by the inducing infection. The phenomenon was first discovered in *Dianthus barbatus* infected by carnation mosaic virus (Gilpatrick and Weintraub, 1952). Systemic acquired resistance (SAR) can be effective against infections by a broad range of pathogens that need not be related to the inducing organism (as reviewed by Madamanchi and Kuc, 1991). For example, in cucumber or tobacco, a first infection with either fungal, bacterial, or viral pathogens protects the plant against subsequent infections by fungal bacterial or viral pathogens both in the infected and uninfected parts of the plant (reviewed by Madamanchi and Kuc, 1991).

In all cases reported so far the induction of resistance is dependent on the production of necrotic lesions by the inducing pathogen. The level of protection is related to the size and number of lesions produced during the first infection (Madamanchi and Kuc, 1991). Protection is not an all or nothing phenomenon as observed in race-specific resistance. Despite this variation in expression, induced resistance can be extremely effective and long lasting. For instance, in cucumber plants infected on the first leaf with tobacco necrosis virus, or the fungus *Colletotrichum lagenarium,* induced resistance lasts for several weeks. A second ''booster'' inoculation following the first inoculation gives protection up to the time of flowering (Madamanchi and Kuc, 1991). The effectiveness of systemic induced resistance was confirmed in field trials for bean (Sutton, 1982), cucumber (Caruso and Kuc, 1977), and tobacco (Tuzun et al., 1986). No deleterious effects on yield were reported. In tobacco plants immunized by subepidermal stem injections with spores of *Peronospora tabacina* the growth of the plants can be enhanced even in the absence of the challenge pathogen (Tuzun and Kuc, 1985).

Since resistance can be induced systemically by localized infections, the production of a signal that activates the resistance mechanism has been hypothesized (Ross, 1966; Kuc, 1982; MacIntyre et al., 1981). Evidence from stem girdling experiments and grafting experiments has indicated that the putative signal moves through the phloem tissue of the vascular system of the plant. (Guedes et al., 1980; Gianninazzi and Ahl, 1983).

Because the inducing infection results in broad protection against multiple pathogens, it is likely that diverse mechanisms are activated in the plant. As already mentioned earlier lignification of epidermal cells occurs more rapidly and to a larger extent in systemically induced cucumber plants than in uninduced controls (Hammerschmidt and Kuc, 1982; Dean and Kuc, 1987). The most thoroughly documented biochemical change associated with the onset of induced resistance is the synthesis of abundant amounts of low molecular weight, mainly acidic, pathogenesis-related (PR) proteins some of which are secreted into the intercellular spaces (Bol and van Kan, 1988; Hosokawa and Ohashi, 1988; Parent and Asselin, 1984; Carr et al., 1987). At present the PR proteins are classified into five serologically and structurally distinct families (Antoniw et al., 1980; van Loon, 1985; van Loon et al., 1987; Bol and van Kan, 1988). The PR proteins have been detected in at least 16 plant species (van Loon, 1985) and can be induced by necrotizing and nonnecrotizing viruses, fungi, and bacteria (van Loon, 1983). The appearance and possible function of PR proteins is correlated with the expression of HR and SAR (Bol and van Kan, 1988; Kassanis et al., 1974). However, the induction of PR proteins alone may not be sufficient for resistance, because the constitutive expression of one or more PR proteins did not affect tobacco susceptibility to TMV (Cutt et al., 1989; Linthorst et al., 1989).

It is well established that resistance to pathogens and the production of at least some PR proteins in tobacco can be induced by salicylic acid (SA) or acetylsalicylic acid (aspirin), even in the absence of pathogenic organisms (White, 1979; Antoniw and White, 1980; van Loon and Antoniw, 1982). For example, in cucumber, extracellular endochitinase is one of the major PR proteins, which is induced systemically following viral, bacterial, or fungal infection (Boller and Métraux, 1986; Métraux et al., 1988). This protein can be effectively induced by SA at the level of RNA accumulation (Métraux et al., 1989). Salicylic acid also reduced the symptoms of tobacco necrosis virus (TNV) infection in asparagus by 90% (Pennazio et al., 1987), and induced PR proteins and resistance to alfalfa mosaic virus in bean and cowpea (Hooft van Huijsduijnen et al., 1986). In Samsun nn tobacco, SA but not virus, was able to induce PR proteins and resistance to subsequent TMV inoculation measured as virus multiplication in the infected leaves (White et al., 1983). In tobacco cultivars containing "N" gene, which confers hypersensitive resistance to TMV, SA decreased the size and number of lesions

(White, 1979). In addition to SA and aspirin, which is rapidly converted to SA in biological tissues (Morgan and Truitt, 1965), only 2,6-dihydroxybenzoic acid was capable of directly inducing PR proteins and resistance to virus (van Loon, 1983). At this time we do not know to what extent SA-induced resistance is based on the induction of PR proteins. It is certainly possible that SA activates additional known and unknown resistance mechanisms.

6. SALICYLIC ACID: A LIKELY ENDOGENOUS SIGNAL IN THE TOBACCO AND CUCUMBER RESISTANCE RESPONSE

The investigation of the role of endogenous SA in disease resistance was undertaken because it was known that exogenous SA applications effectively induce resistance and PR proteins in plants, including tobacco varieties lacking the ''N'' gene. Previous work had also demonstrated that SA is an endogenous signal, which triggers cyanide-insensitive heat production in the inflorescences of thermogenic *Arum* lilies (Raskin et al., 1987; Raskin et al., 1989), suggesting that SA serves as a regulatory molecule in plants.

The hypothesis that SA may be an endogenous messenger that activates host resistance to pathogens, including synthesis of PR proteins was tested by monitoring endogenous levels of SA and PR1 mRNA in TMV-inoculated resistant Xanthic-nc and susceptible Xanthi varieties of tobacco. Salicylic acid levels in the resistant (Xanthi-nc), but not susceptible (Xanthi) tobacco increase almost 50-fold in the inoculated leaves and 10-fold in the uninoculated leaves after TMV inoculation (Figs. 11.1**A** and 11.2**B**) (Malamy et al., 1990). Induction of PR1 genes paralleled the rise in SA levels (Figs. 11.1**B** and 11.2**B**). While TMV induced PR proteins only in Xanthi-nc tobacco, SA was effective in both Xanthi-''nn'' and Xanthi-nc ''NN'' plants (Fig. 11.1**C**). The observed increase in endogenous SA levels is sufficient for the systemic induction of PR-1 proteins (Yalpani et al., 1991).

Another set of experiments has demonstrated that SA increased dramatically in the phloem of cucumber plants inoculated with tobacco necrosis virus or the fungal pathogen *C. lagenarium* (Métraux et al., 1990). The levels of SA increased transiently after inoculation, with a peak reached before SAR was detected (Fig. 11.3).

In tobacco and cucumbers, SA is most likely synthesized in a two-step pathway from *trans*-cinnamic acid. *Trans*-cinnamic acid is, in turn, synthesized from L-phenylalanine by PAL, a central enzyme in the biosynthesis of all phenolic compounds in plants (Chadha and Brown, 1974). The PAL activity increases sharply following infection with pathogens or mechanical wounding (Minamikawa and Uritani, 1964). Similarly, PAL can be induced by fungal elicitors added to the suspension culture of bean cells (Dixon and Harrison, 1990, Lawton and Lamb, 1987).

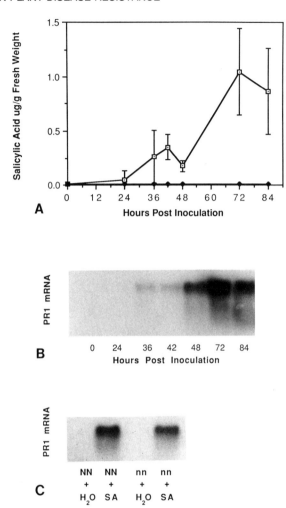

Fig. 11.1. Effect of TMV inoculation on the endogenous salicylic acid and PR1 mRNA in the infected leaves of Xanthi nc (NN) tobacco. **A:** Salicylic acid levels in TMV-infected (☐) and mock-infected leaves (◆). **B:** Northern blot analysis of steady-state levels of PR1 mRNA from TMV-infected leaves of Xanthi nc (NN) tobacco. **C:** Induction of PR1 genes in Xanthi nc (NN) and Xanthi (nn) by exogenous salicylic acid.

The above information is consistent with the proposed role of SA as a signal compound that activates synthesis of PR proteins and the induction of SAR in tobacco and cucumber. Unfortunately the role of endogenous SA in disease resistance in other plants has not been studied and there are some indications that it may not be the ubiquitous signal in disease resistance. In some plants SA induced

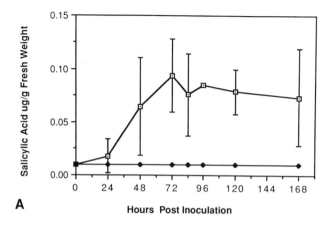

A   Hours Post Inoculation

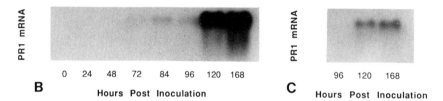

B   Hours Post Inoculation   **C**   Hours Post Inoculation

Fig. 11.2. Effect of TMV inoculation on the endogenous salicylic acid and PR1 mRNA in the uninfected leaves of Xanthi nc (NN) tobacco. Upper, uninfected leaves of mock- or TMV-inoculated plants were harvested and analyzed for SA and PR1 mRNA content. **A:** Salicylic acid levels in uninfected leaves of TMV-inoculated (□) and mock-inoculated (♦) plants. **B:** Northern blot analysis of steady-state levels of PR1 mRNA from TMV-infected leaves of Xanthi nc (NN) tobacco. **C:** Induction of PR1 genes in Xanthi nc (NN) and Xanthi (nn) by exogenous salicylic acid. C: Autoradiogram of the same blot as in (B), exposed for the same time as in Fig. 11.1 B.

only some (Pennazio et al., 1987) or no PR proteins at all (Pennazio and Redolfi, 1980). In soybean SA did not induce resistance to viruses and did not stimulate production of PR proteins (Roggero and Pennazio, 1991).

7. CONCLUSIONS AND OUTLOOK FOR PRACTICAL APPLICATIONS

The growing appreciation of the role of phenolics in resistance to plant pathogens may result in a development of crops with higher levels of resistance. Preformed phenolics that in all likelihood are part of the nonhost resistance mechanisms

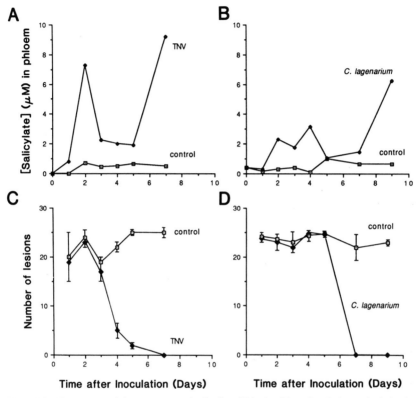

Fig. 11.3. Time course of the appearance of salicylic acid in the phloem in relation to the induction of resistance against *Colletotrichim lagenarium* after initial infection with TNV (**A** and **C**) or *C. lagenarium (***B** and **D**).

might in some cases be useful traits for breeding programs. Further knowledge on their mode and site of action in fungi can provide the agrochemical researcher with useful targets for the design or the screening of new fungicides. Induced lignification might be desirable when occurring at the site of penetration of a pathogen. To influence this process at a local level seems at present an unrealistic task. One solution might be to place the expression of genes involved in lignification under the control of promoters responding to stimuli that are released locally during any plant–pathogen interaction (nonspecific elicitors). The difficulty is to find promoters that have the appropriate specificity. The understanding of phytoalexin metabolism is a prerequisite to develop seeds with improved defense against pathogens. For instance, the phytoalexin pathway might be modified by genetic engineering to change the pattern of phytoalexins produced in a plant upon infection.

This could improve the spectrum of chemical defense and circumvent to some degree the problem of phytoalexin detoxification.

The manipulation of the levels of phenolic compounds, which may serve as signals for activating various defense mechanisms, represents one of the most promising areas for applying biotechnology to crop protection. Salicylic acid might provide a good lead for this research. The biosynthesis of SA is likely to be a two-step enzymatic process (see above). The levels of SA that are sufficient for the activation of PR proteins and some disease resistance are not phytotoxic (Raskin I, unpublished). Increases in endogenous SA may be achieved by enhancing transcription and translation of genes of SA biosynthesis or by blocking the expression of genes involved in SA metabolism. These genes may be of plant or bacterial origin. In either case, transgenic plants with elevated SA levels may be a first step in the engineering of crops with increased resistance to agronomically important pathogens.

Alternatively, chemicals could be found that act like the endogenous signal by triggering resistance mechanisms. Application of such compounds by conventional techniques could be suitable for large scale farming. Such treatments could even be complementary to existing disease control practices, which typically involve treatments designed to directly kill the invading pathogen. Efforts to find such resistance-inducing compounds have yielded some preliminary results: oxalate, phosphate, and 2,6-dichloroisonicotinic acid and its ester derivative were all found to induce resistance, and associated biochemical changes, in much the same way as preinoculations with pathogens (Madamanchi and Kuc, 1991; Métraux et al., 1991).

REFERENCES

Antoniw JF, White RF (1980): The effects of aspirin and polyacrylic acid on soluble leaf proteins and resistance to virus infection in five cultivars of tobacco. Phytopath Z 98:331–341.

Argandona VH, Corcuera LJ, Niemeyer HM, Campbell BC (1983): Toxicity and feeding deterrency of hydroxamic acids from graminae in synthetic diets against the greenbug, *Schizapis graminum*. Ent Exp Appl 34:134–138.

Asada Y, Matsumoto I (1972): The nature of lignin obtained from downy mildew-infected Japanese radish roots. Phytopathology Z 73:208–214.

Bailey JA, Mansfield JW (1982): Phytoalexins. New York: Wiley.

Baltas M, Cazaux L, Gorrichon L, Maroni P, Tisnes P (1988): Sulphinamoylacetates as sulphine precursors. Mechanism of basic hydrolysis and scheme of irreversible inactivation of cinnamoyl alcohol dehydrogenase, an enzyme of the lignification process. J Chem Soc Perkin Trans II 8:1473–1478.

Barz W, Koster J, Weltring K-M, Strack D (1985): Recent advances in the metabolism and degradation of phenolic compounds in plants and animals. Ann Proc Phytochem Soc Eur 25:307–347.

Barz W, Bless W, Borger-Pappendorf G, Gunia W, Mackenbrock O, Meier DE, Otto CH, Super E (1990): Phytoalexins as part of induced defense reaction in plants: their elicitation, function, and metabolism. In van Sumere CF, Leen PJ (eds): Bioactive Compounds From Plants. Chichester: Wiley, 140–156.

Beardmore J, Ride JP, Granger JW (1983): Cellular lignification as a factor in the hypersensitive resistance of wheat to stem rust. Physiol Plant Pathol 22:209–220.

Bell JN, Ryder T, Wingate VPM, Bailey JA, Lamb CJ (1986): Differential accumulation of plant defense gene transcripts in a compatible and an incompatible plant-pathogen interaction. Mol Cell Biol 6:1615–1623.

Bird P (1988): The role of lignification in plant disease. In Singh RS, Singh US, Hess WM, Weber DJ (eds): Experimental Concepts in Plant Pathology. New York: Gordon and Breach, pp 523–535.

Blakeman JP, Atkinson P (1976): Antimicrobial substances associated with the aerial surfaces of plants. In Blakeman JP (ed): Microbial Ecology of the Phylloplane. New York: Academic Press, pp 245–263.

Bol JF, van Kan JAL (1988): The synthesis and possible functions of virus-induced proteins in plants. Microbiol Sci 5:47–52.

Boller T, Métraux JP (1986): Local and systemic induction of chitinase in cucumber plants in response to viral, bacterial and fungal infections. Physiol Mol Plant Pathol 28:161–169.

Carr JP, Dixon DC, Nikolau BJ, Voelkerding KV, Klessig DF (1989): The pathogenesis-related proteins of plants. In Setlow JK (ed): Genetic Engineering. Principles and Methods. New York: Plenum Press, pp 65–109.

Caruso F, Kuc J (1977): Field protection of cucumber, watermelon and muskmelon against *Colletotrichum lagenarium* by *Colletotrichum lagenarium*. Phytopathology 67:1290–1292.

Cassab GI, Varner JE (1988): Cell wall proteins. Annu Rev Plant Physiol Plant Molec Biol 39:321–353.

Chadha KC, Brown SA (1974): Biosynthesis of phenolic acids in tomato plants infected with *Agrobacterium tumefaciens*. Can J Bot 52:2041–2046.

Ciuffetti LM, Weltring KM, Turgeon BG, Yoder OG, Van Etten H (1988): Transformation of *Nectria haematococca* with a gene for pisatin demethylating activity, and the role of pisatin detoxification in virulence. J Cell Biochem 12C:278.

Copaja SV, Niemeyer HM, Wratten SD (1991): Hydroxamic acid levels in Chilean and British wheats. Ann Appl Biol 118:223–227.

Cutt JR, Harpster MH, Dixon DC, Carr JP, Dunsmur P, Klessig DF (1989): Disease response to tobacco mosaic virus in transgenic tobacco plants that constitutively express the pathogenesis-related PR-1b gene. Virology 173:89–97.

Cuypers B, Schmelzer E, Hahlbrock K (1988): In situ localization of rapidly accumulated phenylalanine ammonialyase mRNA around penetration sites of *Phytophthora infestans* in potato leaves. Mol Plant-Microbe Interactions 1:157–160.

Darvill AG, Albersheim P (1984): Phytoalexins and their elicitors—A defense against microbial infection in plants. Annu Rev Plant Physiol 35:243–275.

Dean R, Kuc J (1987): Rapid lignification in response to wounding and infection as a mechanism for induced systemic protection in cucumber. Physiol Mol Plant Pathol 31:69–81.

Dix NJ (1979): Inhibition of fungi by gallic acid in relation to growth on leaves and litter. Trans Br Mycol Soc 73:329–336.

Dixon RA (1986): The phytoalexin response: Elicitation, signalling and control of host gene response. Biol Rev 61:239–291.

Dixon RA, Harrison MJ (1990): Activation, structure and organization of genes involved in microbial defence in plants. Adv Genet 28:165–234.

Ebel J (1986): Phytoalexin synthesis: The biochemical analysis of the induction process. Annu Rev Phytopathol 24:235–264.

Einhellig FA (1986): Mechanism and modes of action of allelochemicals. In Putnam AR, Tang CS (eds): The Science of Allelopathy. New York: John Wiley & Sons, p 317.

Friend J (1981): Plant phenolics, lignification and plant disease. In Reinhold L, Harborne JB, Swain T (eds): Progress in Phytochemistry Vol 7. Oxford: Pergamon Press, pp 197–261.

Friend J (1985): Phenolic substances and plant disease. In van Sumere CF, Lea PJ (eds): Annual Proceedings of the Phytochemical Society of Europe 25. Oxford: Clarendon, pp 367–392.

Gianninazzi S, Ahl P (1983): The genetic and molecular basis of b-proteins in the genus *Nicotiana*. Neth J Plant Pathol 89:275–281.

Giannini JL, Holt JS, Briskin DP (1991): The effect of glyceollin on soybean (*Glycine max* L.) tonoplast and plasma membrane vesicles. Plant Sci 74:203–211.

Gilpatrick JD, Weitstraub M (1952): The effects of aspirin and polyacrylic acid on soluble leaf proteins and resistance to virus infection in five cultivars of tobacco. Science 115:701–701.

Grand C, Sarni F, Boudet AM (1985): Inhibition of cinnamyl-alcohol-dehydrogenase activity and lignin synthesis in poplar (*Populus euramericana* Dode) tissues by two organic compounds. Planta 163:232–237.

Guedes MEM, Richmond S, Kuc J (1980): Induced systemic resistance to anthracnose in cucumber as influenced by the location of the inducer inoculation with *Colletotrichum lagenarium* and the onset of flowering. Physiol Plant Pathol 17:229–235.

Hahlbrock K, Scheel D (1989): Physiology and molecular biology of phenylpropanoid metabolism. Annu Rev Plant Physiol Plant Mol Biol 40:347–369.

Hahn MG, Bonhoff A, Griesebach H (1985): Quantitative localization of phytoalexin glyceollin in relation to fungal hyphae in soybean roots infected with *Phytophthora megasperma* f. sp. glycinea. Plant Physiol 77:591–601.

Hammerschmidt R (1984): Rapid deposition of lignin in potato tuber tissue as a response to fungi non-pathogenic on potato. Physiol Plant Pathol 24:33–42.

Hammerschmidt R, Kuc J (1982): Lignification as a mechanism for induced systemic resistance in cucumber. Physiol Plant Pathol 20:61–71.

Harborne JB (1980): Plant phenolics. In Bell EA, Charlwood BV (eds): Secondary Plant Products. Berlin: Springer Verlag, pp 329–402.

Harborne JB (1990): Role of secondary metabolites in the chemical defence mechanisms in plants. In: Bioactive Compounds From Plants, Ciba Foundation Symposium 154. Chichester: Wiley, pp 126–139.

Harborne JB, Ingham JL, King L, Payne M (1976): The isopentenyl isoflavone luteone as a preinfectional antifungal agent in the genus *Lupinus*. Phytochemistry 15:1485–1487.

Hildebrand DC, Powell CC, Schroth MN (1969): Fire blight resistance in pyrus: Localization of arbutin and β-glucosidase. Phytopathology 70:1534–1539.

Hill G, Stellwaag-Kittler F, Huth G, Schlosser E (1981): Resistance of grapes in different development stages to *Botrytis cinerea*. Phytopathol Z 102:328:338.

Höhl B, Arnemann M, Schwenen L, Stockl D, Bringmann G, Jansen J, Barz W (1989): Degradation of the pterocarpan phytoalexin (—)-maackiain by *Ascochyta rabiei*. Z Naturforsch Sect C 44:771–776.

Hooft van Huijsduijnen RAM, Alblas SW, Ryk RH, Bal JF (1986): Induction by salicylic acid of pathogenesis-related proteins and resistance to alfalfa mosaic virus infection in various plant species. Gen Virol 67:2135–2143.

Hosokawa D, Ohashi Y (1988): Immunochemical localisation of pathogenesis-related proteins secreted into the intercellular spaces of salicylate-treated tobacco leaves. Plant Cell Physiol 29:1035–1040.

Ingham JL, Keen NT, Hymowitz T (1977): A new isoflavone phytoalexin from fungus-inoculated stems of *Glycine wightii*. Phytochemistry 16:1943–1946.

Jersch S, Scherrer C, Huth G, Schlosser E (1989): Protoanthocyanidins as basis for quiescence of *Botrytis cinerea* in immature strawberry fruits. Z Pflanzenkr Pflanzenschutz 96:365–378.

Kassanis B, Gianinazzi S, White RF (1974): A possible explanation of the resistance of virus-infected tobacco plants to second infection. J Gen Virol 23:11–16.

Kirk TK (1971): Effects of microorganisms on lignin. Annu Rev Phytopathol 9:185–210.

Kovats K, Binder A, Hohl HR (1991): Cytology of induced systemic resistance of cucumber to *Colletotrichum lagenarium*. Planta 183:484–490.

Kuc J (1982): Induced immunity to plant disease. Bioscience 32:854–860.

Laks PE, Pruner MS (1989): Flavonoid biocides: structure/activity relations of flavonoid phytoalexin analogues. Phytochemistry 28:87–91.

Lawton MA, Lamb CJ (1987): Transcriptional activation of plant defense genes by fungal elicitor, wounding and infection. Molec Cell Biol 7:335–341.

Legrand M (1983): Phenylpropanoid metabolism and its regulation in disease. In Callow JA (ed): Biochemical Plant Pathology. Chichester: Wiley, pp 367–384.

Legrand M, Fritig B, Hirth L (1976): Enzymes of the phenylpropanoid pathway and the necrotic reaction. Phytochemistry 15:1353–1359.

Legrand M, Fritig B, Hirth L (1978): O-Diphenyl and O-methyltransferase of healthy and tobacco-mosaic-virus-infected hypersensitive tobacco. Planta 144:101–108.

Lewis NG, Yamamoto E (1990): Lignin: Occurrence, biogenesis and biodegradation. Annu Rev Plant Physiol Plant Mol Biol 41:455–496.

Linthorst HJM, Meuwissen RLJ, Kauffmann S, Bol JF (1989): Constitutive expression of pathogenesis-related proteins PR-1, GRP, and PR-S in tobacco has no effect on virus infection. Plant Cell 1:285–291.

Long SR (1989): Rhizobium-legume nodulation: Life together in the underground. Cell 56:203–214.

Lynn DG, Chang M (1990): Phenolic signals in cohabitation: Implications for plant development. Annu Rev Plant Physiol Plant Mol Biol 41:497–526.

MacIntyre JL, Dodds JA, Hare JD (1981): Effects of localized infections of N. tabacum by brome mosaic virus on systemic resistance against diverse pathogens and insects. Phytopathology 71:297–301.

Madamanchi NR, Kuc J (1991): Induced systemic resistance in plants. In Cole GT, Hoch HC (eds): The Fungal Spore and Disease Initiation in Plants and Animals. New York: Plenum Press, pp 347–362.

Malamy J, Carr JP, Klessig DF, Raskin I (1990): Salicylic acid: A likely endogenous signal in the resistance response of tobacco to viral infection. Science 250:1002–1004.

Martin JT, Batt RF, Burchill RT (1957): Fungistatic properties of apple leaf wax. Nature (London) 180:796–797.

Massala R, Legrand M, Fritig B (1987): Comparative effects of two competitive inhibitors of phenyl-alanine ammonia-lyase on the hypersensitive resistance of tobacco to tobacco mosaic virus. Plant Physiol Biochem 25:217–225.

Matern U, Kneusel RE (1988): Phenolic compounds in plant disease resistance. Phytoparasitica 16:153–170.

Maule AJ, Ride J (1976): Ammonia-lyase and O-methyltransferase activities related to lignification in wheat leaves infected with Botrytis. Phytochemistry 15:1661–1664.

Maule AJ, Ride J (1983): Cinnamate 4-hydroxylase and hydroxycinnamate: CoA ligase in wheat leaves infected with Botrytis cinerea. Phytochemistry 22:1113–1116.

Métraux JP, Streit L, Staub T (1988): A pathogenesis-related protein in cucumber is a chitinase. Physiol Mol Plant Pathol 33:1–9.

Métraux JP, Burkhart W, Moyer M, Dincher S, Middlesteadt W, Williams S, Payne G, Carnes M, Ryals J (1989): Isolation of a complementary DNA encoding a chitinase with structural homology to a bifunctional lysozyme/chitinase. Proc Natl Acad Sci USA 86:869–900.

Métraux JP, Signer H, Ryals J, Ward J, Wyss-Benz M, Gaudin J, Raschdorf K, Schmid E, Blum W, Inverardi B (1990): Increase in salicylic acid at the onset of systemic acquired resistance in cucumber. Science 250:1004–1006.

Métraux JP, Ahl-Goy P, Staub TH, Speich J, Steinemann A, Ryals J, Ward E (1991): Induced systemic resistance in cucumber in response to 2,6-dichloro-isonicotinic acid and pathogens. In Hennecke H, Verma DPS (eds): Advances in Molecular Genetics of Plant–Microbe Interactions. Dordrecht: Kluwer, pp 432–439.

Minamikawa T, Uritani I (1964): Arch Biochem Biophys 108:573–574.

Moersbacher BM, Noll UM, Flott BE, Reisener HJ (1988): Lignin biosynthetic enzymes in stem rust infected, resistant and susceptible near-isogenic wheat lines. Physiol Molec Plant Pathol 33:33–46.

Moersbacher BM, Witte U, Konigs D, Reisener HJ (1989): Changes in the level of enzyme activities involved in lignin biosynthesis during the temperature sensitive resistant response of wheat (Sr6) to stem rust (P6). Plant Science 65:183–190.

Moersbacher BM, Noll U, Gorrichon L, Reisener HJ (1990): Specific inhibition of lignification breaks hypersensitive resistance of wheat to stem rust. Plant Physiol 93:465–470.

Morgan AN, Truitt EB Jr (1965): Evaluation of acetylsalicylic acid esterase in aspirin metabolism: Interspecies comparison. J Pharm Sci 54:1640–1646.

Niemeyer HM (1988): Hydroxamic acids (4-hydroxy-1,4-benzoxazin-3ones), defence chemicals in the graminae. Phytochemistry 27:3349–3358.

Overeem JC (1976): Preexisting antimicrobial substances in plants and their role in disease resistance. In Friend J, Threlfall DR (eds): Biochemical Aspects of Plant Parasite Relationships. London: Academic Press, pp 195–206.

Parent J-G, Asselin A (1984): Detection of pathogenesis-related proteins (PR or b) and of other proteins in the intercellular fluid of hypersensitive plants infected with tobacco mosaic virus. Can J Bot 62:564–569.

Paxton JD (1981): Phytoalexins—a working redefinition. Phytopathol Z 101:106–109.

Pennazio S, Redolfi P (1980): Resistance to tomato bushy stunt virus localized infection induced in *Gomphrena globosa* by acetylsalicylic acid. Microbiologia 3:475–479.

Pennazio S, Colariccio D, Roggero P, Lenzi R (1987): Effect of salicylate stress on the hypersensitive reaction of asparagus bean to tobacco necrosis virus. Physiol Mol Plant Pathol 30:347–357.

Raskin I, Ehman A, Melander WR, Meeuse BJD (1987): Salicylic acid—a natural inducer of heat production in *Arum* lilies. Science 237:1601–1602.

Raskin I, Turner IM, Melander WR (1989): Regulation of heat production in the inflorescences of an *Arum* lily by endogenous salicylic acid. Proc Natl Acad Sci USA 86:2214–2218.

Ride JP (1975): Lignification in wounded wheat leaves in response to fungi and its possible role in resistance. Physiol Plant Pathol 5:187–196.

Ride J (1978): The role of cell wall alterations in resistance to fungi. Ann Appl Biol 89:302–306.

Ride J, Pearce RB (1979): Lignification and papilla formation at sites of attempted penetration of wheat leaves by non-pathogenic fungi. Physiol Plant Pathol 15:79–92.

Ride JP (1983): Cell walls and other structural barriers in defence. In Callow JA (ed): Biochemical Plant Pathology. Chichester: Wiley, pp 215–236.

Roggero P, Pennazio S (1991): Salicylate does not induce resistance to plant viruses, or stimulate pathogenesis-related protein production in soybean. Microbiologica 14:65–69.

Ross AG (1966): Systemic effects of local lesion formation. In Beemst ABR, Bijkstra J (eds): Viruses of Plants. Amsterdam: North Holland, pp 127–150.

Schafer W, Straney D, Ciuffetti L, Van Etten HD, Yoder O (1989): One enzyme makes a fungal pathogen, but not a saprophyte, virulent on a new host plant. Science 246:247–249.

Schlosser E (1988): In Singh RS, Singh US, Hess WM, Weber J (eds): Experimental and Conceptual Plant Pathology. New Delhi: Oxford & IBH Publishing Co., pp 465–476.

Schmelzer E, Kruger-Lebus S, Hahlbrock K (1989): Temporal and spatial patterns of gene expression around sites of attempted ungal infection in parsley leaves. Plant Cell 1:993–1001.

Smith DA, Kuhn PJ, Bailey J, Burden RS (1980): Detoxification of phaseollidin by *Fusarium solani* f. sp. phaseoli. Phytochemistry 19:1673–1675.

Somssisch IE, Schmelzer E, Kawalleck P, Hahlbrock K (1988): Gene structure and in situ transcript localization of pathogenesis-related protein 1 in parsley. Mol Gen Genet 213:93–98.

Southerton SG, Deverall BJ (1990): Histochemical and chemical evidence for lignin accumulation during the expression of resistance to leaf rust fungi in wheat. Physiol Mol Plant Pathol 36:483–494.

Stachel SE, Messens E, Montagu MV, Zambryski P (1985): Identification of the signal molecules produced by wounded plant cells that activate T-DNA transfer in *Agrobacterium tumefaciens*. Nature (London) 318:624–629.

Stermer B, Schmid J, Lamb CJ, Dixon RA (1990): Infection and stress activation of bean chalcone synthase promoters in transgenic tobacco. Mol Plant–Microbe Interactions 3:381–388.

Sutton DC (1982): Field protection of bean against Colletotrichum lindemuthianum by *Colletotrichum lindemuthianum*. Aust Plant Pathol 11:50–51.

Tiburzy R, Reisener HJ (1990): Resistance of wheat to *Puccinia graminis* f. sp. tritici: Association of the hypersensitive reaction with the cellular accumulation of ligninlike material and callose. Physiol Mol Plant Pathol 36:109–120.

Thorpe JR, Hall JL (1984): Chronology and elicitation of changes in peroxidase and phenylalanine ammonia-lyase activities in wounded wheat leaves in response to inoculation by *Botrytis cinerea*. Physiol Plant Pathol 25:363–379.

Touze A, Rossignol M (1977): Lignification and the onset of premonition in muskmelon plants. In Solheim B, Raa J (eds): Cell Wall Biochemistry Related to Specificity in Host–Plant Pathogen Interactions. Oslo: Universitetsforlaget, pp 289–293.

Tuzun S, Nesmith W, Ferriss RS, Kuc J (1986): Effects of stem injections with *Peronospora tabacina* on growth of tobacco and protection against blue mold in the field. Phytopathology 70:938–941.

Tuzun S, Kuc J (1984): A modified technique for inducing systemic resistance to blue mold and increasing growth in tobacco. Phytopathology 75:1127–1129.

Van Etten H, Matthews DE, Matthews PS (1989): Phytoalexin detoxification: Importance for pathogenicity and practical implications. Annu Rev Phytopathol 27:143–164.

Van Etten H, Pueppke SG (1976): Isoflavonoid phytoalexins. In Friend J, Threlfall DR (eds): Biochemical Aspects of Plant Parasitic Relationships. London: Academic Press, pp 239–288.

van Loon LC (1983): The induction of pathogenesis-related proteins by pathogens and specific chemicals. Neth J Plant Pathol 89:265–273.

van Loon LC (1985): Pathogenesis-related proteins. Plant Mol Biol 4:111–116.

van Loon LC, Antoniw JF (1982): Comparison of the effects of salicylic acid and ethephon with virus-induced hypersensitivity and acquired resistance in tobacco. Neth J Plant Pathol 89:237–256.

van Loon LC, Gerritsen YAM, Ritter EC (1987): Identification, purification, and characterization of pathogenesis-related proteins from virus-infected Samsun NN tobacco leaves. Plant Mol Biol 9:593–609.

Vance CP, Kirk TK, Sherwood RT (1980): Lignification as a mechanism of disease resistance. Annu Rev Plant Phytopathol 18:259–288.

Walker JC, Stahmann MA (1955): Chemical nature of disease resistance in plants. Annu Rev Plant Physiol 6:351–366.

White RF (1979): Short communications: Acetylsalicylic acid (aspirin) induces resistance to tobacco mosaic virus in tobacco. Virology 99:410–412.

White RF, Antoniw JF, Carr JP, Woods RD (1983): The effects of aspirin and polyacrylic acid on the multiplication and spread of TMV in different cultivars of tobacco with and without the N-gene. Phytopath Z 107:224–232.

Wood RKS (1967): Physiological Plant Pathology. Oxford: Blackwell.

Yalpani N, Silverman P, Wilson TMA, Klier DA, Raskin I (1991): Salicylic acid is a systemic signal and an inducer of pathogenesis-related proteins in virus-infected tobacco. Plant Cell 3:809–818.

Yoshikawa M, Yamauchi K, Masago H (1978): Glyceollin: Its role in restricting fungal growth in resistant soybean hypocotyls infected with *Phytophthora megasperma* var. sojae. Physiol Plant Pathol 12:73–82.

Zang Y, Smith DA (1983): Concurrent metabolism of the phytoalexin phaseollin, kievitone and phaseollin isoflavan by *Fusarium solani* f. sp. phaseoli. Physiol Plant Pathol 23:89–100.

≡ 12

GENETIC ENGINEERING OF MICROORGANISMS FOR IMPROVED BIOCONTROL ACTIVITY

I. CHET
Z. BARAK

The Hebrew University of Jerusalem, Otto Warburg Center for Agricultural Biotechnology, Faculty of Agriculture, Rehovot 76100, Israel

A. OPPENHEIM

The Hebrew University, Hadassa Medical School, Jerusalem 90101, Israel

1. INTRODUCTION

Biological control is becoming an urgently needed component in agriculture. Chemical pesticides have been the object of substantial criticism in recent years, due to the adverse environmental effects causing health hazards to humans and other nontarget organisms, including beneficial natural enemies. Therefore, it is important to develop safer and environmentally feasible control alternatives, mainly by the use of existing living organisms in their natural habitat. These organisms are able to provide protection against a large range of plant pathogenic fungi, without damage to the ecological system.

Potential agents for biocontrol activity are rhizosphere-competent fungi and bacteria which, in addition to their antagonistic activity, are capable of inducing growth responses by either controlling minor pathogens or producing growth-

Biotechnology in Plant Disease Control, pages 211–235
© *1993 Wiley-Liss, Inc.*

stimulating factors (Burr et al., 1978; Harman et al., 1989). The management of rhizosphere bacteria and fungi as biological control agents is much safer, as well as nonpolluting to the environment, when compared to the application of chemical pesticides.

When added as seed treatments, some rhizosphere organisms may protect the subterranean portions of growing plants from attack by plant pathogens throughout the entire growth period (Ahmad and Baker, 1987; Burr et al., 1978; Sivan and Chet, 1989a; Harman et al., 1989). As a consequence, plant growth may be increased (Ahmad and Baker, 1988; Burr et al., 1978; Harman et al., 1989). Biocontrol agents acquire their antagonistic activities by several potential mechanisms: antibiosis, competition, or parasitism using lytic enzymes (Baker, 1989; Chet et al., 1981; Chet, 1987).

The biological control agents *Trichoderma harzianum* (isolate T-95) and *Pythium nunn,* were combined to reduce *Pythium* damping-off of cucumber in greenhouse experiments lasting 3–4 weeks. This work demonstrates that two compatible biological control agents can be combined to give additional control of a soil-borne plant pathogen (Paulitz et al., 1990). Similarly, Nelson et al. (1986) showed that *Trichoderma* biocontrol activity against *Pythium* was improved by the addition of the lectin producing *Enterobacter cloaccae* bacterium. The bacteria bound tightly to *Trichoderma* without any obvious adverse effect on fungal antagonists and yet, the mixture improved biocontrol activity (Lynch, 1990).

Before biocontrol can become an important component of plant disease management, however, it must be effective, reliable, consistent, and economical. To meet these criteria, superior strains, together with delivery systems that enhance biocontrol activity, must be developed (Harman et al., 1989). So far, strains have been selected from the wild, or produced through mutation or protoplast fusion (Harman et al., 1989; Pe'er and Chet, 1990; Harman and Hayes, 1993—Chapter 14 this volume; Hemming, 1990). Existing biological control attributes can be enhanced by improving existing, known biocontrol agents, using genetic manipulation application of methods, such as transformation, protoplast fusion, or preparation of new mutants, using conventional mutagenesis. Genetic manipulations of biocontrol agents can enhance their biocontrol activity and expand its spectrum.

Development of superior strains requires understanding the mechanism by which specific gene products are involved in biocontrol activity, using the right tools to clone the specific gene and, eventually, utilizing the appropriate mechanism to introduce the heterologous gene into the selected host or to change regulation of genes encoding for biocontrol activity. The foreign gene, which is transformed into the cell, can be integrated into the genome or stay as an episomal plasmid. To express heterologous genes in fungi or bacteria, the regulatory regions of the heterologous gene has to be modulated, by promoter and terminator exchange, in order to optimize the activity of the introduced gene in the new system. Nevalainen (1991) used the *cbh*I promoter and *cbh*I terminator to introduce and express the cellulase gene in *Trichoderma reesei*. An interesting investigation of the bean chitinase promoter, triggered in response to fungal attack, was analyzed in transgenic tobacco plants containing the chitinase SB gene promoter fused to the coding region of the *gus*A gene (Roby et al., 1990; Broglie and co-workers, 1993—Chapter 9 in this volume).

The addition of specific genes, which are known to have biocontrol activity, may improve the biocontrol capability of organisms deficient in these particular genes and enhance it, by producing either greater amounts of the enzyme in question or a product with a similar function but complementary properties, in organisms already possessing a similar gene. Another property of an introduced gene for an extracellular enzyme is that it will be constitutively expressed, while native extracellular enzymes are usually inducible. Constitutive expression permits the production of an extracellular barrier that a pathogen would have to transverse, whereas induced enzymes are produced only after the biocontrol agents come into contact with a specific inducer, such as a target pathogen fungus. Thus, a constitutively produced enzyme might be more effective in biocontrol than a similar, induced enzyme.

Inducible enzymes are mobilized in response to stress or pathogen attack (Legrand et al., 1987). Production of specific polysaccharide lyases may depend on the structure of the cell wall of the particular target fungus. Correlation between the production level of a specific enzyme and the control ability of fungi containing particular cell wall constituents has been shown by Elad et al. (1982). These enzymes degrade cell walls of target fungi allowing the mycoparasite to eventually enter the luman of their hyphae (Chet et al., 1981; Chet, 1987).

A new direction for improving biocontrol activity is the introduction of genes directly into plants. A number of plant transformation vectors are available, all of which exploit the natural ability of the soil bacterium *Agrobacterium tumefaciens* to transfer a defined gene into the plant cell genome (Walden et al., 1990a and b). After the genetic manipulations of rhizosphere bacteria and/or fungi, direct genetic manipulations of plants were carried out to test the improved control activity. However, stability, safety, and the possible formulation of the new, genetically engineered system has to be considered (Sundheim et al., 1988).

There is conclusive evidence that the lytic enzyme chitinase plays an important role in the natural control of plant pathogenic fungi. Most of the known rhizobacteria that serve as biocontrol agents produce and excrete chitinase (Inbar and Chet, 1991; Ordentlich et al., 1988). Chitin, a polymer of β-(1–4)-N-acetyl glucosamine and one of the main cell wall constituents of many plant pathogenic fungi, degrades chitin into short oligomers. In higher plants endochitinases have substantial antifungal activity (Schlumbaum et al., 1986; Legrand et al., 1987; Roberts and Selitrennikoff, 1988). An endochitinase complimentary deoxyribonucleic acid (cDNA) clone from barley aleurone cells was isolated (Swegle et al., 1989). The increase in endochitinase mRNA levels in aleurone layers during incubation of barley seeds suggests that this enzyme plays a defensive role against fungi during seed maturation and germination in cereal seeds (Swegle et al., 1989). Samac et al. (1990) reported the isolation and characterization of genes encoding basic and acidic chitinase in *Arabidopsis thaliana*. A report by Laflamme and Roxby (1989) refers to the isolation and nucleotide sequencing of cDNA clones encoding potato chitinase genes. Chitinase has an interesting role in the malaria parasite. There it is known to digest the peritrophic membrane in the mosquito midgut during penetration. Huber et al. (1991) reported the digestibility of this membrane by *Serratia marcescens* chitinase. In addition, the presence of a malaria chitinase that digests 4-methylumbelliferyl chitotriose was reported. A chitinase gene from *Aeromonas hydrophila* was cloned from shrimp-shell enriched soil. The extracted chitinase has a molecular weight of approximately 85,000 and is repressed by glucose (Chen et al., 1991).

It is conceivable that *endo*-chitinases may be required to inhibit certain fungi infection, ascomycetes and basidiomycetous fungi, such as *Rhizoctonia solani* and *Sclerotium rolfsii* would be inhibited by these. Such a range of activity is consistent with the observation that strain T-35 of *T. harzianum* was mycoparasitic on *R. solani* but not on *Fusarium oxysporum,* apparently due to a protein layer coating the latter (Sivan and Chet, 1989a,b). A dichotomy of action of *endo*- versus *exo*-chitinases on different fungal groups is also suggested by Roberts and Selitrennikoff (1988), who found that *S. marcescens* chitinase lacked antifungal activity when tested against *Trichoderma reesii* or *Phycomyces blakesleeanus*. Shapira et al. (1989) found a similar chitinase enzyme from *S. marcescens* to be highly active against *R. solani* and *S. rolfsii*. Chitinases produced by *Trichoderma* have already been implicated in their ability to control plant pathogens (Chet, 1987).

2. PLANT DISEASE CONTROL BY CHITINASE

Serratia marcescens can function in the control of soil plant diseases. We have tested the effectiveness of a partially purified chitinase enzyme, coded by a gene present in *S. marcescens,* and expressed it in *Escherichia coli* as a biological control agent.

Bean seeds, sown in soil artificially infested with *S. rolfsii*, was grown under greenhouse conditions. The pots were irrigated daily with tap water containing a diluted chitinase preparation. Control boxes were irrigated with tap water or tap water containing a protein preparation obtained from equivalent amounts of culture medium of *E. coli* strain A5186 that does not carry the *chi*A gene. It was found that chitinase can effectively reduce the number of diseased plants (Shapira et al., 1989). During the course of these experiments, more than 50% of the plants irrigated with tap water showed severe disease symptoms. The addition of chitinase to the irrigation water significantly reduced the percentage of symptom-bearing plants. The protein preparation isolated from *E. coli* carrying the plasmid ppLchiA showed low levels of chitinase activity and did not reduce disease levels significantly as compared with the control.

The effectiveness of the chitinase preparation in reducing inoculum potential was tested by uprooting all plants used in the first experiments, that is, the first growth cycle, mixing the soil and redividing it into six boxes. Bean seeds were resown and irrigated daily with tap water. In the second growth cycle, disease rate in soil previously treated with chitinase was significantly lower than in the two control soils. Disease incidence after 14 days was 28% in the chitinase treatment, 52% in water control and 47% in soil previously treated with extracellular proteins prepared from *E. coli* that did not produce chitinase (Shapira et al., 1989).

To check the effect of inoculum level on the control of *S. rolfsii,* soil, inoculated with three levels of *S. rolfsii* (40, 80, and 120 mg of sclerotia per kg of soil), was irrigated with chitinase preparation or tap water as control. Inoculum levels of 40 and 80 mg per kilogram of soil significantly reduced disease incidence (42 and 36%, respectively). In another experiment, the incidence of damping-off of cotton, caused by *R. solani* was reduced by the chitinase preparation by 62% at 40°C after 20 days with beans, *S. rolfsii,* and whole viable *E. coli* cells. These cells were also effective in inhibiting *S. rolfsii,* although to a lower degree. Disease incidence after 16 days was 35% in water controls, 37% in soils treated with the control *E. coli* strain A5186, and 19% in soils treated with strain A5187 of *E. coli* carrying the ppLchiA plasmid. Disease incidence in soils treated with strain A5187 of *E. coli* was significantly less than that in the other treatments (p = 0.05). When the same experiment was carried out at 30°C, a temperature at which the pL promoter is repressed, no protection was observed (Shapira et al., 1989).

The genetically engineered *E. coli,* a known soil bacterium, served here as a model system to demonstrate the importance of chitinase in controlling a chitin-containing plant pathogen. Plant protection by the application of a lytic enzyme opens the way for the development of chitinase as an alternative to direct control of soil-borne pathogenic fungi. It also suggests that the introduction of engineered genes into soil bacteria increases control efficacy by combining high expression of a gene coding for a lytic enzyme with rhizosphere competence.

3. CHITINASE EXPRESSION IN *ESCHERICHIA COLI*

Ordered partition of proteins is a central property of biological systems. The translocation of proteins from the sites of synthesis to their final destination is regulated by specific signals. In *E. coli,* a gram-negative bacteria, newly synthesized proteins are directed to specific cellular structures, such as the inner or outer membrane (Cronan et al., 1987; Nikaido and Vaara, 1987; Oliver, 1987). Alternatively, they may end as free proteins in the cytoplasm or periplasmic space.

The laboratory strains of *E. coli*, K12, do not generally excrete proteins into the surrounding growth medium. In a few unique cases one finds proteins that are specifically released to the surrounding media (Luria and Suit, 1987; Mackman et al., 1986; Oropeza-Wekerle et al., 1989; Pugsley and Schwartz, 1984).

Escherichia coli is used extensively for the production at high levels of specific polypeptides (Reznikoff and Gold, 1986). It has recently been shown that in some cases, the expression of foreign genes can lead to their release to the growth medium (Abrahmsen et al., 1986; Ball et al., 1987; d'Enfert et al., 1987; Givskov et al., 1988; Kobayashi et al., 1986; Nagahari et al., 1985; Wandersman et al., 1987; Wong et al., 1989 and Yanagida et al., 1986). Development of secretion systems may yield new means of producing and purifying genetically engineered proteins in bacteria. Moreover, the biological control action of chitinase is dependent on its ability to be secreted to the surrounding environment. This section describes our experiments, demonstrating that a DNA fragment carrying the *chiA* gene from *S. marcescens* leads to high levels of chitinase expression and that a large portion of the enzyme is secreted as an active, processed enzyme to the growth medium.

In order to study the fate of the chitinase enzyme in *E. coli* we constructed a plasmid, ppLchiA, which carries the *chiA* gene under the control of the oLpL operator and promoter of bacteriophage λ (Fig. 12.1; Jones et al., 1986; Shapira

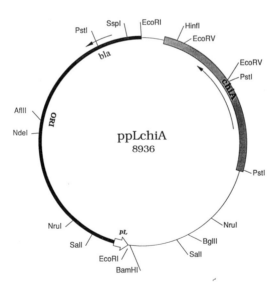

Fig. 12.1. Physical map of the plasmid expressing the *chiA* gene. For detailed steps in the construction of the plasmid carrying the *chiA* gene of *S. marcescens* (see Shapira et al., 1989).

et al., 1989). This plasmid was introduced into strain A2097, to yield strain A5745, which harbors a defective λ prophage carrying a thermosensitive cI repressor.

Chitinase expression was induced by heating to 42°C. In the initial experiments, we found that continuous exposure to this elevated temperature resulted in massive cell lysis. The induction protocol was, therefore, modified to 30 min of induction at 42°C followed by continuous incubation at 40°C. Under these conditions no decrease in optical density was observed during the induction period.

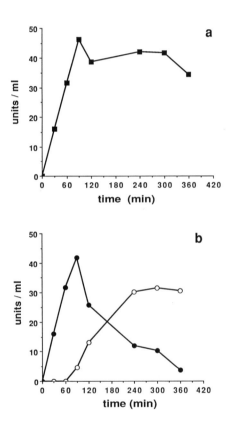

Fig. 12.2. Kinetics of chitinase synthesis following heat induction. Induction was carried out as described in the text. Samples (1 ml) were removed at various times after induction and spun for 5 min at 4000 rpm in a refrigerated centrifuge. The supernatant fraction was removed and respun at 12,000 rpm for 10 min. The cellular pellet was resuspended in 1 ml of water and subjected to cell disruption by sonication. The enzymatic activity was determined by following the cleavage of *p*-nitrophenyldiacetyl-β-D-*N,N'*-chitobioside (Sigma), 0.2 mg/ml, to yield *p*-nitrophenyl (OD410 nm). The reaction (0.5 ml) was carried in 50 m*M* of H$_2$NaO$_4$P buffer, pH 6.3 at 40°C and stopped by the addition of 0.2 ml of 1 *M* NaOH. **(a)** Total enzyme activity. **(b)** Chitinase activity associated with the cellular fraction (●). Chitinase activity found in the supernatant fraction (o).

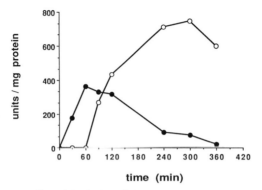

Fig. 12.3. Chitinase specific activity in the cellular and extra-cellular fractions. Chitinase activity associated with the cellular fraction (●). Chitinase activity found in the supernatant fraction (o).

Induction of chitinase expression was found to prevent further cell growth. Chitinase synthesis was initiated at a high rate immediately following heat induction. The increase in the total amount of enzyme was limited to the first 90 min following the heat shock. After 1 h of induction the enzyme was released to the growth medium and the level of chitinase in the medium increased continuously (Figs. 12.2 and 12.3). In time, most of the enzyme was found in the growth medium. Extracellular protein content was measured using sodium decyl sulfate (SDS)-polyacrylamide gel electrophoresis. The cellular fractions were precipitated with trichloroacetic acid (10%) and the proteins were separated by electrophoresis on 10% polyacrylamide gel and stained with Coomassie Blue. We found that chitinase constituted over 10% of the total cellular protein and was the major protein secreted to the growth media.

It is, as yet, unknown how the enzyme is secreted from the cells to the medium. It is possible that chitinase is secreted from the cytoplasm by existing specific pores or ones formed during induction. Alternatively, the leader sequence, present at the amino-terminal end of the chitinase protein, directs the newly formed chitinase to cross the periplasmic membrane, with the aid of the enzymatic system that transports proteins from the cytoplasm to the periplasmic space. The leader peptide is then removed. By determining the amino-terminal end of the chitinase enzyme, we have shown that the secreted enzyme was processed and the amino-terminus leader peptide removed. Preliminary results indicate that chitinase is accumulated in the periplasmic space very early after induction. These findings lead us to favor models in which the outer membrane is modified during the induction process, allowing the diffusion of proteins from the periplasm to the surrounding medium. Such modification may be achieved by the action of the accumulated chitinase on the integrity of the outer membrane. Further experiments are needed to understand the process of chitinase secretion and its importance in biological control.

Southern blot analysis of the *chi*A gene, cloned from *S. marcescens* showed great homology to one of the *Trichoderma* chitinase genes. Based on this finding, we used the *chi*A gene as a probe to isolate a chitinase gene from a cDNA library prepared from *T. harzianum* (T-35) (Barak, Chet, and Oppenheim, unpublished). To prepare the cDNA library, *Trichoderma* (T-35) was grown with 3 g/l of chitin for 7 days and total RNA was isolated. Poly(A) RNA was selected on an oligo dT column and a cDNA library was prepared. The chitinase gene from *T. harzianum* (T-35) has recently been cloned in Bluescript plasmid under the lac promoter. When the transformed *E. coli* was plated on LB + 0.2% chitin plates and induced by 1 mM IPTG, the bacteria showed chitinolytic activity (Fig. 12.4).

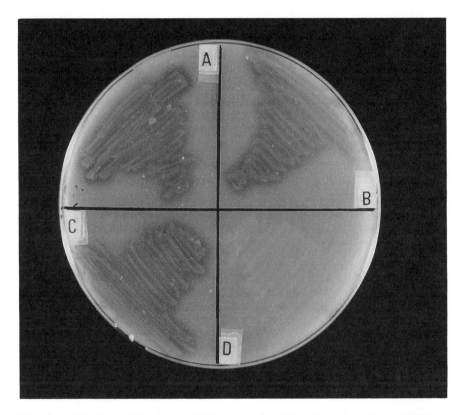

Fig. 12.4. Chitinolytic activity of a cloned chitinase gene from *Trichoderma*. *Escherichia coli* (XLI-Blue) transformed with Bluescript SK plasmid containing the chitinase gene cloned from *T. harzianum*, plated on LB + 0.2% chitin induced by IPTG. Samples A, B, and C show chitinolytic activity, D is a negative control XLI-Blue bacteria.

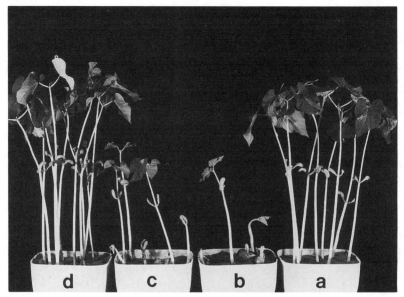

Fig. 12.5. Biocontrol activity of the cloned chitinase gene from *Trichoderma* in a greenhouse experiment. Bean plants, infected by *S. rolfsii,* were irrigated with 10^7 of transformed bacteria in milliliters
per gram of soil for 18 days. From *right* to *left* (**a**) uninfected plants; (**b**) plants infected by *S. rolfsii*
(control plants); (**c**) infected plants treated with XL1-Blue *E. coli*; (**d**) (on the *left* side) infected plants
treated with 10^7 *E. coli* with the cloned *Trichoderma* gene in milliliters per gram of soil.

In greenhouse experiments, bean seedlings were irrigated with *E. coli* XL1-Blue
($10^7/1$ per gr soil per day) transformed with the *Trichoderma* chitinase gene, induced
by 1 mM IPTG. The results presented significant biocontrol activity (Fig. 12.5).
The treatment obviously suppressed the disease caused by *S. rolfsii*. The treated
plants present better growth rate compared to untreated controls, the growth rate
of the plants irrigated with the transformed bacteria after 18 days was even better
than that of the uninfected plants.

4. IMPROVEMENT OF RHIZOSPHERE BACTERIA AS BIOCONTROL
AGENTS

Free-living bacteria, which are associated with plants, have been the focus of
research efforts to support their usage either as soil inoculates to enhance plant
growth (Kloepper et al., 1989) or in biological control of plant pathogens (Weller, 1988). Studies on bacteria that enhance plant growth have concentrated on
the rhizosphere, but studies on bacteria as biological control agents have included

both the rhizosphere and the phylloplane. The rhizosphere is characterized as a region of high microbial diversity and competition for available nutrients. In comparison, the phylloplane is a harsher environment and has less microbiol competition, due to fluctuations in temperature, humidity, and high ultraviolet (UV) radiation. Endophytic bacteria are bacteria isolated from the interior of healthy plants, including ovules (Mundt and Hinkle, 1976; Hollis, 1951). Little is known about the populations and identity of endophytic bacteria compared with bacteria from the rhizosphere or phylloplane.

Several important rhizobacteria, such as *Rhizobium meliloti* and *Pseudomonas putida,* which are excellent root colonizers, lack the ability to synthesize chitinase. Hence, it is assumed that introducing the chitinase gene into these bacteria will enable them to provide protection against plant pathogenic fungi. Sundheim et al. (1988) introduced the chitinase gene, cloned from *S. marcescens,* into *Pseudomonas fluorescens* strains. In *in vitro* experiments, the genetically engineered *fluorescens* suppressed disease caused by *F. oxysporum radolens* but was less effective in controlling *Gaeumannomyces graminis* var. *tritici,* the causal agent of take-all of wheat. Experiments testing the stability of the plasmid showed that it was very unstable in the *Pseudomonas.*

Recently, Sitrit, Barak, Oppenheim, and Chet (unpublished) introduced the *chi*A gene into two rhizobacteria: *R. meliloti* and *P. putida.* Both were chosen for their unique beneficial characteristics. The symbiotic bacterium *Rhizobium* stimulates root nodules inside leguminous plants and fixes atmospheric nitrogen. Because *Rhizobium* colonizes in root nodules, we might expect chitinase synthesis to provide protection against soil fungi. The contribution of *Pseudomonas* strains to the rhizosphere is of great importance. Many *Pseudomonas* strains excrete siderophores that chelate iron ions, thereby improving iron uptake by plant roots. Moreover, iron ion depletion from the soil decreases the growth potential of plant pathogenic fungi (Lopper and Buyer, 1991).

The introduction of the *chi*A gene into these bacteria, in addition to their already existing beneficial properties, makes them superb candidates for biocontrol bacteria. The *chi*A gene was introduced into the bacteria through a broad host range plasmid, carrying the *chi*A gene (pYZ291) (Sitrit, Barak, Chet, and Oppenheim, unpublished) cloned from *S. marcescens* under the P*tac* promoter. The broad host plasmid pYZ291 was transferred from *E. coli* to *P. putida* or *R. meliloti* by conjugation. The use of the P*tac* promoter in *Rhizobium* and *Pseudomonas* allowed constitutive expression of the *chi*A gene in the new hosts, since both bacteria are Lac and so do not encode for a repressor gene. To construct the pYZ291 plasmid the *chi*A gene was first subcloned into pKK223-3, which utilizes the synthetic tac promoter. The resulting pYZ223-3 plasmid has the *chi*A gene cloned under the P*tac*. The next step was to prepare the pYZ291 plasmid, which is the broad host plasmid pRK290 with the *chi*A gene cloned under the P*tac*.

To test the expression of the *chi*A gene in the constructed pYZ291 plasmid, the transformed *E. coli* strain 5039 was grown for 4 h, during which chitinase expression was induced by 1 m*M* IPTG. Samples of total protein were prepared after 1, 2.5, and 4 h of growth and run on 10% SDS-acrylamide gel stained with Coomassie Blue. Results (Fig. 12.6) show the formation of a new protein band of 58 kD, correlating with the cloned chitinase gene from *Serratia* (Sitrit et al., unpublished).

After conjugation between the transformed *E. coli* and *Pseudomonas* or *Rhizobium,* chitinase activity was tested in the transformed *P. putida* or *R. melioti* by detecting chitin degradation on LB plates containing 0.2% chitin and then by 10% SDS-acrylamide gel that showed a new band, 58 kD of chitinase. Moreover, relative chitinase activity was tested by PNP-chitobiase substrate degradation. All experiments showed conclusive high chitinase activity in the transformed bacteria.

The stability of the transformed gene in the rhizobacteria was determined by growing the engineered bacteria without selective pressure. After 20 generations, the transformed *pseudomonads* rapidly lost the plasmid (only 0.5% of the bacte-

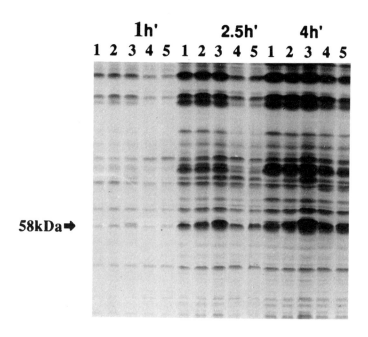

Fig. 12.6. Expression of a cloned chitinase gene in *E. coli* controlled by ptac. Cells were grown for 4 h. Samples of total protein extracted at 0, 1, 2.5, and 4 h, were tested on 10% acrylamide gel. Lane 1: *E. coli* transformed by pKK223-3. Lane 2: *E. coli* transformed by pYZ223-3. Lane 3: *E. coli* transformed by pYZ223-3 induced by IPTG. Lane 4: *E. coli* transformed by pYZ291. Lane 5: *E. coli* transformed by pYZ291 induced by IPTG.

ria grew on selective media). In contrast, the plasmid that was introduced into *Rhizobium* was very stable. More than 95% of the cells grew on LB plates + 50 μg/ml of Km after growing in medium without selective pressure for 100 generations (Sitrit, Barak, Oppenheim, and Chet, unpublished).

Transformed *Pseudomonas* bacteria were tested in greenhouse experiments. Biocontrol activity was tested in bean plants infected by *S. rolfsii* and irrigated with 10^8 bacteria per gram of soil for 18 days. Disease reduction was observed as compared to control plants (Fig. 12.7 **Top**). Similarly, the biocontrol activity was tested in bean plants infected by *R. solani* and irrigated with 10^7 bacteria/1 g of soil for 9 days. Disease reduction was observed as compared to control plants (Fig. 12.7 **Bottom**).

5. IMPROVEMENT OF TRICHODERMA AS A BIOCONTROL AGENT

Trichoderma spp. fungi present in almost every soil, are antagonistic to other fungi. *Trichoderma harzianum* (strain T-35) is a strong rhizosphere colonizer able to parasitize some plant pathogenic fungi. *Trichoderma* strains first grow tropically towards target hyphae, upon contact they coil around the hypha or grow parallel, in tight contact with, the target hypha (Chet et al. 1981). This reaction is mediated by attachment of the *Trichoderma* hypha through lectin binding (Chet, 1987). When physical contact is made, a group of extracellular enzymes are produced, including chitinases, β-1,3-glucanases (Chet, 1987), lipases (Elad et al., 1982) and, possibly, proteinases (Ridout et al., 1988).

To improve the biocontrol activity of *T. harzianum,* the lytic enzyme *chi*A, from *Serratia,* was introduced into it through protoplast transformation, using plasmid DNA. The plasmid used for transformation carried the *chi*A gene, controlled by a promoter, which allowed its constitutive expression, as well as a selectable marker. The selectable marker used was the *amd*S gene, which enables transformed fungi to grow on acetamide or acrylamide as a sole nitrogen source (Tilburn et al., 1983). Protoplasts prepared from filamentous fungi can be transformed by foreign DNA (Fincham, 1989). Protoplasts for transformation were obtained from *T. harzianum* according to the methods of Stasz et al. (1988) followed by Pe'er and Chet (1990). Transformation experiments of *T. reesei* with the p3SR2 plasmid (Hynes et al., 1983) containing the *amd*S gene are reported by Penttila et al. (1987). Similar experiments were repeated in our laboratory (Barak, Pe'er, and Chet, unpublished). In all the above-mentioned experiments the plasmid DNA (p3SR2) was found to integrate into the fungal chromosomes by recombination. This integration was not site specific and was probably accompanied by gene replication and recombination events (Pe'er et al., 1991).

Two strategies were used to transform the *Trichoderma* by the *chi*A gene. The first was transformation by a single type of plasmid, which contained both the

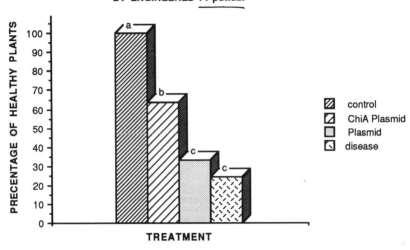

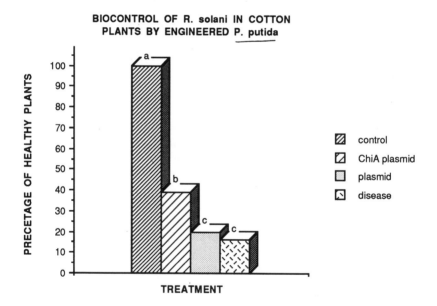

Fig. 12.7. Top: Biocontrol activity of transformed *Pseudomonas* versus *S. rolfsii*. Bean seedlings were infected by *S. rolfsii* and grown for 18 days. The plants were irrigated during 18 days with 10^8 bacteria per 1 gram of soil. **Bottom:** Biocontrol activity of transformed *Pseudomonas* versus *R. solani*. Bean seedlings were infected by *R. solani* and grown for 9 days. The plants were irrigated during 9 days with 10^7 bacteria per 1 gram of soil.

selective marker gene *amd*S and the *chi*A gene from *Serratia.* The second was cotransformation, which is transformation by two types of plasmids: (a) contained the selective marker *amd*S gene (acetamidase) in the p3SR2 plasmid and (b) was the vector containing the *chi*A gene under the appropriate promoter. According to Punt et al. (1987) more than 50% of the transformants that were transformed by two different types of plasmids (cotransformation), contained the selective marker acquired by one type of plasmid and the gene carried by the second vector.

Transformation by a single type of plasmid was carried out by transforming protoplasts with the pp-579 plasmid (Fig. 12.8), which contained both the *amd*S gene and the *chi*A gene, controlled by the chitinase regulatory region of *Serratia.* The plasmid was transformed into *T. harzianum* (T-35) and again, after transformation the DNA integration was not site specific. Deoxyribonucleic acid analysis by hybridization experiments revealed that both the *amd*S and *chi*A genes integrated into the genome. When hybridized to the *amd*S gene, the various transformants (Fig. 12.9**a**) show a different number and intensity of bands. This result indicates that the replication event was different in each transformant. Southern blot analysis of genomic DNA, extracted from the different transformants and hybridized with the *chi*A gene revealed that the *chi*A gene, integrated into the *Trichoderma* genome. As shown in Fig. 12.9**b**, two bands are observed; one of 8.8 kb, representing chitinase from the *Trichoderma,* and the other, of 4.6 kb, representing the new, integrated *chi*A gene (this band does not appear in the control wild type). Although *chi*A gene integration into the chromosome was clear, no expression of the *chi*A gene was detected when cloned under the regulatory region of *chi*A from *Serratia.*

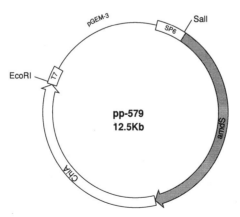

Fig. 12.8. The plasmid pp-579 used for *Trichoderma* transformation. The plasmid pp-579 contains the 5.1-kb fragment encoding the *adm*S gene from *A. nidulans* and the 4.6-kb fragment encoding for the *chi*A gene controlled by the regulatory domains of *Serratia.*

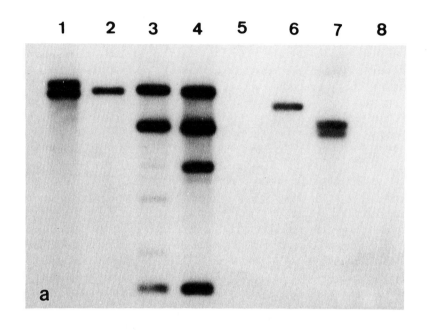

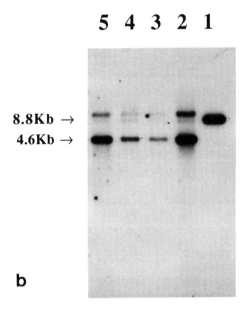

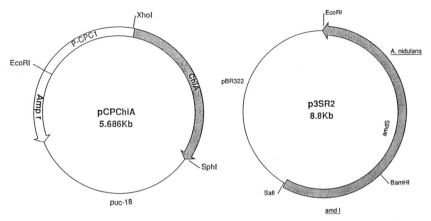

Fig. 12.10. Plasmids used for cotransformation of *Trichoderma*: p3SR2 is the plasmid encoding for acetamidase cloned from *A. nidulans,* and pCPChiA a plasmid encoding for *chi*A regulated by CPC-1 regulatory region (from *Neurospora*).

In cotransformation experiments, two different plasmid types were used (Fig. 12.10) to transform the fungi: The p3SR2 plasmid, together with a plasmid containing the *chi*A gene under the *cpc*-1 regulatory region, from *Neurospora* regulatory region induced by starvation to amino acids (Paulh et al., 1988). More than 90% of the transformants selected on acetamide medium contained the *chi*A gene, when examined by Southern blot (Fig. 12.11). Apparently, the integration of the *chi*A gene into *Trichoderma* DNA occurred in the chitinase gene of the *Trichoderma,* probably by homologous recombination. We can see that the intensity of the band, size 8.8 kb, which represents *Trichoderma* chitinase, is reduced and a new, smaller band appears. The size of this new band is not the same for all transformants, thus indicating integration into several sites.

Specific activity of *Serratia* chitinase was tested by growing the transformants on synthetic medium (Okon et al., 1973). After 6 days the transformants presented a higher expression of chitinase than the wild types (Fig. 12.12). Since the Southern blot analysis showed that the *chi*A gene integrated into the *Trichoderma*

Fig. 12.9a. Southern blot analysis of genomic DNA hybridized with labeled *amd*S gene. Genomic DNA from cotransformed *Trichoderma* cut with EcoRI, and hybridized with labeled *amd*S gene. Lanes: 1, 2, 3, and 4 are different transformants, transformed by p3SR2 plasmid. Lanes 6 and 7 are transformants transformed with the plasmid pp-579. Lanes 5 and 8 are control WT.

Fig. 12.9b. Southern blot analysis of genomic DNA hybridized with labeled chiA gene. Genomic DNA from cotransformed *Trichoderma* cut with EcoRI, and hybridized with nick translated chiA gene. Lane 1: the wt shows only one band of 8.8 kb, which presents *Trichoderma's* chitinase. The different transformants transformed with plasmid pCPChiA are in lanes 2–5; in addition to the 8.8 kb band, a new band of 4.6 kb appears which represents the chiA gene.

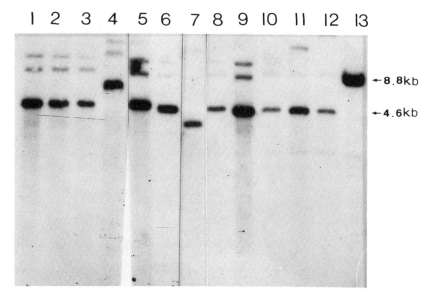

Fig. 12.11. Southern blot analysis of genomic DNA hybridized with labeled *chi*A gene. Genomic DNA from cotransformed *Trichoderma* cut with EcoRI, and hybridized with nick translated *chi*A gene. Lane 13 WT shows only one band of 8.8 kb, which presents chitinase of *Trichoderma*. The different transformants are: 224, 223, 222, 130, 117, 114, 18, 225, 7, 63, 207, 214, and 77 in lanes 1–12, respectively, a new band of 4.6 kb appears which represents the *chi*A gene.

chitinase gene, specific activity of chitinase, induced by 3 g/l of chitin, was tested. The chitinase activity of the transformed *Trichoderma* was not disrupted and was similar to that of the wild types (Pe'er, Haran, Barak, Chet, and Oppenheim, unpublished).

All transformed fungi were tested for their capability to survive in the soil rhizosphere and were found to be stable for at least 4 months. These tests can only be performed with marked isolates. In previous studies of rhizosphere colonization, *Trichoderma* strains resistant to Benomyl (Ahmad and Baker, 1987) or cycloheximide (Sivan and Chet, 1986) were examined. The *amd*S transformants were used as an efficient method for detection of *Trichoderma* in soil. We found *T. harzianum* transformants to be stable through *in vitro* transfer of hyphae or conidia and *in vivo* tests in the greenhouse. *Trichoderma harzianum amd*S transformants developed conidia on acrylamide medium, while *T. reessei* did not (Penttila et al., 1987). About 50% of all *Trichoderma* colonies, selected from rhizosphere and plant roots treated with transformants, were stable on acrylamide (Pe'er et al., 1992). Recovery of marked microorganisms from the soil was shown by Hofte et al. (1990) with the rhizopseudomonas strain 7NSK$_2$ containing a Mu d(*lac*) element. Such a genetically marked fungus can be a very helpful tool in studying

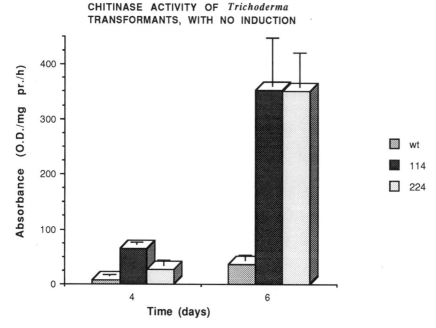

Fig. 12.12. Specific activity of chitinase in *Trichoderma* grown with no induction. Transformants 114 and 224 were grown on minimal medium (synthetic). Chitinase activity was tested after 4 and 6 days. The chitinase specific activity of wt and transformants 114, 224 after 4 days is low. After 6 days the activity of the transformants was raised up to a level of 350 units.

population dynamics of soil-borne fungi, their behavior, survival under various conditions, and rhizosphere competence.

6. CONCLUSION

Selection and isolation of a gene with a high biocontrol activity, such as the chitinase gene, allowed us to test its biocontrol activity at different levels of regulation and expression (Fig. 12.13). Three different chitinase genes from *Serratia, Aeromonas,* and *Trichoderma* were tested by our group and their activity was tested and explored in different systems. The pure enzyme produced was tested directly for its biocontrol activity, as well as its properties as a lytic enzyme. The cloned gene was then expressed in the non-rhizosphere bacteria *E. coli* for biocontrol experiments. The ultimate purpose of this research was to introduce the gene into rhizosphere fungi or bacteria to improve their biocontrol activity. *Rhizobium* and

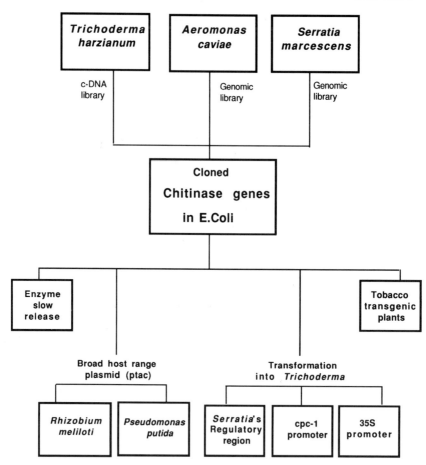

Fig. 12.13. A summary scheme of the different usages of the chitinase gene.

Pseudomonas have long been considered important rhizosphere organisms with beneficial effects on growth, by nitrogen fixation, or chelating iron ions. *Trichoderma* strains provide potential strong rhizosphere competence and the ability to effectively and reliably control a range of plant pathogenic fungi in both the laboratory and greenhouse (Sivan and Chet, 1986, 1989b; Harman et al., 1989). We need to know much more about the mechanisms by which *Trichoderma* strains accomplish biocontrol and what enables them to compete in natural ecosystems. The *Trichoderma* strains we have used are attractive as test organisms for determining biocontrol mechanisms because they are capable of effectively controlling a range of pathogens as well as being rhizosphere competent (Baker, 1989;

Sivan and Chet, 1986, 1989b; Harman et al. 1989). These are the best strains chosen after years of biocontrol research.

Molecular biology provides a tool for producing strains of *fungi and bacteria* that are even more effective than the ones already in use and for elucidating mechanisms of biological control.

By transformation, a new gene, replicated as a stable episomal DNA, has been introduced into the bacteria as a plasmid. The gene is constitutively expressed under the tac promoter, since the host is lac⁻ and no repressor is made. Constitutive expression is at such a level that it does not affect bacterial growth but shows significant biocontrol activity against *S. rolfsii* and *R. solani*. Foreign DNA introduced into *Trichoderma* was found to be integrated into the genome, stable, and expressed when the right promoter was used. These phenomena probably cause changes in the genome, where the DNA was integrated, hence different mutations and phenotypes, which could present a good panel for testing the role of different *Trichoderma* genes may appear. In addition, integration of a new gene downstream a strong promoter of the *Trichoderma* itself, can lead to the strong activation of that gene.

The genetic manipulation of introducing the *chi*A gene into *Trichoderma* or the rhizosphere bacteria did not affect their growth rate. Stability of the transformed *Trichoderma* was high when detected in nonsterile soil at least 6 weeks after being seeded (Pe'er et al., 1992). Moreover, the rhizosphere bacteria *Pseudomonas*, transformed with the broad range host plasmid pRK290 carrying the *chi*A gene controlled by the p*tac*, was unstable, yet *Rhizobium*, with the same plasmid, was very stable. Experiments (Sitrit, Kapulnik, Barak, and Chet unpublished) show that neither the property of nitrogen fixation by these bacteria nor their growth rate were affected. These transformants still have to be tested for their efficacy as biocontrol agents against plant pathogenic fungi. We tested the naturally occurring chitinase activity of *Trichoderma*, induced by chitin, to find whether changes in the chitinase gene from *Trichoderma* was affected. Results show that chitinase activity was not reduced when compared to the wild-type control.

Genetic manipulation has been shown to improve the biocontrol activity of these microorganisms. Some problems remain unsolved, such as choosing the right promoter to produce the right amount of chitinase with no adverse effect on the *Trichoderma* or the bacteria. Recently, upon transforming *Trichoderma* with the *chi*A gene controlled by 35S cauliflower mosaic virus (CaMV), we did, indeed, detect a high constitutive expression of *chi*A gene in the transformants.

We have to consider the safety of these experiments and possible formulations (Baker and Henis, 1990). The use of the *LacZ* genes in biocontrol activity experiments against wheat take-all disease, was recently cleared under regulations of the Federal Insecticide, Fungicide and Rodenticide Act (FIFRA) for field tests (Hemming, 1990). This finding raises the hope that the ptac promoter will also be suitable for field experiments. Such experiments of genetic transformation must be tested for their effect on the ecosystem. We cannot exactly define the nature of the changes that occur in the genomic DNA after transformation takes place and, unless these transformants are tested in the soil, we cannot predict their influence on the biological system. Yet, we may assume that since the introduced gene has a positive influence as a gene with biocontrol activity, and is target oriented, the overall influence of transformants, transformed by the chitinase gene, will be positive with no significant detrimental influence on the environment.

The major advantage of such genetic manipulations is the ability to isolate genes from one strain and introduce them into other varieties of fungi or bacteria, thus enhancing the potency of biocontrol agents and making a single strain effective, stable, and consistent against more than one plant pathogenic fungus, without the hazardous effects of chemical pesticides.

ACKNOWLEDGMENT

This work was supported in part by the United States-Israel Agricultural Research and Development Fund (BARD).

REFERENCES

Abrahmsen L, Moks T, Nilsson B, Uhlen M (1986): Secretion of heterologous gene products to the culture medium of *Escherichia coli*. Nucleic Acids Res 14:7487–7500.

Ahmad J, Baker R (1987): Rhizosphere competence of *Trichoderma harzianum*. Phytopathology 77:182–189.

Ahmad J, Baker R (1988): Growth of rhizosphere-competent mutants of *Trichoderma harzianum* on carbon substrates. Can J Microbiol 34:807–814.

Baker CA, Henis JMS (1990): Commercial production and fermentation of microbial biocontrol agent. In Baker RR, Dunn PE (eds): New Directions in Biological Control, New York: Alan R. Liss, pp 333–344.

Baker R (1989): Some perspectives on the application of molecular approaches to biocontrol problems. In Whipps JM, Lumsden RD (eds): Biotechnology of Fungi for Improving Plant Growth, Cambridge: Cambridge University Press, pp 220–233.

Ball TK, Saurugger PN, MJ Benedik (1987): The extracellular nuclease gene of *Serratia marcescens* and its secretion from *Escherichia coli*. Gene 57:183–192.

Burr TJ, Schroth MN, Suslow T (1978): Increased potato yields by treatment of seed pieces with specific strains of *Pseudomonas fluorescens* and *P. putida*. Phytopathology 68:1377–1388.

Chen JP, Nagayama P, Chang CM (1991): Cloning and expression of chitinase gene from *Aeromonas hydrophila* in *Escherichia coli*. App Environ Microbiol 57:2426–2428.

Chet I (1987): *Trichoderma*—application, mode of action, and potential as a biocontrol agent of soilborne plant pathogens. In Chet I (ed): Innovative Approaches to Plant Disease Control, New York: Wiley, pp 137–160.

Chet I, Harman GE, Baker R (1981): *Trichoderma hamatum*. Its hyphal interactions with *Rhizoctonia solani* and *Pythium* spp. Micob Zool 7:29–38.

Cronan Jr JE, Gennis RB, Maloy SR (1987): Cytoplasmic Membrane, pp 31–55. In Neidhardt FC, Ingraham JL, Low KB, Magasanik B, Schaechter M, Umbarger HE (eds): *Escherichia coli* and *Salmonella typhimurium* Cellular and Molecular Biology, American Society for Microbiology, Washington, DC.

d'Enfert C, Ryter A, Pugsley AP (1987): Cloning and expression in *Escherichia coli* of the *Klebsiella pneumoniae* genes for production, surface localization and secretion of the lipoprotein pullulanase. EMBO J 6:3531–3538.

Elad Y, Hadar Y, Chet I, Henis Y (1982): Prevention with *Trichoderma harzianum* Rifai aggr. of reinfestation by *Sclerotium rolfsii* Sacc. and *Rhizoctonia solani* Kuhn of soil fumigated with methyl bromide and improvement of disease control in tomatoes and peanuts. Crop Protection 1:199–211.

Fincham JRS (1989): Transformation in fungi. Microbiol Rev 53:148–170.

Givskov M, Olsen L, Molin S (1988): Cloning and Expression in *Escherichia coli* of the Gene for Extracellular Phospholipase A1 from *Serratia liquefaciens*. J Bacteriol 170:5855–5862.

Harman GE, Taylor AG, Stasz TE (1989): Combining effective strains of *Trichoderma harzianum* and solid matrix priming to provide improved biological seed treatment systems. Plant Disease 73:631–637.

Hemming BC (1990): Bacteria as antagonists in biological control of plant pathogens. In Alternatives for Suppressing Agricultural Pests and Diseases, New York: Alan R. Liss, pp 223–242.

Hofte M, Mergeay M, Verstraete W (1990): Marking the rhizopseudomonas strain 7NSK$_2$ with a Mu d(lac) element for ecological studies. Appl Environ Microbiol 56:1046–1052.

Hollis JP (1951): Bacteria in healthy potato tissue. Phytopathology 41:350–366.

Huber M, Cabib E, Miller HL (1991): Malaria parasite chitinase and penetration of the mosquito peritrophic membrane. Proc Natl Acad Sci USA 88:2807–2810.

Hynes MJ, Corrick CM, King JA (1983): Isolation of genomic clones containing the *amd*S gene of *Aspergillus nidulans* and their use in the analysis of structural and regulatory mutations. Mol Cell Biol 3:1430–1439.

Inbar J, Chet I (1991): Evidence that chitinase produced by *Aeromonas caviae* is involved in the biological control of soil-borne plant pathogens by this bacterium. Soil Biol Biochem 23:973–978.

Jones JDG, Grady KL, Suslow TV, Bedbrook JR (1986): Isolation and characterization of genes encoding two chitinase enzymes from *Serratia marcescens*. EMBO J 5:467–473.

Kloepper JW, Lifshitz R, Zablotowicz RM (1989): Free-living bacterial inocula for enhancing crop productivity. Trends Biotech 7:39–43.

Kobayashi T, Kato C, Kudo T, Horikoshi K (1986): Excretion of the Penicillinase of an Alkalophilic *Bacillus* sp. through the *Escherichia coli* Outer Membrane Is Caused by Insertional Activation of the *kil* Gene in Plasmid pMB9. J Bacteriol 166:728–732.

Laflamme D, Roxby R (1989): Isolation and nucleotide sequence of cDNA clones encoding potato chitinase. Plant Mol Biol 13:249–250.

Legrand M, Kauffmann S, Geoffroy P, Fritig B (1987): Biological function of pathogenesis-related proteins: four tobacco pathogenesis-related proteins are chitinases. Proc Natl Acad Sci USA 84:6750–6754.

Lopper JE, Buyer JS (1991): Siderophores in microbial interactions on plant surfaces. Mol Plant Microbe Interact 4:5–13.

Luria SE, Suit JL (1987): Colicins and Col Plasmids. In Neidhardt FC, Ingraham JL, Low KB, Magasanik B, Schaechter M, Umbarger HE (eds): *Escherichia coli* and *Salmonella typhimurium* Cellular and Molecular Biology, American Society for Microbiology, Washington, DC, pp 1615–1624.

Lynch JM (1990): Fungi as Antagonists. In Baker R, Dunn PE (eds): New Directions in Biological Control: Alternatives for Suppressing Agricultural Pests and Diseases, New York: Alan R. Liss, pp 243–253.

Mackman N, Nicaud J-M, Gray L, Holland IB (1986): Secretion of Haemolysin by *Escherichia coli*.

Curr. Topics in Microbiol. Immunol. 125:159–181. Magasanik B, Schaechter M, Umbarger HE (eds): *Escherichia coli* and *Salmonella typhimurium* Cellular and Molecular Biology, Washington DC: American Society for Microbiology.

Mundt JO, Hinkle NF (1976): Bacteria within ovules and seeds. Appl Environ Microbiol 32:694–698.

Nagahari K, Kanaya S, Munakata K, Aoyagi Y, Mizushima S (1985): Secretion into the culture medium of a foreign gene product from *Escherichia coli*: use of the *ompF* gene for secretion of human b-endorphin. EMBO J 4:3589–3592.

Nelson EB, Chao WL, Norton JM, Nash GT, Harman GE (1986): Attachment of *Enterobacter cloacae* to hyphae of *Pythium ultimum*: Possible role in the biological control of *Pythium* pre-emergence damping-off. Phytopathology 76:327–335.

Nevalainen KM (1991): The molecular biology of *Trichoderma* and its application to the expression of both homologous and heterologous genes. In Leong SA (ed): Molecular Industrial Mycology Systems and Applications for Filamentous Fungi, New York: Marcel Dekker, pp 129–147.

Nikaido H, Vaara M (1987): Outer Membrane. In Neidhardt FC, Ingraham JL, Low KB, Magasanik B, Schaechter M, Umbarger HE (eds): *Escherichia coli* and *Salmonella typhimurium* Cellular and Molecular Biology, Washington, DC: American Society for Microbiology, pp 7–22.

Okon Y, Chet I, Henis Y (1973): Effect of lactose, ethanol and cycloheximide on translation pattern of radioactive compounds and on sclerotium formation on *Sclerotium rolfsii*. J Gen Microbiol 74:151–158.

Oliver DB (1987): Periplasm and Protein Secretion, In Neidhardt FC, Ingraham JL, Low KB, Magasanik B, Schaechter M, Umbarger HE (eds): *Escherichia coli* and *Salmonella typhimurium* Cellular and Molecular Biology, Washington, DC: American Society for Microbiology, pp 56–69.

Ordentlich A, Elad Y, Chet I (1988): The role of chitinase of *Serratia marcescens* in biocontrol of *Sclerotium rolfsii*. Phytopathology 78:84–88.

Oropeza-Wekerle RL, Muller E, Kern P, Meyermann R, Goebel W (1989): Synthesis, Inactivation, and Localization of Extracellular and Intracellular *Escherichia coli* Hemolysins. J Bacteriol 171:2783–2788.

Paulh JL, Orbach MJ, Legerton TL, Yanofsky C (1988): The cross-pathway control gene of *Nerospora crossa*, cpc-1, encodes a protein similar to GCN4 of yeast and the DNA-binding domain of the oncogene V-*jun*-encoded protein. Proc Natl Acad Sci USA 85:3728–3732.

Paulitz TC, Ahmad JS, Baker R (1990): Integration of *Pythium nunn* and *Trichoderma harzianum* isolate T-95 for the biological control of *Pythium* damping-off of cucumber. Plant Soil 121:243–250.

Pe'er S, Barak Z, Yarden O, Chet I (1991): Stability of *Trichoderma harzianum amdS* transformants in soil and rhizosphere. Soil Biol Biochem (in press).

Pe'er S, Chet I (1990): *Trichoderma* protoplast fusion: a toll for improving biocontrol agents. Can J Microbiol 36:6–9.

Penttila M, Nevalainen H, Ratto M, Salminen E, Knowles J (1987): A versatile transformation system for the cellulolytic filamentous fungus *Trichoderma ressei*. Gen 61:155–164.

Pugsley AP, Schwartz M (1984): Colicin E2 release: lysis, leakage of secretion? Possible role of a phospholipase. EMBO J 3:2393–2397.

Punt PJ, Oliver RP, Dingemanse MA, Pouwels PH, van den Hondel AMJJ (1987): Transformation of *Aspergillus* based on the hygromycin B resistance marker from *Escherichia coli*. Gene 56:117–124.

Reznikoff W, Gold L (eds): 1986. Maximizing Gene Expression. Boston: Butterworths. Ridout MA, Coley-Smith JR, Lynch JM (1988): Fractionating of extracellular enzymes from a mycoparasitic strain of *Trichoderma harzianum*. Enzy Microbiol Technol 10:180–187.

Ridout CJ, Coley-Smith JR, Lynch JM (1988): Fractionating of extracellular enzymes from a mycoparasitic strain of *Trichoderma harzianum*. Enzymol Microbiol Technol 10:180–187.

Roberts WK, Selitrennikoff CP (1988): Plant and bacterial chitinases differ in antifungal activity. J Gen Microbiol 134:169–176.

Roby D, Broglie K, Cressman R, Biddle P, Chet I, Broglie R (1990): Activation of a bean chitinase promoter in transgenic tobacco plants by phytopathogenic fungi. Plant Cell 2:999–1007.

Samac AD, Hironaka CM, Yallaly PE, Shah MD (1990): Isolation and characterization of the genes encoding basic and acidic chitinase in *Arabidopsis thaliana*. Plant Physiol 93:907–914.

Schlumbaum A, Mauch F, Vogeli U, Boller T (1986): Plant chitinases are potent inhibitors of fungal growth. Nature (London) 324:365–367.

Shapira R, Ordentlich A, Chet I, Oppenheim AB (1989): Control of Plant Diseases by Chitinase Expressed from Cloned DNA in *Escherichia coli*. Phytopathology 79:1246–1249.

Sivan A, Chet I (1986): Biological control of *Fusarium* spp. in cotton, wheat and muskmelon by *Trichoderma harzianum*. J Phytopathol 116:38–47.

Sivan A, Chet I (1989a): Degradation of fungal cell walls by lytic enzymes of *Trichoderma harzianum*. J Gen Microbiol 135:675–682.

Sivan A, Chet I (1989b): The possible role of competition between *Trichoderma harzianum* and *Fusarium oxysporum* on rhizosphere colonization. Phytopathology 79:198–203.

Stasz TE, Harman GE, Weeden NF (1988): Protoplast preparation and fusion in two biocontrol strains of *Trichoderma harzianum*. Mycologia 80:141–150.

Sundheim L, Poplawsky A, Ellingboe H (1988): Molecular cloning of two chitinase genes from *Serratia marcescens* and their expression in *Pseudomonas* species. Physio Mol Plant Pathol 33:483–491.

Swegle M, Huang JK, Lee G, Muthukrishnan S (1989): Identification of an endochitinase cDNA clone from barley aleurone cells. Plant Mol Biol 12:403–412.

Tilburn J, Scazzocchio G, Talor GG, Zabicky-Zissman JH, Lockington RA, Davies WR (1983): Transformation by integration in *Aspergillus nidulans*. Gene 26:205–221.

Walden R, Koncz C, Schell J (1990a): The use of gene vectors in plant molecular biology. Methods Mol Cell Biol 1:175–194.

Walden R, Schell J (1990b): Techniques in plant molecular biology—progress and problems. Eur J Biochem 192:563–576.

Wandersman C, Delepelaire P, Letoffe S, Schwartz M (1987): Characterization of *Erwinia chrysanthemi* Extracellular Proteases: Cloning and Expression of the Protease Genes in *Escherichia coli*. J Bacteriol 169:5046–5053.

Weller DM (1988): Biological control of soilborne plant pathogens in the rhizosphere with bacteria. Annu Rev Phytopathol 26:379–407.

Wong KR, Green MJ, Buckley JT (1989): Extracellular Secretion of Cloned Aerolysin and Phospholipase by *Aeromonas salmonicida*. J Bacteriol 171:2523–2527.

Yanagida N, Uozumi T, Beppu T (1986): Specific Excretion of *Serratia marcescens* Protease through the Outer Membrane of *Escherichia coli*. J Bacteriol 166:937–944.

THE GENETIC NATURE AND BIOCONTROL ABILITY OF PROGENY FROM PROTOPLAST FUSION IN *TRICHODERMA*

G. E. HARMAN
C. K. HAYES

Cornell University, Geneva, New York 14456

1. INTRODUCTION

The first requirement for successful biological control is that a highly effective strain be identified and employed. It is also essential that the production system give rise to highly effective, desiccation-tolerant propagules, and that delivery systems and formulations permit high activity of the bioprotectant under conditions of substantial microbial competition (Harman, 1991), but the first requirement still is a highly effective biocontrol strain.

Trichoderma and *Gliocladium spp.* possess great genetic variability. One hundred nine alleles were identified using isozyme electrophoresis at 16 putative loci among 71 strains. This identification permitted ready separation of strains using only these 16 isozyme assays (Stasz et al., 1989). Some strains have a wide spectrum of activity (Harman, 1991) other strains may control only specific pathogens (Lumsden and Locke, 1989) (Smith et al., 1990), while still others may have little or no biocontrol efficacy. Similarly, some strains may grow poorly under some environmental conditions, while others grow well under these same adverse conditions.

Biotechnology in Plant Disease Control, pages 237–255
© 1993 Wiley-Liss, Inc.

Given this wide genetic diversity, several strategies suggest themselves for obtaining strains of *Trichoderma* or *Gliocladium* that are highly effective biological control agents. Of course, one may simply select strains from the wild, and many very useful strains have been obtained in this fashion. It may further be possible to select improved strains from an existing strain. All cells in *Trichoderma* thalli that we have examined are polynucleate, and it seems reasonable that wild strains may be heterokaryotic. If so, isolation of homokaryons from mass isolated strains by single spore isolation (Stasz et al., 1988) may give rise to different and more stable strains. It may also be possible to obtain improved strains by chemical mutagenesis; this procedure has given rise to several effective strains (Papavizas et al., 1982), including the first rhizosphere competent strains known (Ahmad and Baker, 1987). It is also possible to use recombinant DNA methodology for improvement of biocontrol strains and for elucidation of mechanisms; this possibility is discussed elsewhere in this volume (Broglie et al., Suslow, Chet, and Oppenheim).

It would also be useful to use sexual or asexual genetic recombination to obtain new strains. If, for example, two strains were known to possess good ability to control complementary pathogens, a method of combining these desirable attributes into a single strain would be avantageous. The sexual stage for *Trichoderma* and *Gliocladium* is rare, and may be entirely lacking for some biocontrol strains, and so sexual recombination may not be possible. However, the process of parasexuality has been described in fungi. In this process, (1) dissimilar nuclei in a heterokaryotic thallus fuse to give rise to a diploid, which (2) proliferates together with the parental haploid nuclei, (3) eventual sorting out of a diploid nucleus that may become established as a strain, (4) mitotic crossing over during multiplication of diploid nuclei, and (5) vegetative haploidization of diploid nuclei (Pontecorvo, 1956). This process, therefore, results in recombinant progeny strain that combine traits of both parents.

Therefore, if heterokaryons could be readily established, it might be possible to produce progeny strains of *Trichoderma* that possess desirable attributes of both parents. In the presence of appropriate selection procedures, it should be possible to obtain progeny strains that are superior to the parental strains in biocontrol efficacy. Protoplast fusion is a method to efficiently induce heterokaryosis. The purpose of this chapter is to consider efforts in the production of superior strains of *Trichoderma* using protoplast fusion, to consider what is known of the genetic events following fusion (parasexuality seems not to be involved), and to evaluate this process as a method for the genetic improvement of biocontrol efficiency in these fungi.

2. METHODS FOR PROTOPLAST FUSION

2.1. Protoplast Preparation

Protoplast fusion requires an efficient method of producing protoplasts in high numbers (at least 10^8 are required for each fusion). Davis (1985) reviewed general considerations for isolation of fungal protoplasts, while Harman and Stasz (1991) summarized conditions for producing protoplasts of *Trichoderma* spp.

The first consideration is the physiological state of the biomass to be used. Older hyphae or spores generally give poor results (Harman and Stasz, 1991). Toyama et al., (1984) used immature conidia as the starting material, while young hyphae may also work well (Davis, 1985). We place many small (< 3 mm^3) blocks of agar medium colonized by the strain of interest into an appropriate liquid medium and incubate with shaking until each block is covered with hyphae about 2–3 mm in length. These blocks are removed by filtration and aseptically transferred to an enzyme solution for release of protoplasts. Ideally, each protoplast should contain only one nucleus, which limits the level of heterokaryosis that might be obtained. For this reason Toyama et al. (1984) used immature conidia of T. reesei. In our biocontrol strains of *Trichoderma,* we find that all cells are polynucleate (Stasz et al., 1988), so conidia provide no advantage over hyphae, since uninucleate protoplasts cannot be obtained. Also, in our experience, conidia and protoplasts are of similar size, which makes separation difficult.

Additives to the growth medium, especially 2-deoxyglucose (2dG) may permit a more efficient release of protoplasts (Stasz et al., 1989). We routinely grow any strain that we have not used for protoplast isolation without added 2dG, and with concentrations ranging from 150 to 200 mg of 2dG (Stasz et al., 1989). Reducing agents (e.g., dithiothreitol) or ethylenediaminetetraacetic acid (EDTA) may also increase protoplast yield (Davis, 1985). Concentrations of 2dG giving optimal results are strain specific and can only be determined empirically.

Once appropriate material is produced, it must be digested to release viable protoplasts. We examined several enzymes and mixtures of enzymes for this purpose, and found that NovoZym 234 at 13 mg/ml gives good results. Other useful enzymes have been reviewed (Davis, 1985). Our usual procedure is to harvest colonized agar blocks and to suspend these in NovoZym 234 dissolved in 0.7 M NaCl containing 0.6 M sorbitol. Cell wall material and other debris are removed by filtration through cheesecloth or kimwipe, and the protoplasts are recovered by centrifugation and resuspended in a mixture of 600 mM sorbitol, 10 mM Tris-HCl, and 10 mM CaCl$_2$ at pH 7.5 (STC) (Stasz et al., 1988). The sorbitol is an osmoticant that prevents lysis of the protoplasts as they are released; other osmoticants may also be used (Davis, 1985). Calcium is necessary for integrity of the plasma membrane surrounding protoplasts of *Trichoderma.*

2.2. Protoplast Fusion

Once protoplasts have been isolated, fusion is usually accomplished by a procedure similar to that of Turgeon et al. (1985). In general, protoplasts are mixed together in the presence of polyethylene glycol (PEG), which induces fusion between membranes of adjacent cells. In our specific procedure (Stasz et al., 1988), about 10^8 protoplasts, in 1 ml of STC, of each strain to be fused are mixed with 200 µl of a solution containing 60% (w/v) PEG (MW 3350), 10 mM CaCl$_2$, and 10 mM Tris-HCl pH 7.5. This mixture is gently mixed by rolling the tube, and then a 500-µl portion of the fusing solution is added, again mixed gently, and the process repeated with a third 500-µl portion. The protoplast suspension is incubated for 10 min. The mixture is diluted with STC, again in a stepwise fashion. Fused protoplasts are recovered by centrifugation, and the fused protoplast mixture is plated on appropriate media.

Alternatively, protoplasts may be fused in the presence of relatively high pH levels and Ca$^+$, or other procedures (e.g., electroporation) may be used.

In addition, it may be possible to use an alternative to protoplast fusion in order to accomplish cell fusion in *Trichoderma*. Anastomosis occurs when vegetative hyphae of different strains come into contact; this phenomenon can be readily demonstrated. Complementary spore color (e.g., white and yellow) mutants of strains can be prepared and colonies started on opposite sides of a petri dish. When the colonies meet in the center of the plate, a zone of green spores frequently occurs (Fig. 13.1) (Bojnanska et al., 1980). Presumably, the green spores mark a region in which anastomosis has occurred, giving rise to the possibility of complementation of enzymes providing the green pigmentation. Such zones are obvious if two spore color mutants of the same strain are allowed to anastomose, but spore color complementation is rarer if mutants of dissimilar strains are fused.

3. METHODS FOR VISUALIZATION OF FUSED AND REGENERATING PROTOPLASTS

It is useful to be able to visually monitor the process of protoplast fusion and to determine the morphological events accompanying growth and regeneration of fused protoplasts. In order to accomplish this, complementary fluorescent vital (nonlethal) stains are used. Tetramethylrhodamine isothiocyanate (RITC) and fluorescein isothiocyanate (FITC) have been used to label plant and animal cells with yellow-orange and green fluorescence, respectively (Barsby et al., 1984; Junker and Pedersen, 1981). Thus, cells of one line can be stained with one fluorochrome, and the other with the alternative material, and, after fusion, fused cells with one-half of one color and the other of the complementary color can be seen. Using this technique, the frequency of fusion and the numbers of original

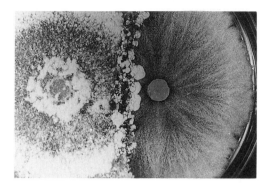

Fig. 13.1. Evidence for hyphal anastomosis and complementation of traits in *Trichoderma*. Two mutants of strain 105 (one largely asporulant, the other producing white conidia) were inoculated on opposite sides of a potato dextrose agar (PDA) plate. At the point of contact, a line of green conidia were produced.

cells making up the fused cell can be observed. However, we found FITC and RITC to be unsuitable labels for *Trichoderma* spp., since only a relatively small proportion of cells were labeled. Therefore, we searched for other stains and found that hydroethidine and rhodamine 6G effectively labeled protoplasts red and green, respectively (Harman and Stasz, 1988). When used at appropriate concentrations and in the dark, theses fluorochromes were nontoxic. They were effective materials for microscopic determination of fusion in *Trichoderma* (Harman and Stasz, 1988).

In addition to visualization, it is also possible to use complementary fluorescent labels to physically sort cells into heterofusant (two color) and nonfused or homofusant cells (one color) using flow cytometry (Shapiro, 1985). Flow cytometry apparatus divides a suspension into droplets, probes each with a laser, and diverts individual droplets into containers with only like particles. This technique also can sort cells by size, so it is possible to sort a population of fused cells into, for example, heterofusants containing more than two cells, heterofusants containing only two cells, and nonfusants. We demonstrated that this was possible using fused *Trichoderma* protoplasts stained with hydroethidine and rhodamine G. However, it probably is necessary to continue selection pressure on heterofusants as they regenerate into thalli, so this technique may not be suited for these fungi.

It also is useful to follow the course of regeneration of protoplasts into thalli. Regenerating cell walls can be stained with Calcofluor, which stains cell wall polymers (Kritzman et al., 1978). We found that *Trichoderma* protoplasts begin to regenerate a fibrillar network over the membrane surface within a few hours after protoplasting. Fused protoplasts frequently are in large aggregates; these appear to combine into giant cells 24–48 h after fusion (Stasz and Harman, unpublished). Thalli probably arise from these polynucleate giant cells.

4. SELECTION AND GENETIC ANALYSIS OF PROGENY

4.1. Methods for Selection of Progeny

Once protoplast fusion is accomplished, some method of selection of heterofusants (i.e., protoplasts fused between dissimilar parents) and homofusants or nonfused protoplasts must be employed. If protoplast fusion mixtures are plated on nonselective media, thalli will grow from the nonfused or self-fused protoplasts more rapidly than from heterofusants, therefore precluding the isolation of useful progeny.

Mellon (1985) summarized five general selection strategies, as follows: (1) nutritional complementation, (2) drug resistance, (3) dead donor, (4) metabolic poisons, and (5) protoplast–nuclear fusion. Of these five, drug resistance and metabolic poisons can be considered as similar, and will be hereafter referred to as toxicant resistance. Of these choices, nutritional complementation has been most frequently used as a selection tool in fungal protoplast fusion (Mellon, 1985). In this method of selection, complementary auxotrophic mutants (i.e., with a requirement for different nutrients, usually a vitamin or amino acid) are prepared. Upon fusion, the fused strain should be able to synthesize all needed nutrients, and the fusion mixture is plated onto basal medium. Strains able to grow on this medium, that is, that are prototrophic, are selected as putative fusant progeny. This selection method has frequently been used in the genus *Trichoderma* (Gracheck and Emert, 1984; Pe'er and Chet, 1990; Stasz et al., 1988). Toxicant resistance would rely upon complementary resistance to toxicants among two strains, the fusants would be expected to possess resistance to both. We have employed this method to select fusants in *Trichoderma* (Hayes and Harman, unpublished), and in other fungi (Bradshaw and Peberdy, 1984). In the dead donor technique, protoplasts of one strain are killed by irradiation, heat or other procedure, and then fused with another strain. Because one strain is not alive, this strain can be a wild strain with no selectable marker (Mellon, 1985). Fusion of strains with isolated nuclei has been accomplished in *Trichoderma* (Sivan et al., 1990). Nuclei were isolated from a wild prototrophic strain and fused with protoplasts of an auxotrophic mutant of another strain.

In addition to these techniques, the use of complementary spore color mutants, as described earlier, may provide a useful selection tool (Bojnanska et al., 1980; Grachek and Emert, 1984; Pe'er and Chet, 1990).

4.2. Methods for Genetic Analysis of Progeny

Once progeny have been selected, it is important to be able to determine their genetic nature. Protoplast fusion gives rise to a mixture of all cellular contents, which adds to the difficulty of analysis. Mitochondria, plasmids, and viruses all contain nucleic acid. These extranuclear sources of genetic material may be combined in fusion progeny, and genetic analysis of progeny ideally will also consider these sources of genetic variation. A substantial body of work demonstrates that protoplast fusion in yeast may give rise to progeny with mixtures of mitochondria from both parents, or to recombinant progeny. This phenomenon has also been demonstrated in *Aspergillus nidulans* (Ferenczy, 1985). Therefore, extranuclear sources of genetic variation may give rise to the phenotype of progeny. A number of general procedures may be employed to determine the genetic nature of protoplast fusion progeny.

4.2.1 Resolution of Heterokaryons

All cells of most *Trichoderma* strains are polynucleate [immature conidia of *Trichoderma reesei* may be an exception (Toyama et al., 1984)]. Therefore, fusion progeny are likely to contain a mixture of nuclei, that is, to be heterokaryons. If parasexuality occurs, diploid and aneuploid nuclei will be formed (Pontecorvo, 1956), which further adds to the complexity of the nuclear mixture. As a consequence, it is impossible to determine whether properties of a fusion progeny derives from the genetic program of a single nuclear type or from the average effect of a mixture of different nuclei.

Complex heterokaryons may be resolved by single sporing. Conidia of *Trichoderma* receive only a single nucleus from the phialide (Stasz et al., 1988). Therefore, isolation of subprogeny from single spores gives rise to homokaryotic thalli. Frequencies of nuclei within a thallus can therefore be examined by determining the numbers of single spore progeny with the character in question.

4.2.2 Phenotypic Markers

Further information concerning the nuclear phenotype can be obtained by examining several of the phenotypic markers used for selection. Spore color, nutritional requirements (e.g., auxotrophy), and toxicant resistance are particularly useful characteristics.

4.2.3. Isozyme Analysis

Since proteins are coded for by specific genes, the distribution of isozyme phenotype should provide useful genetic markers. For this analysis to beneficial, isozymes employed must have specific characteristics. They must be constitutively produced, otherwise variation may occur as a consequence of age or minor environmental fluctuations. They must also be consistently and reliable visualized. The most useful isozyme phenotypes are those with only one or two bands, since these characters are easily interpreted. Stasz et al. (1988, 1989) showed that there was a great deal of isozyme variability within *Trichoderma* and *Gliocladium,* and identified a series of isozymes that give unequivocal differences among strains. One hundred nine alleles were observed at 16 loci among 71 strains tested. In nearly all cases, only a single activity band was detected. This result indicates that strains contained only a single locus for each enzyme, and suggests that all were haploid.

These simple banding patterns are amenable to further analysis. For example, if the parental strains contained different phenotypes for an enzyme, any thallus containing genetic sequences coding for both enzymes should give an isozyme pattern with both banding patterns if the genes are present in roughly equal numbers. Moreover, some enzymes have been shown to be multimeric, that is, composed of subunits assembled after translation. If dimeric enzymes of two separate mobility classes are produced within the same cell, some of the active enzyme will be of the mobility class of one parental strain, some will be of the other parent (i.e., they are homodimeric), and some will be of an intermediate mobility, since they will be composed of one subunit of one strain and one of the other parent. The presence or absence of heterodimeric bands can therefore be used to ascertain whether two separate dimeric enzymes are coded for and produced within the same cell (Micales et al., 1986).

4.2.4. Chromosome Electrophoresis

In addition, the chromosomes of fusants can be examined to determine the genetic nature of progeny. A karyotype comparison of intact chromosomes between the fusants and parentals can be used to determine chromosomal inheritance of the fusants. In classical parasexuality, diploid or aneuploid strains are produced, and strains produced through this process should contain chromosomes of both parents. Standard electrophoresis of DNA cannot be employed to separate chromosomes, since this procedure is not able to resolve large pieces of DNA (larger than about 50 kb, thousand base pairs). Typically, fungal chromosomes range from about 150 kb to 12 Mb (million base pairs) (Mills and McCluskey, 1990). Furthermore, DNA isolated by conventional techniques cannot be used for chromosome electrophoresis, since methods involved in DNA separation and purification invariably shear the chromosomes into fragments.

Techniques for electrophoretic karyotyping have been developed that involve embedding protoplasts or cells in agarose plugs and then gently lysing the cells to release chromosomal DNA. The intact chromosomes are then subjected to pulsed field electrophoresis in which the electrical field is reoriented at specific times

(Schwartz et al., 1982). Large molecules of DNA reorient themselves according to the reorientation of the electrical field (pulses) and migrate through the gels. The interaction of pulse time, voltage, and the influence of the agarose gel result in separation of chromosomal DNA (see Lai et al., 1989, for general review).

We have developed methods permitting electrophoretic karyotyping of *Trichoderma* chromosomes (Hayes et al., 1991). Different strains of *Trichoderma* have different numbers and sizes of chromosomes. For example, strain T95 of *Trichoderma harzianum* has four chromosomes ranging from 2.2 to 5.4 Mb, while strain T12 has only two chromosomes of 4.2 and 5.4 Mb. Therefore, this methodology can be employed to determine the inheritance of chromosomes within progeny strains of *Trichoderma* following protoplast fusion. Other fungi contain variations within chromosomes, for example, *Neurospora crassa* has seven chromosomes that contain minor differences due to translocations (Orbach et al., 1988). *Candida albicans* also has substantial variation in chromosome sizes and numbers (Magee and Magee, 1987). Imperfect fungi may have substantial differences in number and sizes of chromosomes, since chromosomes do not have to undergo pairing during meiosis.

4.2.5. RFLP Analysis

Variation in smaller DNA segments among the parentals and fusants can be visualized by comparison of restriction fragment length polymorphism (RFLP) patterns. The genetic "fingerprint" obtained from these RFLP patterns can be used not only to detect nucleotide differences between the parentals and fusants, but to examine base modifications that may have occurred during protoplast fusion. The RFLP analysis relies upon restriction enzymes that cut isolated DNA at very specific recognition sites. If location of these recognition sites within the genomic DNA differ among strains, specific RFLP patterns will be obtained that can be used to identify a particular strain.

The RFLP analysis can be used to examine total DNA, nuclear DNA, or mitochondrial DNA. Mitochondrial DNA can be separated from nuclear DNA in CsCl gradients containing bis(benzimide) (Garber and Yoder, 1983). In order to speed analysis, we have determined certain restriction enzymes that will give RFLP patterns of only the mitochondrial DNA or the nuclear DNA within a total DNA extract. This separation is possible, since mitochondrial DNA is richer in adenine and thymine dimers than nuclear DNA. Within a total DNA extract, restriction enzymes specifically cleaving at cytosine and guanine rich sites tend to cleave nuclear DNA of *Trichoderma* into very small pieces. The mitochondrial DNA is also cleaved, but to a lesser extent with several bands arising that are useful in strain identification (Hayes et al., unpublished). Other restriction enzymes have been identified that give discernible RFLP patterns of the nuclear genome within a total DNA extract. We have used these principles to determine the nuclear and mitochondrial RFLP patterns of progeny formed by protoplast fusion in *Trichoderma*. This technique has allowed us to identify the parental origin of the fusants (Hayes and Harman, unpublished).

4.2.6. Other Techniques

In fungi other than *Trichoderma,* techniques have been developed for obtaining specific types of protoplast fusion progeny. In some cases, relatively stable progeny strains have been produced that differ from the parentals in various characters. Segregation of these strains has been obtained by growing them in the presence of haploidization agents, for example, benomyl or *p*-fluorophenylalanine (Anne and Peberdy, 1985; Peberdy et al., 1977). Presumably, in such strains, the stable nonparental state is a diploid containing all of the chromosomes of both parental strains. After exposure to the haploidization agent, aneuploid or recombinant haploid strains arise.

Another technique was used with *Cephalosporium* or *Emericellopsis*. When fungi in these genera were fused, heterokaryons grew very slowly, and recombinant nuclei were difficult to detect. Minuth and Esser (1983) therefore prepared "secondary protoplasts" from these slow-growing heterokaryotic thalli. These were plated and aneuploid strains were detected. A similar procedure, but involving grinding slow growing heterokaryotic progeny, was followed in interspecies fusion experiments with *Penicillium* (Anne and Peberdy, 1985). At least in part, this procedure may allow selection of rare genetic types. Hyphae of these fungi may contain many nuclei, and some nuclear types may be more prevalent in some portions of the thallus than in others. If a heterokaryotic thallus is broken apart into individual cells, these rare nuclear types may be easier to detect and isolate than in the original thallus.

5. COMPATIBILITY

When two strains are fused, they may or may not readily give rise to numerous progeny with rapid growth characteristics. If two strains readily give rise to numerous rapid growing progeny they are considered as being compatible, while if they give rise to only a few slow-growing progeny, they have been designated as incompatible, or as being of limited compatibility. We prefer the term limited compatibility, since it seems to us that incompatibility describes an absolute, that is, no progeny maybe expected (Stasz et al., 1989). In other genera, compatibility groups have been identified, with compatible groups occurring among closely related strains (Croft, 1985). The occurrence of compatibility groups in the genus *Aspergillus* and their study using protoplast fusion has been summarized by Croft (1985). In this genus a number of compatibility groups occur, and these appear to be controlled by a group of alleles. The degree of compatibility can be altered by chemical or physical mutation. Similarly, in *Cochliobolus heterostrophus*, asexual compatibility is controlled by specific genes that are inherited independently from those controlling sexual mating type (Leach and Yoder, 1983).

We have examined the degree of compatibility in *Trichoderma* in a number of intra- and interstrain and species fusions (Stasz et al., 1989). If we fused two auxotrophic mutants of the same strain (intrastrain fusion), we obtained numerous, rapidly growing prototrophic progeny similar in all respects to the wild prototrophic progeny similar in all respects to the wild prototrophic parent. Colonies arose on minimal medium within a few days after plating the fusion mixture. The complementation frequency (the number of prototrophic divided by the number of auxotrophic thalli) ranged from 0.2 to 10% (Pe'er and Chet, 1990; Stasz et al., 1989). However, when two mutants of different strains were fused (interstrain fusion) the complementation frequency was lower, ranging from 0.001 to 0.02%. Colonies frequently arose on minimal medium very slowly; some thalli reached visible size only several weeks after plating. These colonies were usually quite different in morphology from the parental auxotrophic strains or from the prototrophic strains from which the auxotrophs were obtained. Results were similar regardless of whether different strains within the same species (intraspecific fusions) or in different species (interspecific fusions) were fused (Stasz et al., 1988, 1989). These data indicate that intrastrain fusions within *Trichoderma* were fully compatible, but that interstrain fusions resulted in limited compatibility.

Limited compatibility restricts the use of protoplast fusion in these fungi. We therefore sought methods to overcome this difficulty. We found in interstrain fusions that nuclei of one parental type are found much more frequently than the other (see Section 6), and we designated the frequent parent the prevalent parent, and the other the nonprevalent parent. It was also possible that limited compatibility was due to factors mediated in the cytoplasm. We therefore isolated nuclei from prototrophic protoplasts of a prevalent strain (T12) and fused these with protoplasts of an auxotrophic strain of a nonprevalent parent (T95-lys-). Eventually, strains were obtained that were prototrophic and isozymically identical to T12 (Sivan et al., 1990).

6. MORPHOLOGICAL AND GENETIC EVENTS FOLLOWING FUSION IN *TRICHODERMA*

We and others followed the morphological and genetic events that follow protoplast fusion in *Trichoderma*, and have found unusual and surprising events.

6.1. Morphological and Nutritional Events

As mentioned earlier, events following intrastrain fusions are quite different from those following interstrain fusions. In intrastrain fusions of two auxotrophs, many progeny are obtained that grow rapidly, and that are similar to the wild prototrophic parent.

Conversely, in interstrain fusions, progeny frequently developed slowly on minimal media. Some also grew slowly on complete media, and, in addition some strains appeared to be only partially prototrophic, that is, they grew more slowly on minimal than on complete medium (Stasz et al., 1988; Stasz and Harman, 1990).

These progeny were often unstable, and sectors frequently occurred (Fig. 13.2) (Stasz and Harman, 1990; Stasz et al., 1988; Pe'er and Chet, 1990). These sectors frequently grew more rapidly than the original thallus, and in many cases were more strongly prototrophic. As a consequence of the substantial variation in the morphology of the strains, and as sectoring occurred, progeny were obtained that were extremely variable in growth rate, pigmentation, nutritional requirements, and level of sporulation (Fig. 13.3) (Stasz et al., 1988; Stasz and Harman, 1990). These events are similar in many respects to those obtained with other fungal genera, including *Penicillium* and *Aspergillus,* particularly when distantly related strains were fused (Anne and Peberdy, 1985; Kevei and Peberdy, 1985).

6.2. Genetic Events Following Protoplast Fusion

The genetic events in intrastrain fusions are relatively simple. A first question to be asked is whether progeny differ from their parents because they are hetero-

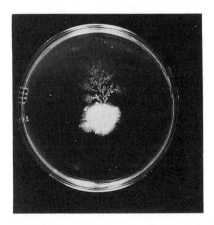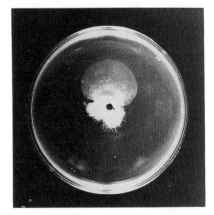

Fig. 13.2. Sectoring of protoplast fusion progeny during vegetative growth on agar medium. Reprinted with permission from Stasz and Harman (1990).

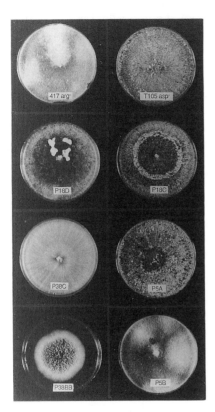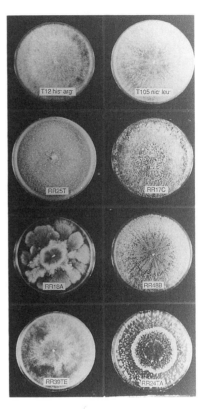

Fig. 13.3. Variation in morphology in progeny produced from protoplast fusion between strain 417 arg- or *Trichoderma koningii* and strain T105 asp- or *Trichoderma viride*, and between strain T12 his-arg- of *T. harzianum* and strain T105 nic-arg- or *T. viride* after 10 days growth on potato dextrose agar. Parental strains are shown at the upper portion of each figure. Similar levels of variation occurred with other interstrain fusions, regardless of taxonomic position. Similar figures have been presented elsewhere (Stasz et al., 1988; Stasz and Harman, 1990; Harman and Stasz, 1991).

karyons, or for some other reason. If conidia were isolated from prototrophic progeny from intrastrain fusions and plated on appropriate diagnostic media, we found that all were of one or the other parental auxotrophic type, and that these parental types were present in approximately equal numbers. The very few progeny that grew on minimal medium probably arose from multinuclear cells, since conidia from these again gave rise only to strains indistinguishable from the auxotrophic parental strains. In our lab, no progeny (or single spored subprogeny) have been obtained from intrastrain fusions that appeared to be anything other than the auxotrophic parental strain. However, Pe'er and Chet (1990) obtained nonparental progeny from an intrastrain fusion.

Primarily, intrastrain protoplast fusion gives rise to balanced heterokaryons. We have detected no evidence for karyogamy or other event in parasexuality (Stasz et al., 1988; Stasz and Harman, 1990); however, Pe'er and Chet (1990) obtained prototrophic colonies from single spore isolations of progeny produced by intrastrain fusion of two different auxotrophs.

Events in interstrain fusions appear quite different. The extreme morphological differences within progeny persist after single sporing (Stasz et al., 1988; Stasz and Harman, 1990), so the observed variation is not simply a consequence of heterokaryosis. In at least some cases, single sporing indicates that progeny are greatly imbalanced heterokaryons, with nuclei representative of one parent outnumbering the other by a ratio of 10,000:1. Other data clearly demonstrate that the variation is also not a consequence of classical parasexuality. If parasexuality occurred, there should be a recombination of genetic characteristics from both parents. Moreover, aneuploid or diploid strains should contain chromosomes of both parents. As described earlier, isozyme analysis, chromosome electrophoresis, and RFLP analysis permit us to determine the genetic nature of progeny.

We have extensively analyzed progeny strains for their isozyme profiles. Soon after progeny are isolated, their isozyme profiles are identical to one parent or the other. Frequently, isozyme patterns within progeny of a given fusion are predominantly of one parent, which we have designated the prevalent parent. Some progeny, when tested soon after fusion, do give banding patterns for some isozymes with double bands indicative of the presence of alleles of both parents, but we never have detected heterodimeric bands. These results indicate that both parental nuclei are present, but that they are not being expressed in the same cell. The simplest explanation of these results is that, in the liquid culture in which thalli are grown to provide biomass for isozyme analysis, sectors occur that contain nuclei of the different parents. This double banding does not persist in progeny, does not occur in sectors, or in single spore progeny (Stasz et al., 1988; Stasz and Harman, 1990). Not only do we not see double bands indicative of two loci of an individual enzyme, but we never see recombination of different isozymes. If the mobility of one isozyme from a particular fusant is that of parent A, the mobility of all the other isozymes of that fusant will also be identical to parent A (Stasz et al., 1988; Stasz and Harman, 1990).

Even though a group of progeny have identical isozyme profiles, they may be very different in nutritional requirements, sporulation ability, color, growth rate, and general colony morphology. Therefore, this variation occurs in the absence of recombination of isozyme phenotype. Since we have examined, in various progeny, 17 separate isozymes, it is unlikely that recombination in isozyme phenotype occurs in protoplast fusion progeny in *Trichoderma*. These data strongly indicate that the variation observed in these progeny occurs as a consequence of factors other than the classical parasexual cycle (Stasz et al., 1988; Harman and Stasz, 1991).

Additional evidence for the absence of the parasexual cycle comes from chromosome electrophoresis. As described earlier, we developed methods for separation and visualization of chromosomes of *Trichoderma*. We have done only a limited number of strains, that is, parental strains T12 and T95 of *T. harzianum*, and a single progeny, 1295-22 that has excellent biocontrol ability (Hayes et al., 1991; Harman, 1991). Strain T12 has two chromosomes, while strain T95 has four chromosomes, with their banding patterns being distinctly different. Strain 1295-22 is isozymically identical to T12, and also has two chromosomes of the same size as T12 (Hayes et al., 1991). If parasexuality had occurred in the production of 1295-22, we would have expected an aneuploid or haploid recombinant in chromosomes arrangement. This result was not seen, which again argues against the occurrence of parasexuality in progeny strain 1295-22.

The RFLP analysis has been used to examine either the mitochondrial or the nuclear genome. We are presently studying several families of fusion progeny, but in general the first data indicate that the RFLP pattern of the nuclear or the mitochondrial genome in fusants agree with the isozyme phenotype (Hayes and Harman, unpublished).

Therefore, the evidence is consistently against the involvement of parasexuality in the production of the very diverse progeny obtained by protoplast fusion in *Trichoderma*. Section 6.3 will examine possible mechanisms by which this paradox may be resolved.

6.3. An Hypothesis To Explain the Postfusion Genetic Events in *Trichoderma*

The data summarized previously clearly indicate that protoplast fusion in *Trichoderma* gives rise to a wide range of progeny characteristics, but that parasexuality is not involved. Indeed, protoplast fusion with these fungi seems not to result in the incorporation of any large portions of DNA from the nonprevalent strain into the progeny. Earlier, we (Harman and Stasz, 1991; Stasz and Harman, 1990) suggested that the genetic events in *Trichoderma* could result either from the interaction of extranuclear DNA, for example, mitochondria, of one strain with nuclear DNA of another, or it could arise by a novel mechanism in which small portions of DNA from one strain are incorporated into the genome of the second strain. Data from RFLP analysis of mitochondrial and nuclear DNA, as well as isozyme profiles, have shown that the fusion progeny always match one or the other parental type (Hayes et al., unpublished). Therefore, variation due to mixtures of nuclear and mitochondrial DNA does not seem to be a major factor in the observed variation within progeny.

Our current hypothesis is that small portions of DNA from one strain are incorporated into the genome of the second strain. While the precise mechanism is unknown, this process would appear to be analogous to the incorporation of plasmids or transposons into genomic DNA. Studies currently underway should prove or disprove this hypothesis.

7. BIOLOGICAL CONTROL EFFICACY OF PROTOPLAST FUSION PROGENY

Regardless of its genetic basis, the diversity of progeny obtained through protoplast fusion in *Trichoderma* can be used as a source of improved strains for biocontrol. Stable strains can be obtained from sectors or from single spore subprogeny. Effective biocontrol strains are rare; in our work only 2–4% of the progeny obtained from a fusion between *T. harzianum* strains T12 his- and T95 lys- were more effective as biological seed treatments than the prototrophic wild parents (Harman and Stasz, 1991), and Pe'er and Chet (1990) found only one progeny strain of the 14 tested from fusion among various mutants of *T. harzianum* ATCC 32173 gave significant biocontrol of *Rhizoctonia solani*. Therefore, screening of progeny sets from protoplast fusion is required. A suitable strategy is to divide the many progeny obtained into sets based upon growth rate, isozyme phenotype, nutritional requirements, and morphology, and to test representatives of each class. Once effective classes are identified, individual progeny within that class can be compared with one another.

Progeny from the fusion between *T. harzianum* strains T95 lys- and T12 his- have been extensively tested and one strain is close to becoming a commercially used biocontrol agent. This progeny grows more rapidly than its prototrophic parental strains and is effective against a range of pathogenic fungi, including *Pythium ultimum, Fusarium graminearum, Rhizoctonia solani, Sclerotium rolfsii, Sclerotinia homoeocarpa,* and *Botrytis cinerea* (Harman et al., 1989, Pearson, Tronsmo, and Harman, unpublished, Nelson, unpublished), but not *Phytophthora* spp. (Smith et al., 1990). This strain is substantially more rhizosphere competent than the parental strains. In field trials, pea roots from plants produced from seeds treated with various strains were removed after plants were at mid-flower. The protoplast fusion strain was found to be present on nearly all root segments sampled, while the parental strains were less frequently found. In laboratory trials, rhizosphere competent strains of *Trichoderma* are distributed in a "u" or "c" shaped curve. Higher numbers of propagules are found at the top and the tip of the root, while lower numbers are observed at the central root portion. The protoplast fusion progeny, however, was found to be equally distributed at high numbers over the entire root surface (Sivan and Harman, 1991) (Table 13.1). Thus, this strain is effective against a range of pathogenic fungi and is able to colonize root surfaces. In addition, it is effective on a wide range of agronomic and vegetable crop species, ranging from cotton to beans to corn. It has been widely tested, and a major corporation recently registered this strain as a microbial pesticide, with the trade

TABLE 13.1. **Mean Levels of the Various Strains of Parental *Trichoderma*
Strains T12 and T95, and Protoplast Fusion Progeny Strain 1295-22 on the
Upper, Middle, and Lower Rhizosphere and Rhizoplane of Corn and Cotton Roots**

| Root Section | Log Colony Forming Units[a] | | | |
	T95	1295-22	T12	MSD[a]
Corn Rhizosphere				
Upper portion	2.5	3.9	3.3	N. S.
Middle portion	1.2	2.9	1.6	1.08
Lower portion	2.8	3.2	1.8	N. S.
MSD[b]	N. S.	N. S.	1.52	
Corn Rhizoplane				
Upper portion	2.8	2.4	2.6	N. S.
Middle portion	1.2	2.9	1.4	N. S.
Lower portion	1.5	3.1	2.4	N. S.
MSD[b]	N. S.	N. S.	N. S.	
Cotton Rhizosphere				
Upper portion	1.3	2.7	2.5	1.13
Middle portion	0.9	2.0	0.6	0.80
Lower portion	0.6	0.7	0	N. S.
MSD[b]	0.45	1.66	1.14	
Cotton Rhizoplane				
Upper portion	1.4	2.6	2.4	0.74
Middle portion	0.8	2.0	1.0	1.07
Lower portion	0.4	1.3	0	N. S.
MSD[b]	N. S.	N. S.	0.12	

[a]Units are in per grams of dry weight of soil or root for rhizosphere or rhizoplane, respectively. In all cases, the upper portion of the roots were considered to be the first 6 cm below the hypocotyl, the lower portions were considered to be the lowest 4 cm. The remainder was considered to be middle portion. Values presented are means across two separate experiments.

[b]The minimum significant difference (MSD) values are as determined by Waller and Duncan's test (GLM procedure, SAS Institute, Cary, NC).

Reprinted with permission from Sivan and Harman (1991).

name "F-Stop®." We expect that this product will be available for use in commercial agriculture within the next few years.

ACKNOWLEDGMENTS

Some of the work described in this chapter was supported in part by grants from the Cornell Biotechnology Program, which is sponsored by the New York State Science and Technology Foundation, a consortium of industries, the US

Army Research Office, and the National Science Foundation; from the US–Israel Binational Agricultural Research and Development Fund (BARD) US-1224-86; and from the Eastman Kodak Co.

REFERENCES

Ahmad JS, Baker R (1987): Rhizosphere competence of *Trichoderma harzianum*. Phytopathology 77:182–189.

Anne J, Peberdy JF (1985): Protoplast fusion and interspecies hybridization in *Penicillium*. In Peberdy JF, Ferenczy, L (eds): Fungal Protoplasts: Applications in Biochemistry and Genetics. New York: Marcel Dekker, pp 259–277.

Barsby TL, Yarrow SA, Shepard JF (1984): Heterokaryon identification through simultaneous fluorescence of tetramethylrhodamine isothiocyanate and fluorescein isothiocyanate labelled protoplasts. Stain Technol 59:217–220.

Bojnanska A, Sipiczki M, Ferenczy L (1980): Characterization of conidiation mutants in *Trichoderma viride* by hyphal anastomosis and protoplast fusion. Acta Microbiol Acad Sci Hung 27:305–307.

Bradshaw RE, Peberdy JF (1984): Protoplast fusion in *Aspergillus*: section of interspecific heterokaryons using antifungal inhibitors. J Microbiol Methods 3:27–32.

Croft JH (1985): Protoplast fusion and incompatibility in *Aspergillus*. In Peberdy JF, Ferenczy L (eds): Fungal Protoplasts: Applications in Biochemistry and Genetics. New York: Marcel Dekker, pp 225–240.

Davis, B (1985): Factors influencing protoplast isolation. In Peberdy JF, Ferenczy L (eds): Fungal Protoplasts: Applications in Biochemistry and Genetics. New York: Marcel Dekker, pp 45–71.

Ferenczy L (1985): Transfer of cytoplasmic genetic elements by protoplast fusion. In Peberdy JF, Ferenczy L (eds): Fungal Protoplasts, Applications in Biochemistry and Genetics. New York: Marcel Dekker, pp 307–321.

Garber RC, Yoder OC (1983): Isolation of DNA from filamentous fungi and separation into nuclear, mitochondrial, ribosomal, and plasmid components. Anal Biochem 135:416–422.

Gracheck SJ, Emert GH (1984): Protoplast formation and fusion using *Trichoderma reesei* mutants. In Nash CH, Underkofler LA (eds): Developments in Industrial Microbiology. Baltimore: Victor Graphics, pp 581–587.

Harman GE (1991): Seed treatments for biological control of plant disease. Crop Protect 10:166–171.

Harman GE, Stasz TE (1988): Fluorescent vital stains for complementary labelling of protoplasts from *Trichoderma* spp. Stain Technol 63:241–247.

Harman GE, Stasz TE (1991): Protoplast fusion for the production of superior biocontrol fungi. In TeBeest DO (ed): Microbial Control of Weeds. New York: Chapman and Hall, pp 171–186.

Hayes CK, Harman GE, Woo SL, Gullino ML (1991): Methods for electrophoretic karyotyping of filamentous fungi in the genus *Trichoderma*. Anal Biochem (Submitted).

Junker S, Pedersen S (1981): A universally applicable method of isolating somatic cell hybrids by two-color flow sorting. Biochem Biophys Res Commun 102:977–984.

Kevei, F, Peberdy JF (1985): Interspecies hybridization after protoplast fusion in *Aspergillus*. In Peberdy, JF and Ferenczy, L (eds): Fungal Protoplasts: Applications in Biochemistry and Genetics. New York: Marcel Dekker, pp 241–257.

Kritzman G, Chet I, Henis Y, Hutterman, A (1978): The use of the brightener "Calcofluor White M2R New" in the study of fungal growth. Isr J Bot 27:138–146.

Lai E, Birren BW, Clark SM, Simon MI, Hood L (1989): Pulsed field gel electrophoresis. Biotechniques 7:34–42.

Leach J, Yoder OC (1983): Heterokaryon incompatibility in the plant-pathogenic fungus, *Cochliobolus heterostrophus*. J Hered 74:149–152.

Lumsden RD, Locke JC (1989): Biological control of damping-off caused by *Pythium ultimum* and *Rhizoctonia solani* with *Gliocladium virens* in soilless mix. Phytopathology 79:361–366.

Magee BB, Magee PT (1987): Electrophoretic karyotypes and chromosome numbers in *Candida* species. J Gen Microbiol 133:425–430.

Mellon FM (1985): Protoplast fusion and hybridization in *Penicillium*. In Timberlake WE (ed): Molecular Genetics of Filamentous Fungi. New York: Alan R. Liss, pp 69–82.

Micales JA, Bonde MR, Peterson, GL (1986): The use of isozyme analysis in fungal taxonomy and genetics. Mycotaxon 27:405–449.

Mills, D, McClluskey, K (1990): Electrophoretic karyotypes of fungi: the new cytology. Mol Plt Micrb Interact 8:351–357.

Minuth W, Esser K (1983): Intraspecific, interspecific, and intergeneric recombination in β-lactam producing fungivia protoplast fusion. Eur J Appl Biotechnol 18:38–46.

Orbach MJ, Vollrath D, Davis RW, Yanofsky C (1988): An electrophoretic karyotype of *Neurospora crassa*. Mol Cell Biol 8:1469–1473.

Papavizas GC, Lewis JA, Abd-El Moity TH (1982): Evaluation of new biotypes of *Trichoderma harzianum* for tolerance to benomyl and enhanced biocontrol capabilities. Phytopathology 72:126–132.

Pe'er S, Chet I (1990): Trichoderma protoplast fusion: a tool for improving biocontrol agents. Can J Microbiol 36:6–9.

Peberdy JF, Eyssen H, Anne J (1977): Interspecific hybridization between *Penicillium chrysogenum* and *Penicillium cyaneo-fulvum* following protoplast fusion. Mol Gen Genet 157:281–284.

Pontecorvo, G (1956): The parasexual cycle in fungi. Annu Rev Microbiol 10:393–400.

Schwartz DC, Saffran W, Welsh J, Haas R, Goldenberg M, Cantor CR (1982): New techniques for purifying large DNAs and studying their properties and packaging. Cold Springs Harbor Symp Quant Biol 47:189–195.

Shapiro HM (1985): Practical Flow Cytometry. New York: Alan R. Liss, 295 pp.

Sivan A, Harman GE (1991): Improved rhizosphere competence in a protoplast fusion progeny of *Trichoderma harzianum*. J Gen Microbiol 137:23–29.

Sivan A, Harman GE, Stasz TE (1990): Transfer of isolated nuclei into protoplasts of *Trichoderma harzianum*. Appl Environ Microbiol 56:2404–2409.

Smith VL, Wilcox WF, Harman GE (1990): Potential for biological control of Phytophthora root and crown rots of apple by *Trichoderma* and *Gliocladium* spp. Phytopathology 80:880–885.

Stasz TE, Harman GE (1990): Nonparental progeny resulting from protoplast fusion in *Trichoderma* in the absence of parasexuality. Exp Mycol 14:145–159.

Stasz TE, Harman GE, Gullino ML (1989): Limited vegetative compatibility following intra- and interspecific protoplast fusion in *Trichoderma*. Exp Mycol 13:364–371.

Stasz TE, Harman GE, Weeden NF (1988): Protoplast preparation and fusion in two biocontrol strains of *Trichoderma harzianum*. Mycologia 80:141–150.

Stasz TE, Nixon K, Harman GE, Weeden NF, Kuter GA (1989): Evaluation of phenetic species and phylogenetic relationships in the genus *Trichoderma* by cladistic analysis of isozyme polymorphism. Mycologia 81:391–403.

Stasz TE, Weeden NF, Harman GE (1988): Methods of isozyme electrophoresis for *Trichoderma* and *Gliocladium* species. Mycologia 80:870–874.

Toyama HY, Yamaguchi A, Shinmyo A, Okada H (1984): Protoplast fusion of *Trichoderma reesei* using immature conidia. Appl Environ Microbiol 47:363–368.

Turgeon BG, Garber RB, Yoder OC (1985): Transformation of the fungal maize pathogen *Cochliobolus heterostrophus* using the *Aspergillus nidulans amdS* gene. Mol Gen Genet 201:450–453.

14

FUNGAL "KILLER" TOXINS AS POTENTIAL AGENTS FOR BIOCONTROL

Y. KOLTIN
I. GINZBERG
A. FINKLER

Department of Molecular Microbiology and Biotechnology, Faculty of Life Sciences, Tel Aviv University, Tel Aviv 69978, Israel

1. INCIDENCE OF "KILLER" TOXINS IN FUNGI

A "killer" phenomenon in fungi, namely, the ability to secrete a toxic agent by some strains of a defined species that is toxic to sensitive individuals of the same species and related species, was initially reported in *Saccharomyces cerevisiae* by Bevan and Makower (1963). A similar phenomenon was earlier reported in bacteria and the secreted agents were referred to as colicins (Fredericq, 1956). The agents described were proteinaceous and generally displayed a narrow range of fungicidal activity. The capacity to secrete an agent that is toxic to sensitive strains of the same species is viewed generally as a factor with a competitive advantage. In fact, a variety of such agents have been known for some time in bacteria (Konisky, 1982). The selective advantage of similar agents in fungi is unknown but this class of agents may constitute an important factor in biocontrol, especially due to their specificity. The ecological effects of such agents are expected to be less severe than those of fungicides.

Information concerning the incidence of "killer" phenomena among the fungi is scarce and most of the available information is related only to yeast forms, both among the Ascomycetes and the Basidiomycetes (Philliskirk and Young, 1975; Stumm et al., 1977; Rogers and Bevan, 1978; Kandel and Stern, 1979;

Biotechnology in Plant Disease Control, pages 257–274
© *1993 Wiley-Liss, Inc.*

Polonelli and Morace, 1986; Puhalla, 1968; Koltin and Day, 1975; Gollubev, 1989; Golubev et al., 1988; Golubev and Churkina, 1990; Golubev and Kuznetsova, 1989). The identification of some of the sources of the toxic agents as being encoded by viruses and plasmids (Koltin and Leibowitz, 1988), along with the purification and characterization of *killer* toxins (Peery et al., 1987; Yamamoto et al., 1986; Savant et al., 1989; Pfeiffer and Radler, 1984) have raised interest in the biotechnological use of some of these agents (Hara et al., 1981; Young, 1981; Bussey et al., 1988; Radler and Knoll, 1988; Tokunaga et al., 1987; Vondrej, 1987; Suzuki et al., 1989; Salek et al., 1990; Cansado et al., 1991). The phenomenon is gradually being detected in a wide range of yeast forms encoded by dsRNA viruses (Schmitt and Tipper, 1990), a DNA plasmid (Gunge and Sakaguchi, 1981), and by nuclear genes (Zekohov et al., 1989; Goto et al., 1990). Yet, in most cases, the source of the information encoding the "killer" factors is not clear.

Although the extent of the *killer* phenomenon in the fungi is not known, the impression is that it is not rare among Ascomycetous yeast forms and among the Basidiomycetes in which the life cycle includes both a mycelial form and a yeast form such as *Ustilago maydis* (reviewed by Schultz et al., 1990). Within those species that are under intensive investigation, such as *S. cerevisiae* and *U. maydis,* the frequency of killers is probably no more than 1–5% in the natural population (Philliskerk and Young, 1975; Day and Dodds, 1979; Day, 1981) although it is significantly higher in some species such as *Hansenula*. Thus far, only one agent isolated from a filamentous form may fit into the category of *killer* factors. Recently, the toxic agent secreted by a strain of *Aspergillus giganteus* was purified and characterized as a small polypeptide of about 6.0 kD (Olson and Groener, 1965; Nakaya et al., 1990) that is toxic to related forms. It is unclear whether the *killer* phenomenon is very rare among filamentous forms, more difficult to detect, or had never received the proper attention. It is suspected that if this phenomenon does confer some selective advantage it should be of broader occurrence among the filamentous forms as well.

In consideration of the *killer* phenomenon for biocontrol, the effect of this phenomenon on pathogenic organisms is of importance. However, since the vast majority of plant pathogens are filamentous forms, contrary to fungal pathogens of mammals, the basic information for assessing the potential of the application of such a phenomenon for biocontrol is lacking. In only one pathogen, *U. maydis,* which is the causative agent of corn smut (Fig. 14.1), was a *killer* phenomenon reported, and at this time it is this system that can form the model for the application of this phenomenon for biocontrol (Nuss and Koltin, 1990). The detailed information derived on the expression of the *killer* toxins in the yeasts can serve as guidelines in the search for similar phenomena in pathogenic forms. Furthermore, survey of yeasts colonizing cactus fruit suggest that in natural populations *killer* phenotype does confer a selective advantage and *killers* can exclude nonkiller strains (Stramer et al., 1987).

Fig. 14.1. Neoplastic development of galls containing the teliospores of *U. maydis*.

2. SOURCE OF THE KILLER TOXIN

The genetic determinants encoding the *killer* toxins have been fully characterized in only a few cases. The mode of inheritance of the *killer* phenotype provides the minimal requirement to distinguish whether the determinants are nuclear
or cytoplasmic. The *killer* phenomenon has been described now in many forms
that lack sexuality and, therefore, non-Mendelian segregation cannot be used in
many cases as an indicator for cytoplasmic inheritance. Other means to distinguish between cytoplasmic determinants and nuclear genes that encode the toxins

can be used. A classical test is seen in the horizontal transmission of the *killer* phenotype within the species by cytoduction or by horizontal transmission to other related species. In some cases the transfer can be performed into species that are amenable to genetic analysis. In both cases, genetically marked strains are necessary to follow the transmission and to verify that in fact only cytoplasmic exchange had taken place and not the exchange of nuclei at the same time. The genetic criteria using formal crosses and cytoduction have been met only in a few cases, such as in *S. cerevisiae* (Mitchell et al., 1973) and *U. maydis* (Puhalla, 1968; Day and Anagnostakis, 1973; Wood and Bozarth, 1973; Koltin and Day, 1976) in which the cytoplasmic nature of the determinants encoding the *killer* factors was demonstrated. Recently, the failure to transmit a *killer* factor by cytoduction led to the detection of a new *killer* factor in *S. cerevisiae,* encoded by a nuclear gene (Goto et al., 1990). Intergeneric transmission was demonstrated with the *killer* phenomenon of the asexual yeast, *Kluyveromyces lactis,* by protoplast fusion (Gunge and Sakguchi, 1981), illegitimate mating (Sugisaki et al., 1985), and by direct transformation with the suspected determinants (Gunge et al., 1982). However, these procedures may encounter numerous obstacles in cases in which a nuclear–cytoplasmic interaction is required to maintain the cytoplasmic determinants; or, to prevent damage to the recipient due to the lack of any immunity to the toxic agents; and also, in those cases in which an internal circuit operates between cytoplasmic genetic elements including organelles (Wickner, 1977, 1989; Gunge, 1988).

The cytoplasmic elements known to encode *killer* toxins are either dsRNA viruses detected in *S. cerevisiae* and *U. maydis* or DNA linear plasmids found in *K. lactis.* In both *S. cerevisiae* and in *U. maydis* most strains contain dsRNA virus particles. The dsRNA contained within these particles encodes the components needed for encapsidation, transcription, and replication (Koltin, 1986; Welsh and Leibowitz, 1980; Bruenn et al., 1980; Icho and Wickner, 1989; Fujimura and Wickner, 1989; Esteban et al., 1989). Some strains secrete a *killer* toxin and these strains contain virus particles with a unique segment of dsRNA that is dependent for its replication and maintenance on the segment encoding the transcriptase and replicase. The toxin-encoding segment, of about 1.0–1.5 kb, is transcribed to a single transcript that encodes the precursor of the *killer* toxin (Bostian et al., 1984; Tao et al., 1990). The processed precursor yields two polypeptides that are secreted and jointly act as the *killer* toxin. The basic organization in the two diverse organisms is similar but the k1 toxin of yeast consists of two covalently linked subunits and the toxin KP6 of *U. maydis* is a binary toxin, consisting of two polypeptides (Peery et al., 1987).

Additional toxins encoded by dsRNA viruses of the same family are known in both *S. cerevisiae* and in *U. maydis* (Young and Yagiu, 1978; Pfeiffer and Radler, 1982; Extremera et al., 1982; Koltin and Day, 1976; Kandel and Koltin, 1978). Those toxins are defined as having a different specificity, since they are toxic to known killers, such as the k1 in *S. cerevisiae* and P6 of *U. maydis*. Furthermore, these toxins are encoded by unique segments of dsRNA that bear no homology to those segments encoding k1 toxin of *S. cerevisiae* and KP6 of *U. maydis* (Field et al., 1983). The basic organization of these toxins is not clear as yet, although some have been partially purified. The most recent toxin associated with a dsRNA virus is the K28 of *S. cerevisiae* that appears to be one glycopolypeptide (16 kD) that is somewhat larger than k1 (Pfeiffer and Radler, 1984). All others have not been purified and characterized to an extent that can allow a structural comparison with the k1 toxin. In *Ustilago* two additional toxins that are encoded by dsRNA (KP1 and KP4) have been identified and partially characterized (Koltin and Day, 1976; Kandel and Koltin, 1978; Shelbourn et al., 1988; Ganesa et al., 1989, 1991). These toxins appear as a single polypeptide in polyacrylamide gel. Recently, the KP4 toxin was characterized in more detail, including the determination of the N-terminal amino acid sequence. The information obtained suggests that it is a single polypeptide of 7.2 kD. Nucleic acid sequencing of the segments encoding the toxins will resolve whether some consist of one subunit or if others are binary as KP6.

Two DNA plasmids encoding a killer toxin were isolated thus far in fungi: one from *K. lactis* (Gunge and Sakaguchi, 1981) and the other from *P. acaciae* (Worsham and Bollen, 1990). The toxin of *K. lactis*, encoded by an 8.9-kb plasmid (pGLK1), is a part of an interdependent complex that includes a second plasmid pGKL2, (13.4 kb) encoding the functions necessary for maintenance of pGKL1 (Gunge and Sakaguchi, 1981). The toxin consists of two subunits encoded from two open reading frames (ORFs) (Stark and Boyd, 1986) and an immunity factor for this toxin is encoded by a separate ORF from the same plasmid (Tokunaga et al., 1987). The unique features of this system are its demonstrated intergeneric transmissability, which makes it attractive in terms of its biotechnological potential. The systems encoded by dsRNA viruses are thought to be restricted to specific species, but this impression is based primarily on the fact that infection by these viruses was impossible until the development of methods of transfection by cotransformation (El-Sherbeini and Bostian, 1987) and the cloning of the dsRNA. Recent studies with the KP6 toxin of *U. maydis* suggest that intergeneric transmission and expression as a cDNA is possible and the expression does not cause damage to the recipient (Tao et al., 1990). However, the expression is contingent on the fidelity of the translation mechanism and the mechanism of protein processing.

The *killer* toxin in *P. acaciae* appears to be associated with a DNA plasmid (7.3 kb). As in *K. lactis*, this system also consists of two components, a 7.3-kb plasmid related to the expression of the toxin and to the immunity to this toxin, and a second plasmid (13.6 kb) that appears to encode functions related to the maintenance of both plasmids. This system is distinct from the one studied in *K. lactis* both in the size of the plasmids and in the host range affected by this toxin.

This result is the first report of a plasmid-encoded *killer* system in *Pichia*. Since the *killer* phenomenon in this genus is not very rare, it is conceivable that some of those cases that were considered as nuclear-encoded toxins will also prove to be of a plasmid origin.

Among the yeasts of both Ascomycetes and Basidiomycetes the list of species in which a killer phenomenon was detected is increasing continuously. Most of the reports do not attribute clearly the source of the toxin to nuclear genes or to viruses and plasmids. These reports include the partially purified 83kD toxin from *P. anomala* (Savant et al., 1989), a 10.7-kD polypeptide of *Hansenula mrakii* (Yamamoto et al., 1986), the 8.5kD polypeptide of *H. saturnus* (Ohta et al., 1984), the 10–30kD mannoprotein of *Sporidiobolus pararoseus* (Golubev et al., 1988), the 18kD *killer* toxin of *Hanseniaspora uvarum* (Radler et al., 1990) and the 36kD broad spectrum glycoprotein from an unidentified *Candida* species (Yokomori et al., 1988). Additional reports suggestive of a *killer* protein have been documented but neither the nature of the toxin nor the source of the information encoding the toxin was determined. These include *Trichosporon capitatum*, a variety of species of *Hansenula* including *H. dimannae, H. canadensis, H. californica* (Morace et al., 1984), *Rhodotorula glutinis* (Golubev, 1989), *Torulopsis glabrata* (Middlebeek et al., 1980), *Candida malodendra, Pichia farinosa*, and *Debaromyces hansenii* (Suzuki et al., 1989), and *Kluyveromyces phaffii* (Rossini and Palpacelli, 1988). Some of the reports may be related to the expression of nuclear genes encoding *killer* toxins.

3. MODE OF ACTION

The fungal *killer* toxins were initially viewed as a group of polypeptides displaying a variety of biological activities similar to the bacterial colicins (Konisky, 1982). The reports included membrane-mediated killing by the k1 toxin of *S. cerevisiae* (de la Pena et al., 1981; Bussey et al., 1990), by the glycoprotein toxin of *Torulopsis glabrata* (Bussey and Skipper, 1975), and by the *killer* toxin of *Pichia kluyveri* (Kagan 1983), inhibition of adenylyl cyclase by the *K. lactis* toxin (Sugisaki et al., 1983) and nucleolytic activity by the *U. maydis* killer proteins (Levine et al., 1979). The killer toxin of *S. cerevisiae* K28 was found recently to inhibit DNA synthesis (Schmitt et al., 1989).

Genetic and biochemical studies of the interaction of k1 toxin with the sensitive cells indicate that this interaction involves a set of cell-surface specific interaction that include binding to the 1,6-β-D-glucan component of the cell wall (Hutchins and Bussey, 1983). Following the binding to the cell wall an energy-dependent process leads to the lethal effect. Ion leakage appears to be the primary effect. The exposure to the toxin causes a rapid inhibition of the proton pump, K^+, and amino acid uptake (de la Pena, 1980) leading to a reduced proton gradient across the plasma membrane accompanied by acidification of the cytoplasm and K$^+$ efflux. Leakage of small metabolites ensues including adenosine triphosphate (ATP). Similar effects were also noted with the killers of *T. glabrata* and *P. kluyveri* Middelbeek et al., 1980b). Based on these data it was proposed that these toxins perturb the membrane. Whether the effect is directly on a component of the proton pump or directly by the formation of transmembrane channels was unknown for some time. Using a planar phospholipid bilayer membrane Kagan (1983) has shown that the toxin of *P. kluyveri* forms *in vitro* ion permeable channels. Similarly, k1 toxin of *S. cerevisiae* was shown to form channels in isolectin liposomes in which the toxin was incorporated (Martinac et al., 1990). Furthermore, patch-clamping experiments have shown that k1 forms ion-channels *in vivo* in spheroplasts from sensitive cells of *S. cerevisiae*.

The mode of action of the *U. maydis* KP6 toxin was reexamined using affinity purified and high-performance liquid chromatography (HPLC) purified toxin. The nuclease activity that was found previously to be associated with the toxin is not displayed by the purified toxin (Ginzberg, 1990). Most of the information available suggests that the binary toxin affects the membranes of sensitive cells. Spheroplasts prepared from sensitive cells lyse following exposure to the toxin whereas those from resistant cells are unaffected. Both K$^+$ and Ca^{2+} can lead to rescue of the treated cell if provided shortly after exposure to the toxin. Although current data suggest that the mode of action is similar to the yeast k1, sequence data indicate that the two toxins are distinct and only some similarity can be found based on the hydrophobicity profile of these toxins.

As for the other toxins, the information related to the mode of action of *S. cerevisiae* K28 is very recent and appears to act as an inhibitor of DNA synthesis (Schmitt et al., 1989). The mode of action of the *K. lactis* toxin was reexamined and found not to inhibit the yeast adenylyl cyclase (White et al., 1989).

4. APPLICATION OF "KILLER" TOXIN FOR PLANT PROTECTION

To date the only model for the application of a *killer* toxin for plant protection can be suggested by the information accumulated on the virus-encoded toxin of *U. maydis*. As mentioned previously, most, if not all, strains of *U. maydis* contain dsRNA, some of which is encapsidated in virus particles of about 41 nm in diameter (Fig. 14.2) consisting primarily of one major coat protein. Other dsRNA molecules are not encapsidated (Seroussi et al., 1989). The encapsidated dsRNA consists of a unique pattern of segments ranging from 6.7 kb to 300 bp and the copy number of these segments ranges from 200 to 1000 (Fig. 14.3). The copy number of the unencapsidated segments is cell cycle dependent and the titer of some can decline to a point that is not discernible by conventional means. These segments are not associated with the *killer* system. The viral associated dsRNA

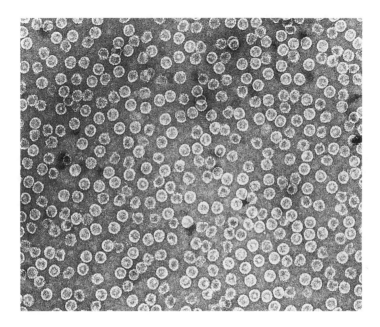

Fig. 14.2. Purified virus particles (41 nm) obtained from cells of *U. maydis*.

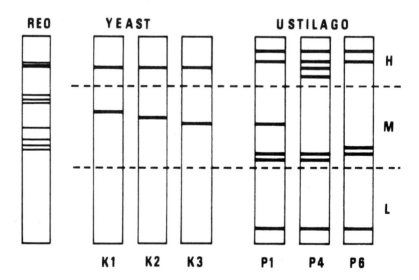

Fig. 14.3. From *left* to *right,* pattern of segmented dsRNA molecules from Reo virus type 3, from the *S. cerevisiae* viruses k1, k2, and k3 and the *U. maydis* viruses P1, P4, and P6.

segments can be detected throughout all growth phases although their replication occurs during log phase and the most active stage of transcription is in late log early stationary phase (Ben-Zvi et al., 1984).

Through a series of experimental procedures including formal genetic crosses, cytoplasmic transmission of virions, infection as a cotransformation (Finkler and Koltin, unpublished), nucleic acid characterization, and virus purification three virus subtypes, designated P1, P4, and P6, have been implicated in the *killer* phenomenon (reviewed by Koltin, 1988). The three virus subtypes are distinguished by a distinct pattern of dsRNA segments that are typical to each subtype (Fig. 14.3). In addition, all three subtypes secrete a toxin, but the fungal cells harboring a specific subtype are sensitive to the two toxins associated with the other subtypes. Three distinct specificities designated KP1, KP4, and KP6 are associated with these toxins (Fig. 14.4). The way in which the cells secreting the toxin are resistsant to this toxin is not completely clear. Nuclear genes conferring resistance to each of the toxins are known and in fact every strain secreting the toxin that was isolated from a natural population displayed resistance determined by a nuclear gene. The allele conferring resistance is recessive and it is suspected that this resistance is due to the loss of a cell wall or membrane receptor. In addition, for one of the toxins, KP1, resistance determined by a dsRNA molecule and referred

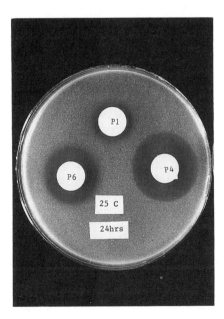

Fig. 14.4. Killer expression by cells containing viruses P1, P4, and P6. The lawn consists of cells from a sensitive strain of *U. maydis*.

to as immunity, was also detected (Peery et al., 1982). The immunity to k1 of *S. cerevisiae* is determined by a part of the pretoxin (reviewed by Bussey et al., 1990). However, in P6 of *U. maydis,* the best characterized of the three viruses, a similar mechanism does not function (Finkler et al., 1992). Nuclear resistance genes are not uncommon in the natural population of *U. maydis* but were not detected thus far in other Ustilaginales.

The viral system encoding the toxins consists of a complex set of dsRNA segments that are typical to each virus type. However, the complexity reflects the basic interactions that involve the maintenance of the different dsRNA segments, including the replication and encapsidation of these segments and the expression of the toxin. Genetic studies and *in vitro* translation and transcription studies (reviewed by Koltin, 1988) have shown that these functions reside on different segments of dsRNA and, as was shown with the P6 subtype, the system can be reduced to three basic segments that play a role in the maintenance and expression of the toxin. The largest segment encodes all the maintenance functions, and thus far involvement in maintenance by host genes was not detected; the medium size segment encodes the toxin and the smallest segment of 350 bps is homologous to the 3′ terminal end of the toxin coding segment (Field et al., 1983). Some of the other segments appear to be unnecessary for maintenance and toxin expression, although those are found encapsidated in virus particles, antigenically indistinguishable from those containing the toxin encoding information and the large segment that encodes the maintenance functions. These accessory segments are still considered a part of the P6 virus complex, since they differ from other dsRNA molecules that are found in *U. maydis* and are not encapsidated. In fact, these accessory segments, as well as the segment encoding the toxin and the 350 bp segment, are considered as satellites of the large segment that constitutes the prototype of the major virus genome with replication, transcription, and encapsidation capacity (Buck et al., 1984).

The viral information encoding the KP1 and KP4 toxins was also identified by similar procedures through the study of variants, both natural and induced, in which one or more of the dsRNA segments was lacking. In addition, through directed crosses, specific segments were exchanged and the *killer* toxin specificity was shown to be linked to a specific segment. One segment in each subtype encodes the toxin, although for an unknown reason the small 350 bp dsRNA segment, that is homologous to the 3′ terminal portion of the toxin-encoding segment, is always present in strains that secrete the toxin. It is not known whether this is an essential component for the expression or secretion of the toxin, but in recent studies complementary deoxyribonucleic acid (cDNA) clones of the viral information were expressed and secreted both in a heterologous system devoid of this segment and in strains of *U. maydis* that do not contain this segment. This result suggests that this segment is not crucial for expression (Tao et al., 1990). However, the expression in all these cases is less efficient than in virus-infected cells and many transformed strains do not express the cloned information (Finkler and Koltin, unpublished). It is not known at present whether efficient expression of the toxin is dependent in any way on the small segment.

The detailed organization, as determined by sequence analysis of the toxin-encoding segment, is known for one of the three virus subtypes (Tao et al., 1990), namely, for P6. The information indicates that this segment is 1200 bp with one ORF of 219 aas. The 3' end of the molecule has no ORF, and it is this part that is homologous to the small 350 bp segment that is always found in secreting strains. One mRNA transcript is transcribed from this segment and one polypeptide of about 27 kD is translated from this transcript. The active toxin is known to be constituted of two polypeptides, that are not covalently linked, and jointly constitute a binary toxin (Peery et al., 1987). The two components were purified and their terminal amino acids determined. A comparison of their known amino acid sequence and the predicted sequence from the cloned segment indicated, in fact, that a 27-kD protein can be encoded from the cloned segment and that this polypeptide contains all the landmarks, as a secreted protein with a signal sequence and as a precursor for smaller polypeptides, since within this precursor known protease clevage sequences were identified. These suspected processing sites are precisely at points that are expected to yield two polypeptides with the exact amino termini as determined experimentally with the purified polypeptides. Furthermore, comparison of the actual cyanogen bromide cleavage sites in the polypeptides with those predicted based on the sequence added another dimension to the confirmation that, in fact, the cloned segment is the correct segment, that a preprotoxin is encoded by the virus, and that the active toxin is a processed product of this 27-kD polypeptide. Full confirmation was obtained by a biological and immunological test of the cloned segment in an inducible expression vector in *S. cerevisiae*. Cells of *S. cerevisiae* expressed a toxin effective against cells of *U. maydis* that were sensitive to KP6 and ineffective against cells resistant to KP6. The secreted toxin was immunologically cross reactive with the antibodies known to react with one of the two components of the binary toxin. The results indicate that the binary toxin consists of two distinct polypeptides consisting of 79 and 81 amino acids, designated α and β, and both are needed to kill the cell (Fig. 14.5). Stoichiometric estimations suggest that the interaction with the cell requires a ratio of 1:5 of α to β. The polypeptides are very stable in a wide range of pH and the toxin activity is retained within a range of pH 5–9. The abundance of cysteine in the molecules may offer this high degree of stability. The molecules contain intramolecular S-S bonds but not intermolecular S-S bonds (Peery et al., 1987).

The toxin interacts with the target cell at two levels. Both subunits bind to the cell wall. The α polypeptide appears to bind with a somewhat higher affinity than the β polypeptide, since the binding of this subunit appears to be irreversible. The β subunit also binds to the cell walls but at a somewhat slower rate. The

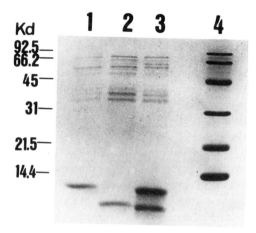

Fig. 14.5. From *left* to *right,* each of the purified subunits of the *U. maydis* KP6 toxin as expressed by mutants, the wild-type KP6 toxin, and molecular weight markers.

binding of both subunits is rapid and the cells cannot be rescued after 15 min of exposure. Both α and β polypeptides are required to affect cells and spheroplasts. Whereas intact cells of both resistant and sensitive strains bind the toxin, spheroplasts of resistant mutants are unaffected by the toxin. Since resistance is a recessive trait it appears that the loss of some component, such as receptors, lead to the refractory effect of the spheroplasts. Furthermore, the toxin is highly specific for the Ustilaginales. All tests with various fungal specimens, both yeast and filamentous forms, as well as with bacterial forms have indicated a high degree of specificity for various forms within the Ustilaginales (Fig. 14.6) (Koltin and Day, 1975; Koltin, 1986). None of the specimens, tested including a collection of marine yeast forms of Basidiomycetes, responded to the toxins of *U. maydis*. Therefore, it appears that a specific membrane receptor may be required for the interaction with the toxin. The polysaccharide preference in binding is not known as yet.

The sensitivity of a broad range of Ustilaginales to the toxin, including such species as *U. hordei, U. nigra, U. tritici, U. avenae,* and others, and the absence of resistance within these species, as known thus far, suggest the use of the viral information as a potential source of resistance against each of the pathogens. However, numerous problems remain to be resolved. It is unknown to what extent the plant cells can tolerate the expression of such a toxin. Just because the cells are insensitive when they are exposed to the toxin says nothing about their response once these proteins are expressed by these cells. It is unknown whether the plants will utilize the signal sequence to secrete the toxin or whether it is preferable to

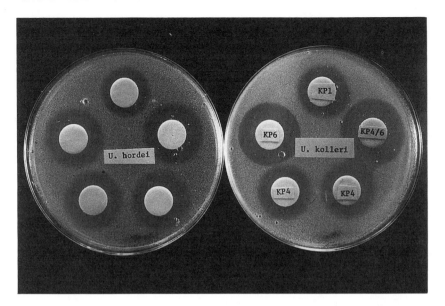

Fig. 14.6. Sensitivity of *U. kolleri* and *U. hordei* to the KP6 toxin of *U. maydis*.

delete the signal sequence and ensure the retention of the polypeptides. How the plant cells will manage such a foreign protein is unknown. It may be sequestered in a form that will not be accessible for inhibition of the fungal development. The affinity to different polysaccharides had not been characterized sufficiently as yet. However, it is known that the toxin binds efficiently to cellulose and, therefore, only very efficient expression of the toxin may be necessary for effective protection.

The fungal processing mechanism utilizes recognition sequences, such as the Lys Arg sequence at the junctions. It is not known whether these signals can be recognized by the plant processing mechanism or if new junctions will have to be implanted if the entire precursor is expressed in the plant cells. Although this problem can be circumvented by treating each subunit as an individual gene, other alternatives, such as gene fusions of α and β, have not been examined as yet. Finally, the prospect for the use of the viral information is primarily in the protection of wheat, barley, and oats, since within the pathogen of corn, resistance is already not uncommon in wild populations. Even though vectors for transformation are available for corn (Gordon-Kamm et al., 1990), a way to introduce DNA into other grain crops is still unavailable.

The advent of molecular genetics opens new horizons, which allow us to attempt to use heterologous genes as new sources of resistance to plant pathogens. With years of emphasis on plant breeding based on genetic reservoirs, the use of heterologous genes appears as a logical extension, and in some instances as the only resource beyond chemical protection. However, this exercise, which appears preferable to the use of chemical means for plant protection, may be no better than conventional breeding and the development of new fungicides. The use these new sources of resistance will clearly create some selective pressure. With conventional breeding and the development of new fungicides one has learned the time frame in which one can anticipate that the specific approach will provide protection. But, with the use of heterologous genes the lesson is yet to be learned. Since resistance to the toxin does not appear to confer a disadvantage under laboratory conditions, the same may be true in the field. However, it is molecular genetics and recombinant DNA technology that has opened the way for the use of heterologous genes. The same basic tools may be used to circumvent resistance mechanisms by genetic engineering of the cloned sequences.

REFERENCES

Bevan EA, Makower M (1963): The physiological basis of the killer character in yeast. 21st International Congress on Genetics. Vol. 1, pp 202–203.

Ben-Zvi BS, Koltin Y, Mevarech M, Tamarkin A (1984): RNA polymerase activity in virions from *Ustilago maydis*. Mol Cell Biol 4:188–194.

Bostian KA , Elliott Q, Bussey H, Burn V, Smith A, Tipper DJ (1984): Sequence of the pretoxin dsRNA gene of type 1 killer yeast: multiple processing events produce a two-component toxin. Cell 36:741–751.

Bruenn J, Bobek L, Brennan V, Held W (1980): Yeast viral RNA polymerase is a transcriptase. Nucleic Acids Res 8:2985–2997.

Buck KW, Ackermann HW, Bozarth RF, Bruenn JA, Koltin Y, Rawlinson CJ, Ushiyama R, Wood HA (1984): Six groups of double-stranded RNA mycoviruses. Intervirology 22:17–23.

Bussey H, Skipper N (1975): Membrane-mediated killing of *Saccharomyces cerevisiae* by glycoproteins from *Torulopsis glabrata*. J Bacteriol 124:476–483.

Bussey H, Vernet T, Sdicu A-M (1988): Mutual antagonism among killer yeasts: competition between K1 and K2 killers and a novel cDNA-based K1–K2 killer strain of *Saccharomyces cerevisiae*. Can J Microbiol 34:38–44.

Bussey HC, Boone C, Zhu H, Vernet T, Whiteway M, Thomas DY (1990): Genetic and molecular approaches to synthesis and action of the yeast killer toxin. Experientia 46:193–200.

Cansado J, Longo E, Calo P, Sieiro C, Velazquez JB, Villa TG (1991): Role of killer character in spontaneous fermentations from NW Spain—Ecology, distribution and significance. Appl Microbiol Biotechnol 34:643–647.

Day PR (1981): Fungal virus populations in corn smut from Connecticut. Mycologia 73:379–391.

Day PR, Anagnostakis SL (1971): Corn smut dikaryon in culture. Nature New Biol 231:19–20.

Day PR, Anagnostakis SL (1973): The killer system in *Ustilago maydis*: heterokaryon transfer and loss of determinants. Phytopathology 63:1017–1018.

Day PR, Dodds JA (1979): Viruses of plant pathogenic fungi in Viruses and Plasmids in Fungi. Lemke PA (ed): New York: Marcel Dekker, pp 202–235.

de la Pena P, Barros F, Gascón S, Laso PS, Ramos S (1981): Effect of yeast killer toxin on sensitive cells of *S. cerevisiae*. J Biol Chem 256:10420–10425.

de la Pena P, Barros F, Gascón S, Ramos S, Lazo PS (1980): Primary effects of the yeast killer toxin. Biochem Biophys Res Commun 96:544–550.

El-Sherbeini M, Bostian KA (1987): Viruses in fungi: infection of yeast with the K1 and K2 killer virus. Proc Natl Acad Sci USA 84:4293–4297.

Esteban R, Fujimura T, Wickner RB (1989): Internal and terminal cis-acting sites are necessary for *in vitro* replication of the L-A double-stranded RNA virus of yeast. EMBO J 8:947–954.

Extremera AL, Martin I, Montoya E (1982): A new killer toxin produced by *Saccharomyces cerevisiae*. Curr Genet 5:17–19.

Field LJ, Bruenn JA, Chang TH, Pinchasi O, Koltin Y. (1983): The *Ustilago maydis* viral dsRNAs of different size code for the same product. Nucleic Acids Res 11:2765–2778.

Finkler A, Perry T, Tao J, Bruenn J, Koltin Y (1992): Immunity and resistance to the KP6 toxin of *Ustilago maydis*. Molec Gen Genetics (in press).

Fredericq P. (1956): Resistance st immunité aux colicine. C R Seances Soc Biol Paris 150:1514–1517.

Fujimura T, Wickner RB (1989): Reconstitution of template dependent *in vitro* transcriptase activity of a yeast double-stranded RNA virus. J Biol Chem 264:10872–10877.

Ganesa C, Chang Y-J, Flurkey WH, Randhawa ZI, Bozarth RF (1989): Purification and molecular properties of the toxin coded by *Ustilago maydis* virus P4. Biochem Biophys Res Commun 162:651–657.

Ganesa C, Flurkey WH, Rondhawa ZI, Bozarth RF (1991): *Ustilago maydis* virus P4 killer toxin: characterization, partial amino terminus sequence, and evidence for glycosylation. Arch Biochem Biophys 286:195–200.

Ginzberg I (1990): The *Ustilago maydis* killer system: The organization of the toxin encoding genes and characterization of their products. Ph.D. Dissertation, Tel Aviv University.

Golubev WI (1989): The action spectrum of killer toxins produced by *Rhodotorula glutinis* and its taxonomic significance. Mikrobiologia 57:99–104.

Golubev WI, Churkina LG (1990): Wide distribution of killer strains in *Rhodotorula mucilaginosa* yeast. Ser Biol 6:854–861.

Golubev WI, Kuznetsova LB (1989): Formation and spectrum of action of mycocins of Basidiomycete yeast *Cryptococcus laurentii* (Kufferath) Skinner. Mikrobiologia 58:980–984.

Golubev WI, Tsiomenko AB, Tickomirova LP (1988): Plasmid-free killer strains of the yeast *Sporidiobolus pararoseus*. Mikrobiologia 57:805–810.

Gordon-Kamm WJ, Spencer TM, Mangano ML, Adams TR, Daines RJ, Start WG, O'Brien JV, Chambers SA, Adams Jr WR, Willets NG, Rice TB, Mackey CJ, Krueger RW, Kausch AP, Lemaux PG (1990): Transformation of maize cells and regeneration of fertile transgenic plants. Plant Cell 2:603–618.

Goto K, Iwase T, Kichise K, Kitano K, Totuka A, Obata T, Hara S (1990): Isolation and properties of a chromosome-dependent KHR killer toxin in *Saccharomyces cerevisiae*. Agric Biol Chem 54:505–509.

Gunge N (1988): *Kluyveromyces* linear DNA plasmids. In Koltin Y, Leibowitz J (eds): Viruses of Fungi and Simple Eukaryotes. New York: Marcel Dekker, pp 265–282.

Gunge N, Murata K, Sakaguchi K (1982): Transformation of *Saccharomyces cerevisiae* with linear DNA killer plasmids from *Kluyveromyces lactis*. J Bacteriol 151:462–464.

Gunge N, Sakaguchi K (1981): Intergeneric transfer of deoxyribonucleic acid killer plasmids pGKL1 and pGKL2 from *Kluyveromyces lactis* by cell fusion. J Bacteriol 147:155–160.

Hara S, Iimura Y, Oyama H, Kozeki T, Kitano K, Otsuka K (1981): The breeding of cryophilic killer wine yeasts. Agric Biol Chem 45:1327–1334.

Hutchins K, Bussey H (1983): Cell wall receptor for yeast killer toxin: involvement of (1,6)-β-D-glucan. J Bacteriol 154:161–169.

Icho T, Wickner RB (1989): The double-stranded RNA genome of yeast virus L-A encodes its own putative RNA polymerase by fusing two open reading frames. J Biol Chem 264:6716–6723.

Kagan BL (1983): Model of action of yeast killer toxins: Channel formation in lipid bilayer membranes. Nature (London) 302:709–711.

Kandel J, Koltin Y (1978): Killer phenomenon in *Ustilago maydis*: comparison of the killer proteins. Exp Mycol 2:270–278.

Kandel JS, Stern TA (1979): Killer phenomenon in pathogenic yeast. Antimicrob Agents Chemother 15:568–571.

Koltin Y (1986): The killer systems of *Ustilago maydis*. In Buck KW (ed): Fungal Virology. Boca Raton: CRC Press, pp 109–143.

Koltin Y (1988): The killer systems of *Ustilago maydis*: Secreted polypeptides encoded by viruses. In Koltin Y, Leibowitz J (eds): Viruses of Fungi and Simple Eukaryotes, New York: Marcel Dekker, p 209–242.

Koltin Y, Day PR (1975): Specificity of *Ustilago maydis* killer proteins. Appl Microbiol 30:694–696.

Koltin Y, Day PR (1976): Inheritance of killer phenotype and double-stranded RNA in *Ustilago maydis*. Proc Natl Acad Sci USA 73:594–598.

Koltin Y, Leibowitz M (1988): Viruses of Fungi and Simple Eukaryotes. New York: Marcel Dekker, pp 434.

Konisky J (1982): Colicins and other bacteriocins with established modes of action. Annu Rev Microbiol 36:125–144.

Levine R, Koltin Y, Kandel JS (1979): Nuclease activity associated with the *Ustilago maydis* virus-induced killer proteins. Nucleic Acids Res 6:3717–3731.

Martinac B, Zhu H, Kubalski A, Zhou X, Culbertson M, Bussey H, Kung C (1990): Yeast k1 killer toxin forms ion channels in sensitive yeast spheroplasts and in artificial liposomes. Proc Natl Acad Sci USA 87:6228–6232.

Middlebeek EJ, Hermans JMH, Stumm C, Mutjens HL (1980a): High incidence of sensitivity to yeast killer toxins among *Candida* and *Torulopsis* isolates of human origin. Antimicrob Agents Chemother 17:350–354.

Middelbeek EJ, Stumm C, Vogels GD (1980b): Effects of a *Pichia kluyveri* killer toxin on sensitive cells. Anto van Leeuwen 46:205–220.

Mitchell DJ, Bevan EA, Herring AJ (1973): The correlation between dsRNA in yeast and the killer character. Heredity 31:133–134.

Morace G, Archibusacci M, Sestila M, Polonelli L (1984): Strain differentiation of pathogenic yeasts by the killer system. Mycopathologia 84:81–85.

Nakaya K, Omata K, Okahashi I, Nakamura Y, Kolkenbrock H, Ullrich N (1990): Amino acid sequence and disulphide bridges of an antifungal protein isolated from *Aspergillus giganteus*. Eur J Biochem 193:31–38.

Nuss DL, Koltin Y (1990): Significance of dsRNA genetic elements in plant pathogenic fungi. Annu Rev Phytopathol 28:37–58.

Ohta Y, Tsukada Y, Sugimori T (1984): Production, purification and characterization of HYI, an anti-yeast substance produced by *Hansenula saturnus*. Agric Biol Chem 48:903–908.

Olson BM, Groener GL (1965): Alpha sarcin, a new antitumor agent. I. Appl Microbiol 13:314–321.

Peery T, Koltin Y, Tamarkin A (1982): Mapping the immunity function of the *Ustilago maydis* P1 virus. Plasmid 7:52–58.

Peery T, Shabat-Brand T, Steinlauf R, Koltin Y, Bruenn J (1987): Virus-encoded toxin of *Ustilago maydis*: Two polypeptides are essential for activity. Mol Cell Biol 7:470–477.

Pfeiffer P, Radler R (1982): Purification and characterization of extracellular and intracellular killer toxin of *Saccharomyces cerevisiae* strain 28. J Gen Microbiol 128:2699–2706.

Pfeiffer P, Radler R (1984): Comparison of the killer toxin of several yeasts and the purification of a toxin of type K2. Arch Microbiol 137:357–361.

Philliskirk G, Young TW (1975): The occurrence of killer character in yeasts of various genera. Antonie van Leeuwenhoek. J Microbiol 41:147–151.

Polonelli L, Morace G (1986): Reevaluation of the yeast killer phenomenon. J Clin Microbiol 24:866–869.

Puhalla JE (1968): Compatibility reactions on solid medium and interstrain inhibition in *Ustilago maydis*. Genetics 60:461–474.

Radler F, Knoll C (1988): Formation of killer toxin by apiculate yeasts and interference with fermentation. Vitis 27:111–132.

Radler F, Schmitt MJ, Meyer B (1990): Killer toxin of *Hanseniaspora uvarum*. Arch Microbiol 154:175–178.

Rossini G, Palpacelli V (1988): Effect of freezing and substratum composition on the production and the stability of the killer toxin secreted by *Kluyveromyces phaffii*. Ann Fac Agrar Univ Stud Perugia 42:473–478.

Rogers D, Bevan EA (1978): Group classification of killer yeasts based on cross-reactions between strains of different species and origin. J Gen Microbiol 105:199–202.

Salek A, Schnettler R, Zimmermann U (1990): Transmission of killer activity into laboratory and industrial strains of *Saccharomyces cerevisiae* by electroinjection. FEMS Microbiol Lett 70:67–72.

Savant AD, Abdelar AT, Ahearn DG (1989): Purification and characterization of the anti-*Candida* toxin of *Pichia anomala* WC65. Antimicrobiol Agents Chemother 33:48–52.

Schmitt M, Brendel M, Schwartz R, Radler F (1989): Inhibition of DNA synthesis in *Saccharomyces cerevisiae* by yeast killer toxin K28. J Gen Microbiol 135:1529–1535.

Schmitt MJ, Tipper DJ (1990): K28, a unique double-stranded RNA killer virus of *Saccharomyces cerevisiae*. Mol Cell Biol 10:4807–4815.

Schultz B, Bannuett F, Dahl M, Schleisinger R, Schafer W, Martin J, Herskowitz I, Kahmann R (1990): The b alleles of *U. maydis,* whose combinations program pathogenic development, code for polypeptides containing homeodomain-related motif. Cell 60:295–306.

Seroussi E, Peery T, Ginzberg I, Koltin Y (1989): Detection of killer-independent ds-RNA plasmids in *Ustilago maydis* by a simple method of extraction of dsRNA. Plasmid 21:216–225.

Shelbourn SL, Day PR, Buck KW (1988): Relationships and function of virus double-stranded RNA in a P4 killer strain of *Ustilago maydis*. J Gen Virol 69:975–982.

Stark MJR, Boyd A (1986): The killer toxin of *Kluyveromyces lactis*: characterization of the toxin subunits and identification of the genes which encode them. EMBO J 5:1995–2002.

Stramer WT, Ganter P, Aberdeen V, Lachance MA, Phaff HJ (1987): The ecological role of killer yeasts in natural communities of yeasts. Can J Microbiol 33:783–796.

Stumm C, Hermans JHM, Middelbeek EJ, Croes AF, de Vries GJML (1977): Killer sensitive relationships in natural habitats. Antonie van Leeuwenhoek. J Microbiol 41:125–128.

Sugisaki Y, Gunge N, Sagakuchi K, Yamasaki M, Tamura G (1983): *Kluyveromyces lactis* killer toxin inhibitor adenylate cyclase of sensitive yeast. Nature (London) 304:464–466.

Sugisaki Y, Gunge N, Sagakuchi K, Yamasaki M, Tamura G (1985): Transfer of DNA killer plasmids from *Kluyveromyces lactis* to *Kluyveromyces fragilis* and *Candida pseudotropicalis*. J Bacteriol 164:1373–1375.

Suzuki C, Yamada K, Okada N, Nikkuni S (1989): Isolation and characterization of halotolerant killer yeasts for fermented foods. Agric Biol Chem 53:2593–2597.

Tao J, Ginsberg I, Banerjee N, Held W, Koltin Y, Bruenn JA (1990): *Ustilago maydis* KP6 killer toxin: Structure, expression in *Saccharomyces cerevisiae,* and relationship to other cellular toxins. Mol Cell Biol 10:1373–1381.

Tokunaga T, Wada N, Hishinuma F (1987): Expression and identification of immunity determinants on linear DNA killer plasmids pGKL1 and pGKL2 in *Kluyveromyces lactis*. Nucleic Acids Res 15:1031–1046.

Vondrej A (1987): A killer system in yeasts: Applications to genetics and industry. Microbiol Sci 4:313–316.

Welsh DJ, Leibowitz MJ (1980): Transcription of killer virion double-stranded RNA in vitro. Nucleic Acids Res 8:2365–2375.

White JH, Butler AR, Stark MJR (1989): *Kluveromyces lactis* toxin does not inhibit yeast adenylyl cyclase. Nature (London) 341:666–668.

Wickner RB (1977): Deletion of mitochondrial DNA bypassing of chromosomal gene needed for maintenance of the killer plasmid of yeast. Genetics 87:441–452.

Wickner RB (1989): Yeast virology. FASEB J 3:2257–2265.

Wood HA, Bozarth RF (1973): Heterokaryon transfer of virus-like particles associated with a cyto-plasmically inherited determinant in *Ustilago maydis*. Phytopathology 63:1019–1021.

Worsham PL, Bollen PL (1990): Killer toxin production in *Pichia acacia* is associated with linear DNA plasmids. Curr Genet 18:77–80.

Yamamoto T, Imai M, Tochibana K, Mayumi M (1986): Application of monoclonal antibodies to the isolation and characterization of a killer toxin secreted by *Hansenulla mrakii*. FEBS Lett 195: 253–257.

Yokomori Y, Akiyama H, Shimuzu K (1988): Toxins of a wild *Candida* killer yeast with a novel killer property. Agric Biol Chem 52:2797–2801.

Young TW (1981): The genetic manipulation of killer character into brewing yeast. J Inst Brew 87:292–295.

Young TW, Yagiu M (1978): A comparison of the killer character in different yeasts and its classifi-cation. Antonie van Leeuwenhoek, J Microbiol Serol 44:59–77.

Zekhov AM, Soom YO, Nesterova GF (1989): New type of killer activity of the yeast *Saccharomyces cerevisiae* and its mode of inheritance. Genetika 35:1365–1372.

Zhu H, Bussey H (1991): Mutational analysis of the functional domains of yeast k1 killer toxin. Mol Cell Biol 11:175–181.

15

BIOLOGICAL CONTROL OF PLANT PESTS AND PATHOGENS: ALTERNATIVE APPROACHES

JENNIFER A. THOMSON

Department of Microbiology, University of Cape Town, Private Bag Rondebosch, 7700 South Africa

1. INTRODUCTION

There is a worldwide swing to the use of ecologically safe, ''environmentally friendly'' methods of protecting crops from pests and pathogens. The use of potentially harmful chemical sprays and pesticides is viewed with disfavor in many countries, and some compounds have even been deregistered. Biological control methods using naturally occurring bacteria, fungi, or viruses have in the past received limited approval from users as their ability to protect plants has often been inferior to results obtained by chemical means. However, this situation can now be changed with the use of recombinant DNA technology by means of which scientists are able to design organisms with the desired activities.

The most effective way of achieving crop protection is to introduce the desired resistance into the plant by the addition of one or more genes. For instance, viral resistant plants have been obtained by the introduction of the relevant coat protein genes (Nejidat and Beachy, 1990; Stark and Beachy, 1989) or antisense RNA (Powell et al., 1989), insect resistance by the introduction of a *Bacillus thuringiensis* toxin gene (Perlak et al., 1990; Vaeck et al., 1987) or the cowpea

Biotechnology in Plant Disease Control, pages 275–290
© 1993 Wiley-Liss, Inc.

trypsin inhibitor (Hilder et al., 1989), and fungal resistance by the introduction of a plant or bacterial chitinase gene (Neuhaus et al., 1991; Logemann, Jach, Lipphardt, Leah, Mundy, Oppenheim, Chet, and Schell, unpublished).

This approach has a few drawbacks however. First, not all commercially important plant cultivars are susceptible to transformation. The genes of interest have to be introduced into a transformable variety and then crossed into the desired plant. This process can lead to the loss of other important traits. Although it is likely that the difficulties encountered in transforming plants that are presently refractory to the technique will be overcome in time, at the moment they can present problems. Second, every cultivar has to be made individually transgenic, which is a major undertaking in terms of time and expense. Third, codon usage in plants is very different from that in bacteria (Murray et al., 1989; Konigsberg and Godson, 1983). Therefore, if the gene is of bacterial origin, such as the *B. thuringiensis* toxin gene, it is unlikely to be expressed efficiently in plants. Indeed, Monsanto C. was forced to synthesize a totally redesigned toxin gene conforming to plant codon usage before they achieved acceptable insect protection in transgenic plants (Perlak et al., 1991). Finally, if the gene is expressed in a part of the plant that is consumed by humans or animals, it will have to be disclosed to the consumer. Public reaction to, for instance, a tomato carrying a *B. thuringiensis* toxin gene is unlikely to be favorable without a massive publicity campaign explaining that the toxin is harmless to all living creatures apart from a few herbivorous insects. Therefore, although this approach will undoubtedly be the most effective in the long term, other approaches are also needed.

This chapter discusses the development of plant-associated bacteria as delivery organisms of gene products that can protect plants against pests and pathogens. There are three main approaches to the development of such biocontrol agents. The first is the isolation of naturally occurring bacteria with disease suppressing capabilities. Such strains are, however, rare. One of the best examples is *Agrobacterium radiobacter* strain 84, which was isolated by New and Kerr in 1972. This strain has proven itself over the years to be an effective biocontrol strain against crown gall disease caused by *Agrobacterium tumefaciens* (Shim et al., 1987).

The second approach is to isolate a strain with some protective capabilities but to improve it by genetic engineering. An example of this is the development of *A. tumefaciens* strain J73, which also protects against crown gall disease (Webster et al., 1986) but required some genetic modification before it was safe to use. Another example is the deletion of the ice-nucleation gene from *Pseudomonas syringae*. The resultant bacteria, when sprayed onto plants, compete with the wild-type bacteria, which cause frost damage to crops (Lindeman and Suslow, 1987).

The third approach is to design a biocontrol strain *ab initio* by inserting into a bacterium, which is capable of harmlessly colonizing a particular plant, a gene that codes for a toxin active against a pest or pathogen. A variation of this has been pioneered by Mycogen who kill their genetically modified bacteria prior to use (Twombly, 1990).

In all these approaches two major features of the controlling strain are important. First, it must produce sufficient amounts of the toxin to effectively kill the target and second, it must colonize the relevant plant efficiently. This chapter discusses ways in which both these important characteristics can be improved by genetic engineering.

2. NATURALLY OCCURRING BIOLOGICAL CONTROL STRAINS

Agrobacterium radiobacter strain 84 protects plants against crown gall produced by strains of *A. tumefaciens* carrying a Ti plasmid of the nopaline or agrocinopine type (Engler et al., 1975). It does so by producing an adenine nucleotide derivative, agrocin 84, which is specifically taken up by strains carrying the relevant types of Ti plasmid (Tate et al., 1979; Holsters et al., 1980). Once inside the cells the agrocin effectively inhibits DNA synthesis and the cells die (Das et al., 1978). This apparently anomalous suicidal behavior of sensitive strains was explained by Ellis and Murphy (1981) who showed that the agrocin "pirates" an existing uptake system for a group of sugar phosphodiesters, the agrocinopines, which are the "legitimate" substrates for the agrocin permease.

In addition to the effectiveness of the agrocin, strain 84 colonizes plants efficiently. A comparison was made between strains 84 and J73. Both produce agrocins active against *A. tumefaciens* carrying nopaline Ti plasmids, but only strain 84 protects tomato plants against crown gall. The ability of strain 84 to colonize the plants was shown to be superior to that of J73 over a period of 65 days (Fig. 15.1; MacRae et al., 1988). This colonization efficiency was found to occur in the rhizosphere, the rhizoplane, and on the stems of the plants tested.

The importance of a biocontrol strain being able to successfully colonize plants was supported by the findings of Shim et al. (1987). They transferred the agrocin 84 plasmid to strain C58 and found that the transconjugants were less effective at controlling tumor formation. They carried out colonization studies on almond seedlings and found the strain to be a poor colonizer.

Strain 84 has a somewhat limited host range in that strains of *A. tumefaciens* carrying other types of Ti plasmids that do not encode the specific permease are resistant. In addition, the biotype 3 strains that have a narrow host range restricted to grapevines are resistant to agrocin 84 (Kerr and Panagopoulos, 1977). Scientists in many countries have therefore been screening strains for new biocontrol capabilities. Panagopoulos et al. (1978) tested 11 avirulent biotype 1 and 2 strains that formed bacteriocins active *in vitro* against pathogenic biotype 3 strains for control. However, none proved effective, and they concluded that the isolation

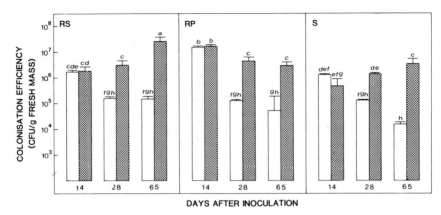

Fig. 15.1. In vivo colonization of tomato seedlings by J73 (☐) and K84 (▨). The bars above the histograms indicate standard deviations, and the letters indicate statistically significant differences; bars bearing the same letter indicate that there is no statistical difference between those values. RS, Rhizosphere; RP, rhizoplane; S, stem. Reprinted with permission from MacRae et al., 1988.

of a strain similar to K84 but inhibiting biotype 3 strains was unlikely. In fact, they considered the prospect of biological control of crown gall on grapevines an unlikely occurrence.

In support of this result, it was found that although the agrocin produced by J73 inhibits biotype 3 strains *in vitro* (Webster and Thomson, 1986), it does not control the disease on grapevines (Table 15.1; MacRae, Webster, Whillier, and Thomson, unpublished). This result may occur because the agrocin does not col-

TABLE 15.1. Biological Control on Grapevines

Biocontrol Strain	Number of Grapevines with Galls (Gall Index[a]) after Inoculation with Grapevine Pathogens			
	At2	A2/1	G2/4	None
None	3 (12)	3 (12)	2 (8)	0
K84	2 (6)	2 (6)	2 (8)	0
J73	3 (12)	3 (9)	2 (6)	0
H6	0	0	0	0
H6D17	0	0	0	0

[a]Gall index: number of plants with galls × average gall size. Gall size scored from 1 to 4. 1 = 1–5-mm diameter; 2 = 5–10-mm diameter; 3 = 10–15-mm diameter; 4 = >15-mm diameter.

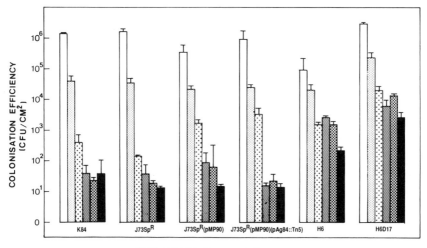

Fig. 15.2. Colonization of the grapevine rhizosphere by agrocin-producing strains of A. tumefaciens. Days after inoculation:☐ , 0; ▦ , 30; ▨ , 64; ▨ , 120; ▨ , 191; ■ , 248. (MacRae, Webster, Whillier, and Thomson, unpublished.)

onize grapevines efficiently (Fig. 15.2; MacRae, Webster, Whillier, and Thomson unpublished). Interestingly enough, strain K84 is also a poor colonizer of grapevines.

These findings emphasize the fact that bacteria, which both produce toxic compounds and colonize plants efficiently, are rare. Indeed it has been stated ''Few if any biological control measures have ever been tested and used to the extent that *Agrobacterium radiobacter* strain K84 has been used for control of crown gall. . . . It is remarkable that a single strain of bacteria could be so effective in such diverse geographic habitats.'' (Moore and Warren, 1979). It is a rare strain indeed.

Staphorst et al. (1985) used an alternative approach for the isolation of a biocontrol strain for grapevine pathogens. They isolated Agrobacterium strains from gall tissue of grapevines and identified the nonpathogenic biotype 3 strains that showed strong *in vitro* agrocinogenic activity against pathogenic biotype 3 strains. They found that some of them successfully reduced crown gall formation on grapevines in greenhouse experiments. We tested one of these strains, H6, further and found very good control (Table 15.1; MacRae, Webster, Whillier, and Thomson, unpublished). A related strain H6D17 also controlled the disease, but the latter colonized grapevines more efficiently (Fig. 15.2; MacRae, Webster, Whillier, and Thomson, unpublished). It may therefore have more potential for long-term biocontrol.

It is clear that nature is unlikely to provide us with many ready-made biocontrol agents. The next approach is, therefore, to find a potential biocontrol strain and improve it.

3. GENETIC IMPROVEMENT OF BIOCONTROL STRAINS

3.1. Protection Against Crown Gall Disease

Although strain K84 has been used commercially for the control of crown gall since 1973 without mishap, a report from Greece indicated that the plasmid encoding the agrocin, pAgK84, could have been transferred to a pathogenic strain of *A. tumefaciens*. Such strains would no longer be subject to biological control. To overcome this problem Jones et al. (1988) deleted the region on the agrocin-encoding plasmid that is responsible for conjugal transfer. The new strain, K1026, is identical to strain K84 except for a 5.9-kb deletion of pAgK84 covering the Tra region.

Strain K1026 was subjected to field testing in Australia and found to be safe and effective. It is now marketed under the trade name No Gall® and, thus, is the first genetically engineered organism to be used as a pesticide (Kerr, 1989).

In a search for a broader spectrum biocontrol strain for crown gall, strain J73 was found, which produced a bacteriocin active against a range of *A. tumefaciens* strains, including all three biotypes and all types of Ti plasmids (Webster et al., 1986). As strain J73 was itself a pathogen it had to be cured of its Ti plasmid, pTiJ73. This procedure was done by the introduction of selectable plasmids carrying the origins of replication of either the nopaline Ti plasmid, pTiC58, or the octopine Ti plasmids, pTi15955 (Webster and Thomson, 1988). Although the resulting strains were nonpathogenic (Fig. 15.3) they still possessed a plasmid migrating in agarose gels to the same position as pTiJ73, about 200-kb. The problem was resolved when Tn5 mutagenesis was used to inactivate agrocin production. During these experiments it was found that some of the genes involved in agrocin production were located on a 200-kb plasmid that comigrated on gels with pTiJ73. One of the agrocin⁻ mutants isolated had undergone a major rearrangement that enabled the separation of the Ti and agrocin plasmids on 0.4% agarose gels. Thus the 200-kb plasmid present in J73, cured of its Ti plasmid, was the agrocin plasmid.

Strain J73, cured of its Ti-plasmid, can be used as a biocontrol agent to protect plants against strains of *A. tumefaciens* resistant to strain 84. However, when their host ranges overlap strain 84 is superior, possibly due to its ability to colonize plants efficiently as mentioned earlier. What causes this colonization is unknown. It is certainly not due to the inability of strain J73 to produce attachment strands to anchor it to plant surfaces (Fig. 15.4; MacRae et al., 1988).

Fig. 15.3. Galls produced on tobacco plants inoculated with various strains of *A. tumefaciens*. Front row, *left* to *right*, J73; J73 Sp^R; C58C1 (pMP90); J73 (pMP90) Sp^R. Back row, *left* to *right*, *E. coli*; J73 (pK1) Sp^R.

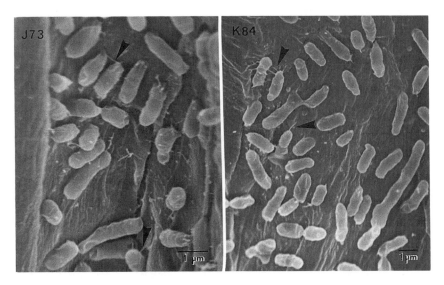

Fig. 15.4. Scanning electron micrographs showing attachment strands between bacteria and tomato roots. Reprinted with permission from MacRae et al., 1988.

3.2. Protection Against Frost Damage

Many pathovars of *P. syringae* are ice-nucleation active (INA) and cause frost damage to plants by the initiation of ice crystal formation at temperatures as high as $-1.5°C$ (Lindow, 1983). The ice formation spreads inter- and intracellularly and results in the disruption of cell membranes (Burke et al., 1976). The extent of frost injury is directly related to the log of the number of INA bacteria on leaf surfaces (Hirano et al., 1985). Therefore, any means of reducing the INA bacteria on leaf surfaces decreases damage (Lindow and Connell, 1984). In the absence of any INA bacteria plants can supercool to below $-5°C$ before suffering frost damage (Lindow, 1983). Strains of *P. syringae* deficient in ice-nucleation have been constructed by deletion of a sequence of chromosomal DNA coding for a membrane protein essential for ice-nucleation (Lindow, 1985). These strains are being evaluated for use in frost control on agricultural crops.

4. ARTIFICIALLY DEVELOPED BIOCONTROL STRAINS

4.1. Live Strains

The gram-positive bacterium *B. thuringiensis* produces proteins that are specifically toxic to a variety of insect pests but have no activity against humans, other vertebrates, and beneficial insects (Hofte and Whiteley, 1989). The commercial use of these proteins is, however, limited by high production costs and the instability of the crystal proteins when exposed in the field. One alternative approach is to isolate a strain of *B. thuringiensis* active against a particular insect pest and clone the gene into a bacterium that colonizes the plant of interest.

Eldana saccharina is a lepidopteran whose larvae bore into sugarcane with devastating effects (Fig. 15.5). Commercial formulations of *B. thuringiensis* have some ameliorating effect, which is limited due to their action on the surface of the canes. Soil samples in affected areas were screened for strains of *B. thuringiensis*. One isolate was found that produced a toxin extremely active against *E. saccharina* larvae. The *tox* gene was cloned and sequenced and found to be homologous to the toxin produced by strain HD-73.

Bacteria were isolated from sugarcane and tested for their ability to colonize the plants. An isolate of *Pseudomonas fluorescens* was identified as an efficient colonizer. To facilitate the cloning of foreign genes into this strain, a number of vectors were tested for their stability. The strain pKT240 was chosen, both due to its stability and to its relatively high copy number (Bagdasarian et al., 1983). The *tox* gene was subcloned onto this vector and bioassays have shown it to be toxic to *E. saccharina* larvae.

Fig. 15.5. Damage caused to sugarcane by the larval borer, *E. saccharina*.

We wished to introduce into strain 14 the capability of growing on sucrose to improve its ability to colonize sugarcane. The sucrase operon of *Vibrio alginolyticus* consists of genes encoding a sucrase enzyme, an enzyme II^{Sucrose} protein of the phosphoenolpyruvate dependent phosphotransferase system (PTS), a fructokinase, and a repressor (Blatch et al., 1990). As it was not known whether *P. fluorescens* encodes a compatible PTS, the sucrase operon was cloned onto pKT240 and intro-

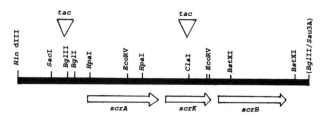

Fig. 15.6. Restriction enzyme map of the *V. alginolyticus* sucrase operon showing the two positions where the *tac* promoter was introduced.

duced into the isolated strain. The sucrase gene was not expressed and, therefore, the *tac* promoter, known to cause efficient expression in *P. putida* (Bagdasarian et al., 1983), was introduced immediately upstream of *scr*B, which encodes the sucrase enzyme (Fig. 15.6). This procedure allowed expression of the gene and enabled the strain to degrade sucrose. The *tac* promoter was then cloned upstream of the *scr*A gene that encodes the enzyme II^{Sucrose} protein. Although this construct still expressed the sucrase gene it did not allow the strain to grow on sucrose. Therefore, this strain does not possess a PTS system compatible with the *V. alginolyticus* sucrase operon.

The levanase gene from *Bacteroides fragilis,* which also cleaves sucrose (Blatch and Woods, unpublished), was then cloned onto pKT240 and introduced into strain 14. This gene was expressed even without the introduction of the *tac* promoter and allowed the strain to grow on sucrose as a sole carbon source. This method enhances its ability to grow in the sucrose rich interior of sugarcane.

A *B. thuringiensis* toxin gene has also been cloned into the endophytic bacterium *Clavibacter xyli* and injected into maize seed by scientists at the biotechnology company Crop Genetics. From there it spreads to all parts of the growing plant and protects it from the European corn borer, *Ostrinia nubilalis* (Twombly, 1990). As the bacteria cannot live outside the plant it is considered to be relatively safe to the environment.

4.2. Nonviable Strains

Before genetically engineered organisms can be subjected to field trials and ultimately released into the environment they have to be approved by various regulatory authorities that are only just beginning to come to terms with the way they should deal with this new category of products. To overcome this problem, which will obviously lead to long delays before permission is granted for field trials, scientists at the biotechnology company Mycogen kill their bacteria before releasing them. These scientists cloned a *B. thuringiensis tox* gene into a strain of *Pseudomonas,* but before they are used the cells are killed and then treated so that the cell wall remains intact. This procedure effectively encapsulates the toxic biopesticide in a long-lived protective coat and the product can be sprayed onto crops. As the bacteria are killed before use they cannot spread in the environment (Twombly, 1990).

4.3. The Development of Broad Host Range and Integration Vectors

A number of vectors are available for cloning genes into plant-associated bacteria. Most are based on broad host range plasmids belonging to incompatibility groups IncP1, IncP4 (IncQ), and IncW (Schmidhauser et al., 1988). The vector pKT240 constructed by Bagdasarian et al. (1983) is mobilizable to, and stably maintained in a number of gram-negative species. To make it more convenient a multiple cloning site carrying a *tac* promoter, which results in high level gene expression in many gram-negative species, was introduced (Fig. 15.7).

It is frequently preferable to insert cloned genes into the bacterial chromosome where they are stably maintained at a more natural copy number. A number of chromosomal integration vectors have been developed based on transposons (Barry, 1986). The stability of the insert is improved if the transposase gene is located on the donor plasmid but outside the inverted repeats that act as its substrate. Thus the transposase enzyme is not integrated into the chromosome during transposition (Obukowicz et al., 1987; De Lorenzo et al., 1990).

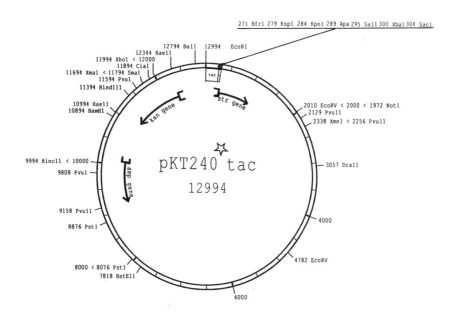

Fig. 15.7. Restriction enzyme map of pKT240*tac*.

There is some debate, however, over the desirability of releasing bacteria carrying genes coding for antibiotic resistance into the environment. Herrero et al. (1990) therefore developed vectors to insert foreign genes into the chromosome without the use of antibiotics. These results are based on the use of resistances to the herbicide biolaphos, mercuric salts and organomercuric compounds, and arsenite.

4.4. The Use of Rhizosphere, Phylloplane, or Endophytic Bacteria

The next question that arises is the choice of bacteria to be used as delivery organisms for the gene product(s). Different bacteria are found on and around roots (the rhizoplane and rhizosphere), on leaves (the phylloplane), or in the interior regions of plants (endophytic bacteria). Therefore, the choice will depend largely on the part of the plant at risk from the pest or pathogen.

Growth of bacteria in the rhizosphere and rhizoplane is much higher than in the surrounding soil. This finding is attributed to the stimulatory effects of matter released by plant roots, commonly referred to as root exudates. These compounds include sugars, amino acids, hormones, metabolic intermediates, and vitamins. These root exudates are released as a result of leakage, active secretion, and cell autolysis. Gram-negative rods, particularly members of the genera *Pseudomonas, Flavobacterium,* and *Alcaligenes* predominate in this environment (Alexander, 1977).

Less is known about bacteria that colonize the leaf surface or phylloplane. The surfaces of leaves are usually hydrophobic due to the presence of cutin and wax, which vary under different environmental conditions. This layer reduces the amount of nutrients that are leached from the leaf. In addition, the phylloplane is a harsh environment for microbial survival due to fluctuations in temperature and humidity and high ultraviolet (UV) radiation. Therefore, the growth of microorganisms on leaves is normally restricted and any slight benefit that can be obtained is advantageous (Campbell, 1989). An approach to achieving such benefits is outlined in the next section.

Even less is known about endophytic bacteria that are found in the interior regions of plants including the cortex and the xylem of roots, stems, and leaves. In a study on the bacteria in xylem of citrus tree roots, Gardner et al. (1982) identified 556 strains from 13 genera. The most frequent were *Pseudomonas* (40%), *Enterobacter* (18%), and *Bacillus, Corynebacterium*, and other gram-positive bacteria (16%). Bacterial populations as high as 10^4 cfu/g of xylem were found. It is likely that microbial colonization of the rhizoplane could lead to invasion into the cortex and xylem.

As mentioned earlier, which of these types of plane colonizers should be used in the development of a biocontrol system will depend largely on the pest or pathogen targeted. Herbivorous insects, and bacteria, such as *Erwinia* spp., mainly attack leaves and, thus, phylloplane bacteria would be appropriate. Bacteria such as *A. tumefaciens* attack at the crown, the junction between the stem and the root and, therefore, rhizosphere bacteria could be used. Stalk borers burrow into stems, and fungi invade a variety of locations, especially roots. Therefore, endophytes and rhizoplane bacteria would be the strains of choice.

5. IMPROVEMENT OF BACTERIAL COLONIZATION OF PLANTS

The two most important factors in a biocontrol bacterial strain are the synthesis of sufficient quantities of a potent antipest or antipathogen agent and the effective colonization of the target plant. Ways in which the former strain may be improved, by the use of efficient promoters, vectors, and so on, have already been discussed. How to improve the latter feature is less obvious, partly because so little is known about the ways in which nonpathogenic bacteria colonize plants. It is not likely that we can learn much from studies of colonization by *A. tumefaciens* or *Rhizobium* species, since the former species is a pathogen and the latter species has a unique interaction with plants resulting in nodule formation.

The growth of bacteria in the rhizosphere is usually attributed to their utilization of root exudates. It is unlikely that the phylloplane exudes many nutrients, but it is possible that sufficient release occurs to support the bacterial populations found in this environment. As mentioned previously, the growth of this bacteria is so restricted that any slight benefit that can be obtained is advantageous. What causes endophytic bacteria to grow inside the plant has received little, if any, attention.

The approach that we are taking in an attempt to understand and improve colonization is to introduce into bacteria genes encoding enzymes that could allow the recipient to utilize plant components. Our first venture, as described earlier, was to introduce into a sugarcane-colonizing pseudomonad a levanase gene that confers on the strain the ability to grow on sucrose as a sole carbon source. We are now investigating the role of other cloned genes whose products degrade plant components such as cellulose, xylan, pectin, and protein.

6. CONCLUSION

This chapter has tried to highlight alternative approaches to the biological control of plant pests and pathogens. As efficient, naturally occurring biocontrol agents are unlikely to be readily obtainable, scientists must depend on the use of recombinant DNA technology to develop such strains. This approach can either take the form of the improvement of a potential biocontrol strain, or the design of a totally novel organism. Whichever approach is used, cognizance must be taken of two most important factors, namely, stable production of sufficient quantities of the toxic compound and effective long-term colonization of the targetted region of the plant. Optimization of the former approach can be achieved in a number of ways, including the use of efficient promoters, integration vectors, and so on. However, the latter approach is less amenable to improvement, as so little is known about the features of a bacterium that enable it to thrive in a given niche on a plant. Therefore, the development of biocontrol strains as discussed here will require a far greater understanding of plant–bacterial interactions.

REFERENCES

Alexander M (1977): Introduction to soil microbiology. New York: Wiley, 2nd ed.

Bagdasarian MM, Amann E, Lurz R, Ruckert B, Bagdasarian M (1983): Activity of the hybrid trp-lac (tac) promoter of *Escherichia coli* in *Pseudomonas putida*. Construction of broad-host-range, controlled-expression vectors. Gene 26: 273–282.

Barry GF (1986): Permanent insertion of foreign genes into the chromosomes of soil bacteria. Bio/Technol 4:446–449.

Blatch GL, Scholle RR, Woods DR (1990): Nucleotide sequence and analysis of the *Vibrio alginolyticus* sucrose uptake-encoding region. Gene 95:17–23.

Burke MJ, Gustsa LA, Quamme HA, Weiser CJ, Li PH (1976): Freezing and injury to plants. Annu Rev Plant Physiol 27:507–528.

Campbell R (1989): Biocontrol on leaf surfaces. Cambridge: Cambridge University Press. Ch 3, pp 67–94; 195–196.

Das PK, Basu M, Chatterjee GC (1978): Studies on the mode of action of agrocin 84. J Antibiotics 31:490–492.

De Lorenzo V, Herrero M, Jakubzik U, Timmis KN (1990): Mini-Tn5 transposon derivatives for insertion mutagenesis, promoter probing, and chromosomal insertion of cloned DNA in Gram-Negative Eubacteria. J Bacteriol 172:6568–6572.

Ellis JG, Murphy PJ (1981): Four new opines from crown gall tumors—their detection and properties. Mol Gen Genet 181:36–43.

Engler G, Holsters M, Van Montagu M, Schell J, Hernalsteens JP, Schilperoort R (1975): Agrocin 84 sensitivity: A plasmid determined property in *Agrobacterium tumefaciens*. Mol Gen Genet 138:345–349.

Gardner JM, Feldman AW, Zablotowicz RM (1982): Identity and behavior of xylem-residing bacteria in rough lemon roots of Florida citrus trees. Appl Environ Microbiol 43:1335–1342.

Herrero M, De Lorenzo V, Timmis KN (1990): Transposon vectors containing non-antibiotic resistance selection markers for cloning and stable chromosomal insertion of foreign genes in gram-negative bacteria. J Bacteriol 172:6557–6567.

Hilder VA, Barker RF, Samour RA, Gatehouse AMR, Gatehouse JA, Boulter D (1989): Protein and cDNA sequences of Bowman–Birk inhibitors from the cowpea (*Vigna unguiculata Walp.*). Plant Mol Biol 13:701–710.

Hirano SS, Baker LS, Upper CD (1985): Ice nucleation temperature of individual leaves in relation to population sizes of ice nucleation active bacteria and frost injury. Plant Physiol 77: 259–265.

Hofte H, Whiteley HR (1989): Insecticidal crystal proteins of *Bacillus thuringiensis*. Microbiol Rev 53:242–255.

Holsters M, Silva B, van Vliet F, Genetello C, de Block M, Dhaese P, de Picker A, Inze D, Engler G, Villarroel R, van Montagu M, Schell J (1980): The functional organization of the nopaline *A. tumefaciens* plasmid pTiC58. Plasmid 3:212–230.

Jones DA, Ryder MH, Clare BG, Farrand SK, Kerr A (1988): Construction of a Tra⁻ deletion mutant of pAgK84 to safeguard the biological control of crown gall. Mol Gen Genet 212:207–214.

Kerr A (1989): Commercial release of a genetically engineered bacterium for the control of crown gall. Agric Sci (Nov) 41–44.

Kerr A, Panagopoulos CG (1977): Biotypes of *A. radiobacter* var. *tumefaciens* and their biological control. Phytopath Z 90:172–179.

Konigsberg W, Godson GN (1983): Evidence for use of rare codons in the *dna*G gene and other regulatory genes of *Escherichia coli*. Proc Natl Acad Sci USA 80:687–691.

Lindemann J, Suslow TV (1987): Competition between ice nucleation-active wild type and ice nucleation-deficient deletion strains of *Pseudomonas syringae* and *P. fluorescens* biovar I and biological control of frost injury on strawberry blossoms. Phytopathology 77:882–886.

Lindow SE (1983): The role of bacterial ice nucleation in frost injury to plants. Annu Rev Phytopathol 21:363–384.

Lindow SE (1985): Ecology of *Pseudomonas syringae* relevant to the field use of Ice⁻ deletion mutants constructed *in vitro* for plant frost control. In Halvorson HO, Pramer D, Rogul M (eds): Engineered organisms in the environment: scientific issues. Washington, DC: American Society for Microbiology, pp 23–35.

Lindow SE, Connell JH (1984): Reduction of frost injury to almond by control of ice nucleation active bacteria. J Am Soc Hortic Sci 109:48–53.

MacRae S, Thomson JA, van Staden J (1988): Colonization of Tomato Plants by two Agrocin-Producing Strains of *Agrobacterium tumefaciens*. Appl Environ Microbiol 54:3133–3137.

Moore LW, Warren G (1979): *Agrobacterium radiobacter* strain 84 and biological control of crown gall. Annu Rev Phytopathol 17:163–179.

Murray EE, Lotzer J, Eberle M (1989): Codon usage in plant genes. Nucleic Acids Res 17:477–498.

Nejidat A, Beachy RN (1990): Transgenic tobacco plants expressing a coat protein gene of tobacco mosaic virus are resistant to some other tobacco viruses. Mol Plant–Microbe Interact 3:247–251.

Neuhaus JM, Ahl-Goy P, Hinz U, Flores S, Meins F (1991): High-level expression of a tobacco chitinase gene in *Nicotiana sylvestris*. Susceptibility of transgenic plants to *Cercospora nicotianae* infection. Plant Mol Biol 16:141–151.

New PB, Kerr A (1972): Biological control of crown gall: field measurements and glasshouse experiments. J Appl Bact 35:279–287.

Obukowicz MG, Perlak FJ, Bolten SL, Kusano-Kretzmer K, Mayer EJ, Watrud LS (1987): IS50L as a non-self transposable vector used to integrate the *B. thuringiensis* δ-endotoxin gene into the chromosome of root-colonizing pseudomonads. Gene 51:91–96.

Panagopoulos CG, Psallidas PG, Allivizatos AS (1978): Studies on 3 of *A. radiobacter* var. *tumefaciens*. Proceedings of the 4th International Conference on Plant Bacteriology, Angers, pp 221–228.

Perlak FJ, Deaton RW, Armstrong TA, Fuchs RL, Sims SR, Greenplate JT, Fischhoff DA (1990): Insect resistant cotton plants. Bio/Technol 8:939–943.

Perlak FJ, Fuchs RL, Dean D, McPherson SL, Fischhoff DA (1991): Modification of the coding sequence enhances plant expression of insect control protein genes. Proc Natl Acad Sci USA 88: 3324–3328.

Powell PA, Stark DM, Sanders PR, Beachy RN (1989): Protection against tobacco mosaic virus in transgenic plants that express tobacco mosaic virus antisense RNA. Proc Natl Acad Sci USA 86:6949–6952.

Schmidhauser TJ, Ditta G, Helinski DR (1988): Broad-host-range plasmid cloning vectors for Gram-negative bacteria. In Rodriguez RL, Denhardt DT (eds): Vectors: A survey of Molecular Cloning Vectors and their uses. Boston: Butterworth, pp 287–332.

Shim JS, Farrand SK, Kerr A (1987): Biological control of crown gall: construction and testing of new biocontrol agents. Phytopathology 77:463–466.

Staphorst JL, van Zyl FGH, Strijdom BW, Groenewold ZE (1985): Agrocin-producing pathogenic and nonpathogenic biotype-3 strains of *A. tumefaciens* active against biotype-3 pathogens. Curr Microbiol 12:45–52.

Stark DM, Beachy RN (1989): Protection against potyvirus infection in transgenic plants: evidence for broad spectrum resistance. Bio/Technol 7:1257–1262.

Tate ME, Murphy PJ, Roberts WP, Kerr A (1979): Adenine N6-substituent of agrocin 84 determines its bacteriocin-like specificity. Nature (London) 280:697–699.

Twombly R (1990): Genetic engineers at Crop Genetics and Mycogen place hopes on two different strategies for pest control. The Scientist (July 9) pp 1, 8, 9, 28.

Vaeck M, Reynaerts A, Hofte H, Jansens S, De Beuckeleer M, Dean C, Zabeau M, van Montagu M, Leemans J (1987): Transgenic plants protected from insect attack. Nature (London) 328:33–37.

Webster J, Dos Santos M, Thomson JA (1986): Agrocin-producing *Agrobacterium tumefaciens* strain active against grapevine isolates. Appl Environ Microbiol 52:217–219.

Webster J, Thomson JA (1988): Genetic analysis of an *Agrobacterium tumefaciens* strain producing an agrocin active against biotype 3 pathogens. Mol Gen Genet 214:142–147.

BIOLOGICAL ACTIVITIES OF BACTERIA USED IN PLANT PATHOGEN CONTROL

STEPHEN T. LAM
THOMAS D. GAFFNEY

Agricultural Biotechnology Research Unit, CIBA-Geigy Corporation, Research Triangle Park, North Carolina 27709

1. INTRODUCTION

Attempts to use bacteria for plant disease control or yield improvement were started in the early part of this century (see Cook and Baker, 1983). Interest in the area has increased steadily since the mid-1960s (Baker, 1987), and experimental applications in many crop–pathogen systems have now been reported (see Weller, 1988 and references cited therein). More recently, research efforts have begun to move beyond system descriptions and towards mechanistic studies. A number of biological activities have been postulated to be involved in successful disease control. These include the abilities to colonize the appropriate host plant parts and to produce antagonistic compounds, such as antibiotics, siderophores, ammonia, cyanide, and hydrolytic enzymes. The biocontrol bacteria also may provide plant protection through induction of host plant defense mechanisms, elimination of plant signals that trigger pathogen development, or competition for nutrients. In this chapter we review evidence that implicate involvement of these activities, with the emphasis on the more recent results. We also discuss emerging methods and approaches that should be useful for furthering our understanding of the involvement of these activities in plant protection and yield enhancement. In light of the current focus of the field, most of the material is drawn from research

Biotechnology in Plant Disease Control, pages 291–320
© 1993 Wiley-Liss, Inc.

on control of soil-borne pathogens. The discussion on colonization is restricted to the rhizosphere for the same reason. A number of reviews on related topics are available (Weller, 1988; Handelsman and Parke, 1989; de Weger and Lugtenberg, 1990; Défago and Haas, 1990; Gutterson, 1990; Lugtenberg et al., 1991).

2. BIOLOGICAL ACTIVITIES

2.1. Rhizosphere Colonization

Colonization of the host plant rhizosphere by the biocontrol bacteria is generally believed to be necessary for disease suppression (see, e.g., Suslow, 1982; Weller, 1988; Bowen, 1991; Parke, 1991). Perhaps because of the apparent self-evidence of the concept, few studies can be found that were devoted to examining the relationship between the population level of the biocontrol bacteria and the level of disease control. Nevertheless, data that are available do support the notion. Xu and Gross (1986) applied different levels (from 10^4 to 10^{10} cfu) of the *Pseudomonas putida* strain W4P63 to potato seed pieces, which were then inoculated with *Erwinina carotovora* subsp. *atroseptica* and planted. They found that increasing the levels of W4P63 applied correlated with decreasing preemergence seed piece decay (as measured by plant emergence). The population levels of W4P63 on the roots after planting were not examined. The effect of bacterial concentration on control of *Pythium* damping-off in cucumbers was studied by Elad and Chet (1987). With the *Pseudomonas cepacia* isolate 808, an effect of applied bacterial concentration on disease control efficacy was observed. On the other hand, no apparent effect was found with the *P. putida* isolate 805. But, since all levels applied, including the lowest, gave good control (~70% healthy plants compared to <10% healthy plants in the untreated control), it is conceivable that had even lower levels of bacteria been applied, a concentration effect might have been observed. Here again, the population level of the biocontrol bacteria on plant roots after planting was not monitored.

In a study on the effect of *Pseudomonas fluorescens* strain M-4 on potato plant growth in soil infested with *Verticillium dahliae* (Leben et al., 1987), the bacteria were applied on potato seed pieces at two different levels. A significant linear relationship was observed between plant growth and level of M-4 applied. There was a 10-fold difference in population levels of M-4 on the potato plant roots at harvest. The *P. fluorescens* strain 2-79 has been shown to control take-all of wheat caused by *Gaeumannomyces graminis* var. *tritici* (Weller and Cook, 1983). Using antibiotic resistant derivatives of 2-79, Bull et al. (1991) demonstrated a direct relationship between the population levels of bacteria applied to wheat seeds and those recovered from plant roots 14 days after planting. The experiments were carried out in natural soil in the greenhouse. Similar results were obtained whether

TABLE 16.1. Relationship Between Population Level of
P. fluorescens Strain BL915 and Disease Control

Input	Population of BL915[a]			Disease Rating[b]
	Day 1	Day 2	Day 3	
4.3×10^6	2.0×10^8	2.3×10^8	1.3×10^7	1.0
2.3×10^5	3.4×10^7	7.4×10^7	1.0×10^7	3.3
3.6×10^4	8.3×10^6	1.8×10^7	3.2×10^5	3.0
Controls				
No bacteria				3.0
No pathogen				1.0

[a]cfu/(seed + root) (input through day 2) or cfu/root (day 3).

[b]Disease symptoms were rated from 1 to 5. 1 = healthy. The numbers represent the average of eight plants.

the soil was infested with *G. graminis* var. *tritici* or not. Moreover, an inverse relationship was observed between the population level of bacteria on roots and the number of lesions formed by *G. graminis* var. *tritici* on seminal roots.

In autoclaved soil, bacteria applied to seeds quickly colonize the emerging plant roots and reach a level characteristic of the particular bacterial–plant combination, regardless of the inoculum size (Bennett and Lynch, 1981). This level is normally not reached in natural soil, presumably because of competition from the established microflora, and a direct effect of inoculum size on root population level is observed as discussed earlier. Torkewitz and Lam (1991) showed that such competition could be simulated under gnotobiotic conditions by coinoculating two competing bacterial strains. Population levels of the biocontrol *P. fluorescens* strain BL915 on cotton seedling roots were controlled by coinoculating seeds at different ratios with strain BL924, which was ineffective in disease control, and were monitored during the first 3 days after planting (Table 16.1); the critical period when the seedlings were highly susceptible to infection by *Rhizoctonia solani* (Torkewitz and Lam, 1991). Only seeds inoculated with the highest level of biocontrol bacteria were protected against damping-off by *R. solani*. A similar threshold population level requirement for biological activity was observed by Iswandi et al. (1987) in their study on the effect of the rhizopseudomonad strain 7NSK2 on maize and barley growth.

We know very little about the details of root colonization, though most would agree that the process includes at least the following: bacterial migration towards plant root, attachment, distribution along the elongating root, and proliferation on the root. A number of bacterial traits have been examined with respect to their possible role in these subprocesses.

Pseudomonas strains have been shown to migrate actively towards soybean seeds in soil (Scher et al., 1985). No correlation between chemotactic ability and root colonization was observed when the bacterial strains were directly inoculated onto seeds (Scher et al., 1988), but the inoculation procedure might have circumvented the need for chemotaxis.

Bacterial cell surface molecules may aid in attachment to plant root and enhance colonization. Agglutination has been correlated with colonization and disease suppression (Chao et al., 1988; Anderson et al., 1988; Tari and Anderson, 1988). The Tn*5* mutants that were no longer agglutinable (Agg⁻) adhered to cucumber roots at lower levels, colonized to a lesser extent, and provided less protection against *Fusarium oxysporum* (Tari and Anderson, 1988). A genomic fragment corresponding to the site of Tn*5* insertion in Agg⁻ mutant 5123 has been isolated (Buell and Anderson, 1991). This fragment complemented the Agg⁻ phenotype in 5123 but not other Agg⁻ mutants, indicating the likely involvement of more than one genetic locus in agglutination. Lipopolysaccharide (LPS) may be another class of cell surface molecules that has a role in colonization. Twenty-three out of twenty-four *Pseudomonas* strains isolated from potato root surface produced LPS with O-antigenic side chains (de Weger et al., 1987a). Mutants with altered O-antigenic side chains were impaired in colonization (de Weger et al., 1989).

The role of motility in colonization was examined by several groups. In each case nonmotile mutants were compared with their motile wild-type parents. In two studies (Howie et al., 1987; Scher et al., 1988) the mutants colonized as well as the parents, while in a third study (de Weger et al., 1987b) the nonmotile mutants were impaired in their ability to colonize lower parts of the plant roots. The bacteria–plant systems studied and test conditions were all different. It is certainly conceivable that motility may be required only under some conditions and not others.

Brand et al. (1991) identified a superior potato root-colonizing *Pseudomonas* strain, WCS365, by inoculating seed potatoes with a mixture of equal amounts of 11 root-colonizing *Pseudomonas* strains (marked with antibiotic resistance), planting the seed potatoes in the field, then recovering bacteria from the potato plant roots after 3.5 months of growth. Among the 11 inoculated strains, WCS365 was the predominant strain recovered from the upper parts of the roots, and the only strain recovered from the lower parts of the roots. When compared with the other strains, WCS365 did not stand out with respect to motility or antagonistic activities. It did, however, grow faster than the other strains in culture medium (more efficient utilization of nutrients?) and attach more firmly to roots. The authors suggested that these properties may be relevant for competitive colonization.

The impact of single substrate utilization on colonization was examined by Lam et al. (1991), using mannopine utilization (Mut) as the model system. The relative competitiveness of two *Pseudomonas* strains that differed only in Mut (one strain was a Tn*5* generated Mut⁻ mutant of the other) was determined. Near isogenic tobacco lines differing in mannopine production (Mop) were obtained by trans-formation with *Agrobacterium rhizogenes*. Tobacco seeds were inoculated with a mixture of the two bacterial strains and the ratio between the two was determined (input ratio). Four weeks after planting, bacteria were recovered from the seed-ling roots and the ratio between the Mut$^+$ and Mut$^-$ bacteria were again deter-mined (output ratio). On Mop$^-$ plants (no mannopine in the environment), the input and output ratios were essentially the same (Fig. 16.1), indicating that the two bacterial strains were equally competitive in the absence of mannopine. On Mop$^+$ plants, however, Mut$^+$ bacteria increased relative to the Mut$^-$ bacteria, indicating that, in the presence of mannopine, the ability to utilize mannopine conferred a competitive advantage.

The production of antibiotics and siderophores are two of the most visible activ-ities of biocontrol bacteria. Because of their known antagonistic properties towards other organisms, they have often been proposed to have a role in disease control (see Sections 2.2 and 2.3). For the same reason, they may be expected to have a role in the natural ecology of the producing organisms. The impacts of antibiotics and siderophores on colonization and disease control were examined in parallel in a number of studies (e.g., Howie and Suslow, 1986; Bakker et al., 1987; Loper, 1988; Thomashow and Weller, 1988; Keel et al., 1992). In general, mutants defec-tive in the production of antibiotics or siderophores colonized the plant host roots at least as well as the wild-type parent within the time frames of the studies. One study (cited in Thomashow, 1991) did show that the ability to produce phenazine contributed to the long-term survival of the producing organism in soil habitats.

2.2. Antibiotics

The suspicion that antibiosis may be involved in disease suppression by a biocontrol bacterium usually starts with evidence of inhibition of the target patho-gen *in vitro*. A number of approaches can be used to substantiate such suspi-cions. If the chemical structure of the inhibitory compound has been identified, the purified or synthetic compound can be applied in place of the biocontrol bac-terium in plant protection experiments and compared to the effect of the bio-control bacterium. Isolation of the inhibitory compound from rhizospheres of plants to which the biocontrol bacterium has been applied provides further sup-porting evidence that the compound may be involved. Mutants that are defective in the production of the inhibitor can be isolated and their disease control efficacy compared to the wild-type parent. This genetic approach is very versatile, since it can be applied regardless of the nature of the inhibitory compound, or even when the compound is unknown. Each approach has limitations. For example, the inhibitory compound in question may be adsorbed or inactivated in soil, and it would not be possible to mimic the effect of the biocontrol organism using the purified com-pound. The genetic approach, on the other hand, relies on the assumption that the

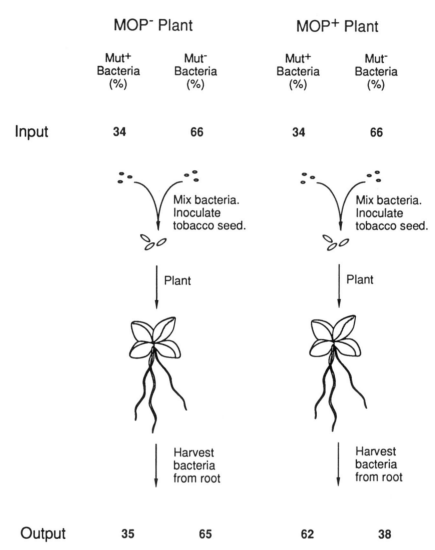

Fig. 16.1. Competition between a Mut bacterium and its nonutilizing mutant.

mutant examined is altered only in the production of the compound in question, which is not always easy to prove. Ideally, the potential role of a compound in disease control should be examined using as many approaches as are available. At present, only a few systems have been examined at such level of detail. These include the production of phenazine in the control of take-all of wheat (caused by

G. graminis var. *tritici*) and 2,4-diacetylphloroglucinol in the control of black root rot of tobacco (caused by *Thielaviopsis basicola*) and take-all of wheat.

The *P. fluorescens* strain 2-79, which has been shown to control take-all of wheat (Weller and Cook, 1983), produces the antibiotic phenazine-1-carboxylic acid (Gurusiddaiah et al., 1986; Brisbane et al., 1987). This compound inhibits *G. graminis* var. *tritici in vitro* at less than 1 µg/ml, and its production in the rhizosphere of wheat plants inoculated with strain 2-79 has been demonstrated (Thomashow et al., 1991). Thomashow and Weller (1988) isolated mutants of strain 2-79 that were defective in phenazine production (Phz$^-$) following Tn*5* transposon mutagenesis. Six independent, prototrophic Phz$^-$ mutants were tested. Each was no longer inhibitory to *G. graminis* var. *tritici* and provided significantly less protection against take-all (though colonizing the host plant roots at levels comparable to the wild-type parent). Phenazine could be isolated from roots of wheat inoculated with Phz$^+$ strains but not from those inoculated with Phz$^-$ strains. Roots with antibiotic were healthier than roots without (Thomashow et al., 1991), indicating a direct correlation between the presence of antibiotic and disease suppression. Two of the mutants, 2-79-B46 and 2-79-782, have since been shown to be pleiotropic; they were also defective in the production of anthranilic acid, and apparently altered in motility and cell surface properties (Thomashow and Pierson, 1991). Complementation of these mutants to Phz$^+$ by the plasmid pPHZ49-6 also restored the ability to produce anthranilic acid. Three other mutants, 2-79-99, 2-79-2510, and 2-79-42355, were complemented to Phz$^+$ by the plasmid pDWE108A. This plasmid, when transferred into other *Pseudomonas* strains that normally did not produce phenazine and did not inhibit *G. graminis* var. *tritici in vitro*, conferred upon the transconjugants the ability to do both (Thomashow, 1991). Similar results were obtained with the *P. aureofaciens* strain 30-84 (Thomashow and Pierson, 1991).

The *P. fluorescens* strain CHA0 suppresses black root of tobacco (caused by *T. basicola*), take-all of wheat (caused by *G. graminis* var. *tritici*), and other root diseases (Stutz et al., 1986; Défago et al., 1990). This strain produces a number of secondary metabolites, including a siderophore, hydrogen cyanide, and the antibiotics pyoluteorin and the acetylphloroglucinols. To examine the role of 2,4-diacetylphloroglucinol (Phl$^-$) in disease control, Keel et al. (1990–1992) isolated a Tn*5* mutant of strain CHA0 that no longer produced the compound Phl$^-$. This mutant showed less inhibition of *T. basicola* and *G. graminis* var. *tritici in vitro* and, though colonizing the rhizosphere as well as the wild-type parent, provided less disease control. Complementation of the mutant to Phl$^+$ also restored fungal inhibition *in vitro* and disease control. Similar results were obtained with *P. aureofaciens* strain Q2-87, another take-all control strain that produces 2,4-diacetylphloroglucinol (Vincent et al., 1991). In this case, complementing cosmids were isolated which, when transferred into the *Pseudomonas* strains 2-79 and 5097 (neither of which were normally Phl producers), resulted in Phl-producing transconjugants. 2,4-Diacetylphloroglucinol was shown to have broad antibacterial, antifungal, and phytotoxic activity (Keel et al., 1992). One of the target pathogens, *G. graminis* var. *tritici,* was shown to be the most sensitive fungus to the compound. The compound was recovered from the rhizosphere of wheat inoculated with the wild-type and the complemented strains (Phl$^+$) but not from that inoculated with the Phl$^-$ mutant. Correlation was also observed between the presence of antibiotic and root health.

The nonpathogenic *Agrobacterium radiobacter* strain K84, which produces the antibiotic agrocin 84, is used successfully to control crown gall (see review by Ryder and Jones, 1990). Several lines of evidence implicate the involvement of agrocin 84 in the disease control process. The biosynthetic genes of agrocin 84 are located on the plasmid pAgK84. Transfer of pAgK84 to another *Agrobacterium* strain results in the acquisition by the recipient of the abilities to produce agrocin 84 and control crown gall simultaneously (Ellis and Kerr, 1978), and curing of the plasmid in a strain results in loss of the ability to produce agrocin 84 and reduction in disease control efficacy (Cooksey and Moore, 1982). Moreover, only agrocin sensitive strains of *Agrobacterium tumefaciens* are controlled by strain K84. Finally, addition of partially purified agrocin 84 to the infection court (plant wounds) mimics the effect of application of the strain itself (Moore, 1988).

The role of oomycin A in the control of *Pythium* by *P. fluorescens* strain HV37a and the regulation of its biosynthesis have been discussed recently in an excellent review (Gutterson, 1990; see also Howie and Suslow, 1991, and Section 3.2). The structure of oomycin A has not yet been determined. Accordingly, most of the results have come from genetic studies. The production of oomycin A in the rhizosphere, for example, has been monitored indirectly through the use of *lux* (Howie et al., 1988; see Section 3.2) and β-galactosidase (Howie and Suslow, 1991) gene fusions in genes involved in the biosynthesis of the compound. Other compounds that have been implicated in disease control by mutant analysis include pseudomycin (Lam et al., 1987; Harrison et al., 1988), herbicolin A (Kempf and Wolf, 1989; Kempf and Schroth, 1991), pyocyanine (Flaishman et al., 1990), pyrrolnitrin (Homma et al., 1991), and pyoluteorin (Kraus and Loper, 1991). Purified pyrrolnitrin and pyoluteorin were shown to suppress disease symptoms when applied to seeds (Howell and Stipanovic, 1979, 1980; Homma and Suzui, 1989). Pyrrolnitrin (Homma et al., 1991) and herbicolin A (Kempf and Schroth, 1991) have been recovered from the rhizosphere of plants inoculated with the producing organisms.

2.3. Siderophores

The role if siderophore production by biological control strains in the antagonism of phytopathogens has been the subject of excellent recent in-depth reviews (Loper and Buyer, 1991; Loper and Ishimaru, 1991). Iron depletion of competing microbes, including phytopathogens, resulting from their lack of receptors for a given biocontrol strain's ferric siderophore is thought to be the mechanism of this type of antagonism (Loper and Ishimaru, 1991). Since both siderophore production itself and the ability of any available siderophore to contribute to plant disease suppression can be influenced by a variety of physical, chemical, and biological factors (Loper and Buyer, 1991; Loper and Ishimaru, 1991), it is perhaps not surprising to find it impossible to make broad generalizations about the importance of siderophores in biological control. For example, instances where the pyoverdine class of siderophore has been implicated in either plant growth promotion or disease suppression have been described (Kloepper and Schroth, 1981; Bakker et al., 1986; Schippers et al., 1987; Becker and Cook, 1988; Loper, 1988), as have instances where pyoverdine production had little or no observable effect (Ahl et al., 1986, Weller et al., 1988; Gutterson, 1990).

Recent literature supports the contention that the contribution of siderophore biosynthesis to the efficacy of biocontrol strains requires examination, and in some instances reexamination, on a case-by-case basis. A pyoverdine-negative mutant of *P. fluorescens* strain CHA0 has been described (Keel et al., 1989; Haas et al., 1991; Keel and Défago, 1991), which is not impaired in its ability to protect wheat from diseases caused by *Pythium ultimum* or *G. graminis* var. *tritici*. This was true in the wheat/*Pythium* system even under conditions of iron deficiency. The authors noted that iron competition could still be playing some role in disease suppression if *P. fluorescens* CHA0 synthesizes siderophore(s) in addition to pyoverdine. Paulitz and Loper (1991) obtained Tn5 mutants of *P. putida* strain N1R deficient in pyoverdine production (Pvd-phenotype). The Pvd$^-$ derivatives showed no reduction in ability to protect cucumber from *Pythium* damping-off in three soil types. Hamdan et al. (1991) generated mutants of *P. fluorescens* strain 2-79, which were either unable to produce the antibiotic phenazine-1-carboxylic acid (Phz$^-$ phenotype) or unable to produce a fluorescent pyoverdine siderophore (Flu$^-$ phenotype). The Phz$^+$Flu$^-$ derivatives showed no reduction in *in vitro* inhibition of *G. graminis* var. *tritici,* and little or no reduction of suppressiveness to wheat take-all *in situ*. In contrast, Phz$^-$Flu$^+$ derivatives provided significantly less disease suppression than did the wild-type strain, supporting the idea that phenazine antibiotic production plays a major role in the biological control of take-all by strain 2-79. A Phz$^-$Flu$^-$ strain was constructed, and it demonstrated some residual biocontrol activity, which proved to be due in part to an antifungal factor thought to be anthranilate. The authors indicated that this antifungal factor apparently was responsible for the fungal inhibition previously credited to Pvd in a Phz$^-$ derivative of strain 2-79 (Weller et al., 1988), and cautioned that the ability of fluorescent pseudomonads to synthesize a variety of antifungal compounds under iron-limited conditions can lead to confusion in accurately assessing the role of pyoverdines in biocontrol.

Additional recent siderophore-related research involving rhizosphere bacteria has focused on outer membrane receptors for ferric siderophores and the identification of regulatory genes required for the expression of siderophore biosynthetic genes. Leong et al. (1991) described the identification of the *pupB* gene of *P. putida* WCS358, which encodes an 87-kD outer-membrane protein involved in iron transport by heterologous siderophores. They also identified two putative positive regulatory loci whose products can activate transcription of genes coding for synthesis of the strain's native siderophore, pseudobactin 358 (see also Section 3.3). Morris et al. (1992) constructed a mutant derivative of *Pseudomonas* sp. strain M114 lacking an outer membrane receptor capable of transporting two heterologous siderophores in addition to the native ferric pseudobactin. The mutant demonstrated a total inability to utilize ferric pseudobactin M114, and its growth was retarded significantly. Since ferric pseudobactin M114 produced by the receptor mutant was apparently responsible for the observed growth inhibition, and since wild-type strains are subject to similar growth inhibition in the presence of siderophores that they cannot utilize, the authors suggested that there may be evolutionary pressure for bacteria to acquire receptors for heterologous siderophores. Consistent with this idea, strain M114 was found to contain an additional receptor for pseudobactin MT3A, a siderophore that the strain does not produce.

2.4. Volatile Compounds

A number of *Enterobacter cloacae* strains have been shown to be effective biocontrol agents against damping-off diseases caused by *Pythium* in cucumbers, peas, beets, and cotton (Hadar et al., 1983; Nelson, 1988). Nelson et al. (1986) showed that the *E. cloacae* strains reduced *Pythium* growth *in vitro,* but no antibiotics, toxins, or cell wall degrading enzymes were detected in culture filtrates. The bacteria were also found to reduce seed colonization by *Pythium* during the first 20 h of germination, to bind *Pythium* hyphae, and to agglutinate *Pythium* cell wall fragments. The various phenomena (disease suppression, inhibition of seed colonization, fungal growth inhibition, fungal hyphae binding, and fungal cell wall agglutination) all appeared to be related. All were reduced or eliminated in the presence of certain sugars such as D-glucose or sucrose. Sugars that did not interfere with hyphae binding also did not interfere with the other processes. Moreover, *E. cloacae* was effective as a biocontrol agent only on seeds with low sugar exudation. All the evidence together, though circumstantial, built a good case that hyphae binding may be involved in disease suppression. It was, however, not clear how hyphae binding could lead to inhibition of further fungal development and infection.

Howell et al. (1988) found that *Pythium* growth was inhibited by *E. cloacae* in partitioned agar plates (with the organisms in separate quadrants), indicating the involvement of a volatile inhibitor. The inhibitor was identified as ammonia, and the inhibitory effect of ammonia on *Pythium* was confirmed. Interestingly, the production of ammonia by *E. cloacae* was influenced by various sugars in the same way as the other phenomena discussed earlier, that is, there was a positive correlation between ammonia production and disease control. The authors suggested that ammonia was produced as a byproduct when amino acids in seed exudates were utilized by *E. cloacae* as carbon sources in the absence of readily metabolized sugars. They also suggested that hyphal attachment, placing the bacteria in close juxtaposition to the pathogen, might lower the level of ammonia production required for effective disease control. More recently, high level ammonia production in the spermosphere of cotton seeds inoculated with *E. cloacae* was detected (Howell and Stipanovic, personal communication). Little ammonia was produced with either the seed or the bacteria omitted. Addition of glucose to the soil also reduced ammonia production to the background level. These results further lend support to the involvement of ammonia in disease suppression *in vivo*.

The *P. fluorescens* strain CHA0 produces a number of secondary metabolites (discussed earlier), one of which is hydrogen cyanide (HCN). Strain CHA0 provides more effective plant protection in iron-rich soil than in iron-poor soil, and iron stimulates HCN production in this strain (Keel et al., 1989). To determine the role of HCN in disease suppression, Voisard et al. (1989) constructed a cyanide-negative (hcn^-) mutant, CHA5 (with an Ω-Hg insertion in the *hcn* genes), and tested its disease control efficacy against black root rot. Strain CHA5 provided reduced disease suppression when compared to strain CHA0, but was better than the no bacteria control. Introduction of a plasmid carrying the hcn^+ genes (pME3013) into strain CHA5 restored HCN production (to higher than the wild-type level) and disease control efficacy. The hcn^+ genes, when inserted into strain P3, a *P. fluorescens* strain that is naturally hcn^- and provides poor plant protection, conferred upon the recombinant the ability to produce HCN and provide improved plant protection (Voisard et al., 1989). Strain CHA5 did not protect wheat against *G. graminis* var. *tritici* (Défago et al., 1990). Interestingly, another hcn^- mutant, CHA77 (a deletion mutant), protected wheat against *G. graminis* var. *tritici* at wild-type levels, though it behaved similar to strain CHA5 in the tobacco system (Haas et al., 1991). These results raise some interesting questions about the potentially different role of HCN in different host–pathogen systems.

Strain CHA0 caused root stunting and proliferation of long root hairs in tobacco (Défago et al., 1990). As in disease suppression, strains CHA5 and P3 had little effect on root hair formation, while CHA5(pME3013) and the P3 derivative containing the hcn^+ genes caused increased root hair formation (Défago et al., 1990). Thus, HCN appeared to have a direct effect on plant roots and led the authors to raise the possibility that HCN may induce plant defense mechanisms (Voisard et al., 1989; Défago et al., 1990).

2.5. Hydrolytic Enzymes

It has long been recognized that extracellular hydrolytic enzymes synthesized and released by soil microbes might play a role in the antagonism of phytopathogenic fungi. In particular, numerous correlations between fungal antagonism and bacterial production of chitinases and/or β-1,3,-glucanases have been noted (e.g., Mitchell and Alexander, 1961; Mitchell and Alexander, 1963; Skujins et al., 1965; Sneh, 1981; Inbar and Chet, 1991b; Lim et al., 1991). Chitin and β-1,3-glucans are major constituents of many fungal cell walls (Sietsma and Wessels, 1979), and various workers have demonstrated *in vitro* lysis of fungal cell walls either by bacterial chitinase or β-1,3-glucanase alone or by a combination of both enzymes (Skujins et al., 1965; Kobayashi et al., 1974; Ordentlich et al., 1988, Shapira et al., 1989, Fiske et al., 1990). Such studies have lent support to the hypothesis that these hydrolytic enzymes contribute to biocontrol efficacy. Recently, several attempts to test this hypothesis with genetic and molecular biology approaches have been made.

The cloning of chitinase genes from *Serratia marcescens* QMB1466 by Jones et al. (1986) allowed the disruption of the *chiA* locus and the recombinational exchange of the disrupted gene into the *Serratia* chromosome. The ChiA deficient mutant demonstrated reduced chitinase production, reduced inhibition of fungal germ tube elongation, and reduced biological control of *F. oxysporum* f.sp. *pisi* infection of pea seedlings in a greenhouse assay. It was proposed that chitinase most likely was acting in concert with an array of other compounds produced by *Serratia* to slow fungal growth. Sundheim et al. (1988) cloned chitinase genes from *Serratia marcescens* BJL 200, and introduced one such chitinase gene into a *P. fluorescens* strain on a broad-host-range plasmid. Culture filtrates of a *Pseudomonas* transconjugant that expressed chitinase activity were reported to inhibit *in vitro* the germ tube growth of two *F. oxysporum* strains. Enhanced biocontrol of *F. oxysporum* f.sp. *redolens* by the *Pseudomonas* transconjugant on radish in a growth chamber experiment was also observed, although in this instance the effect was slight and the plant population tested was small. A *Pseudomonas stutzeri* isolate with native chitinase and β-1,3-glucanase activities was identified recently (Lim et al., 1991) and demonstrated to inhibit growth of *Fusarium solani in vitro*. Chitinase under-producing and over-producing mutants were recovered by mutagenesis either with nitrosoguanidine or ultraviolet (UV) light, and fungal inhibition was positively correlated with chitinase production in *in vitro* assays. It was not determined in this study whether the *P. stutzeri* strain, or any of the mutant derivatives, exhibited biocontrol activity in greenhouse experiments. Shapira et al. (1989) constructed an *Escherichia coli* strain harboring the *Serratia chiA* gene on a multicopy plasmid. Optimal expression of *chiA* in this system was temperature dependent and was driven by the strong bacteriophage lambda P_L promoter. Greenhouse experiments indicated that daily application of the chitinase-producing *E. coli* strain reduced disease incidence of *Sclerotium rolfsii* on bean plants.

Interest also has increased in testing the antifungal efficacy of bacterial chitinases and glucanases by expressing them in plants, where it is hoped that their activity will complement existing antifungal plant defenses, which often already include chitinases and glucanases (reviewed in Boller, 1985; Carr and Klessig, 1989; and Bol et al., 1990). Since plant and bacterial chitinases often differ in their mechanism of action (Roberts and Selitrennikoff, 1988), it is possible that bacterial and plant hydrolytic enzymes will have a synergistic effect that might inhibit the growth of fungal pathogens that might otherwise overcome the activity of the plant enzyme(s) alone. Jones et al. (1988) expressed the *chiA* gene of *S. marcescens* in tobacco, while Fiske et al. (1990) cloned two genes with β-1,3-glucanase activity from *Bacillus circulans* as a first step toward expressing them in plants. Success in obtaining increased resistance to *R. solani* in tobacco plants expressing the *S. marcescens* chitinase has been reported recently by Dunsmuir et al. (1991), and Suslow et al. (1988) observed enhanced resistance to *Alternaria* in such plants.

As noted earlier, numerous soil bacteria with biological control activity have been identified that possess chitinase and/or β-1,3-glucanase activities. With few exceptions, though, little is known about whether possession of such enzymes plays a role in biocontrol. Do these enzymes inhibit fungal growth directly *in vivo* (e.g., by lysis of hyphal cell walls as has often been proposed), or indirectly [by releasing plant or fungal wall fragments, which might elicit a plant defense response (Keen and Yoshikawa, 1983)]? On the other hand, do these enzymes only function in many instances to allow the bacterium to utilize as carbon source the fungal cell wall material of organisms killed by other mechanisms (e.g., antibiotics)? It will be necessary to extend the research of Jones et al. (1986) with the *S. marcescens* system to other biocontrol microorganisms by cloning the genes encoding various hydrolytic enzymes in other microbes, inactivating them, reintroducing the mutant versions into the host microbes' genomes, and assessing the impact in biocontrol assays as a step toward answering such questions.

2.6. Induction of Systemic Resistance

The acquisition of systemic resistance to a variety of pathogens by plants previously inoculated either with a pathogen or with inducing chemical compounds has been documented in numerous systems (reviewed in Sequeira, 1983; Kuc, 1987; Ryals et al., 1991), and progress has been made at understanding plant responses at the molecular level (e.g., Ward et al., 1991). An intriguing question (and an exciting area for research now being pursued by several groups) is whether any of the molecular events that occur when a plant "recognizes" a pathogen and mounts a defense to it also are triggered when plants are colonized by so-called beneficial microorganisms. For example, one may speculate that aggressive colonization of host plant tissue by a microbe that is not itself a pathogen may, as a result of metabolic activity or other means, generate a "signal" that induces plant defense responses. Prior inoculation of a plant with such a microbe would be expected to enhance resistance to later challenges by pathogenic microbes.

Kempe and Sequeira (1983) conducted a study in which they attempted to induce resistance in potato to a highly virulent strain of *Pseudomonas solanacearum* by prior inoculation of potato tubers with various pseudomonads, including two *P. fluorescens* strains. Interestingly, dipping potato eyepieces in suspensions of some of the pseudomonads, including the *P. fluorescens* isolates, reduced disease severity when plants were challenged 3–4 weeks later with stem inoculations of the virulent *P. solanacearum* strain. Anderson and Guerra (1985) suggested that colonization with beneficial isolates of *P. putida* might increase the defensive potential of bean plants against *F. solani* f. sp. *phaseoli* when it was noted that lignin increased in root tissue in response to *P. putida* colonization. This notion was further supported by the observation that colonization of bean roots by *P. putida* enhanced root surface peroxidase activity (Albert and Anderson, 1987). The production of HCN by *P. fluorescens* CHAO has been suggested as another potential inducer of plant defense mechanisms by the generation of cyanide stress in root tissue (Défago et al., 1990; see Section 2.4). The induction of plant defenses through herbicidal activity has been proposed as one possible mechanism by which the production of the secondary metabolite 2,4-diacetylphloroglucinol by this same strain might provide tobacco with protection from *T. basicola* Keel et al., 1992).

Van Peer et al. (1991) examined whether the biocontrol of *Fusarium* wilt of carnation by *Pseudomonas* sp. strain WCS417r involved an induce resistance mechanism. When root-bacterized carnations were challenged 1 week later by stem inoculation with *F. oxysporum* (i.e., pathogen and antagonist were separated spatially to rule out competition for nutrients), a significant reduction in *Fusarium* wilt was noted relative to that seen either when plants were not bacterized with the *Pseudomonas* strain or when bacterization was simultaneous with the *Fusarium* stem inoculation. Phytoalexin accumulation in the stem was implicated, but did not occur in the absence of the pathogen. The authors suggested that bacterization of the roots with the pseudomonad generated signals that induced sensitization of the stem, ultimately leading to higher phytoalexin accumulation when the fungal challenge occurred.

A classic demonstration of induced systemic resistance in plants involves a cucumber/*Colletotrichum* system (Kuc, 1987). Treatment of a lower leaf with pathogen induces plant defense responses that protect other portions of the plant from the full effects of a later challenge with *Colletotrichum* (or a variety of other fungal, viral, and bacterial pathogens) (Kuc, 1987). The cucumber/*Colletotrichum* model was utilized to screen a collection of 94 plant growth-promoting rhizobacteria (PGPR) for the presence of bacterial strains which, when applied as a seed treatment, could induce resistance in a highly susceptible cucumber cultivar to *Col*-

letotrichum orbiculare applied as a leaf inoculum (Wei et al., 1991). Six such strains were identified. None of the six could be recovered from petiole samples on the day of inoculation with *C. orbiculare,* indicating that the bacteria and the fungal pathogen were spatially separated and could not be interacting directly, thereby ruling out such explanations as competition or antagonism for the protective effect. Necrosis of root tissue did not appear to be required for the induction of resistance.

The mounting evidence that microbial root colonizers that are not themselves pathogens (and which, in the study of Wei et al. (1991) above, actually led to enhanced plant growth relative to nonbacterized controls) can induce resistance to plant pathogens raises a variety of questions which, as they are answered, may lead to the development of biological control strains with more effective combinations of activities. For instance, what is the nature of the inducing factor(s)? Are particular cell surface components recognized by responsive host plants? Do phytotoxic bacterial secondary metabolites trigger a defense response (Défago et al., 1990; Keel et al., 1992)? It is quite possible that multiple means of inducing resistance exist, and some of these may prove to be amenable to manipulation by recombinant DNA techniques such that inducing ability might be transferred to other potent biocontrol strains as an additional complementary mode of action.

2.7. Competition for Nutrients or Elimination of Signals

Competition for nutrients, like colonization, is generally believed to have a role in biological control probably more because the concept seems to make sense than because of conclusive evidence. The concept is extremely difficult to prove. It lacks the simplicity of a central element, like antibiotics, which one can perturb and observe the consequences. One exception is the competition for a specific, limiting nutrient, such as iron, where specific molecules, such as siderophores (discussed in Section 2.3), are elaborated for its acquisition.

Most nutrients in the rhizosphere come from plant exudates. Some compounds serve not only as nutrients, but also as signals that initiate interactions between plants and microorganisms. Most fungal pathogens exist in a state of dormancy in the soil (Lockwood, 1977). Infection occurs only after dormancy is broken in the presence of signals or stimulants from the plant host. A biocontrol agent can provide plant protection by efficient removal of such signals or stimulants from the rhizosphere (Elad and Chet, 1987; Nelson and Craft, 1991; Paulitz, 1991).

Elad and Chet (1987) identified 6 bacterial strains out of 130 examined to be effective biocontrol agents against *Pythium* spp. No lytic enzymes were detected in these strains. No correlation was found between disease suppression and production of inhibitory subtsances or competition against fungal mycelial growth in liquid culture. Significant correlation was found between disease suppression and the ability to inhibit oospore germination. The inhibition required the presence of growing bacterial cells; bacterial culture supernatant did not slow inhibition. They proposed that bacterial competition for nutrients in the rhizosphere might have led to the reduction of oospore germination. More recently, Nelson and Craft (1991) showed that the growth of disease suppressive strains of *E. cloacae* in cotton seed exudate reduced the stimulating activity of the exudate on *Pythium* sporangium germination. Addition of concentrates of the treated exudate to fresh untreated exudate did not reduce the stimulating activity of the untreated exudate, indicating the absence of stimulating substances rather than the presence of inhibitory substances in the treated exudate. Treatment with nonsuppressive bacterial strains did not reduce the stimulating activity of seed exudates. It was suggested that metabolism of propagule germination stimulants may be an important trait in the biological control of *Pythium* diseases.

Besides soluble stimulants, germinating seeds also evolve volatile substances that stimulate fungal spore germination (Catska et al., 1975; Harman et al., 1978; Norton and Harman, 1985). Ethanol and acetaldehyde were identified as major components of cotton and pea seed volatiles (Vancura and Stotzky, 1976; Gorecki et al., 1985; Nelson, 1987), and both were shown to stimulate fungal spore germination (Harman et al., 1980; Nelson, 1987; Paulitz, 1991). Paulitz (1991) found that treatment of soybean and pea seeds with *P. putida* strain N1R increased plant emergence in *Pythium* infested soil. The bacterial strain did not produce antibiotics *in vitro,* and its siderophore did not affect biocontrol activity (Paulitz and Loper, 1991). Seed treatment with the bacteria reduced both the stimulating effect of seed volatiles on fungal growth and the concentrations of ethanol and acetaldehyde produced by the germinating seeds. He suggested that the disease suppressiveness of strain N1R might be the result of metabolism of the seed volatiles (Paulitz, 1991). With the identification of the specific stimulants, it should be possible to determine the role of bacterial metabolism of these stimulants in biocontrol by examining mutants that can no longer metabolize them.

3. EMERGING METHODS AND APPROACHES

3.1. Monitoring Biocontrol Bacteria in the Rhizosphere

A variety of approaches exist to detect the presence and persistence of particular biocontrol strains once they have been applied to seeds, roots, or soil. Depending on the particular requirements of the researcher, one or more of these techniques might have particular advantages. Some of the methods useful in monitoring the presence of particular rhizobacteria are discussed below with representative examples of how such methods have been applied.

Strains marked either with spontaneously arising resistances to such antibiotics as rifampicin and nalidixic acid or with antibiotic resistance genes introduced on transposable elements often have been used in experiments that estimate population size by dilution plating on selective media (e.g., Kloepper et al., 1980). The introduction of genes encoding β-galactosidase and lactose permease into fluorescent pseudomonads that otherwise lack these activities (Drahos et al., 1986, Lam et al., 1990) is another means by which root-colonizing populations can be estimated. Using a Tn5-*Lac* transposon, Lam et al. (1990) isolated a collection of Lac[+] mutants of *P. fluorescens* and developed a competition assay to identify a subset of marked mutants, which were competition-defective relative to the wild-type strain (see Section 3.3).

The use of *Vibrio* bioluminescent genes by Shaw and Kado (1986) to monitor the presence and movement of phytopathogenic bacteria during the disease process also has been extended to the tracking of root-colonizing microbes. Mahaffee et al. (1991) introduced the *lux* operon into several rhizobacteria and were able to detect the presence of a *P. putida* isolate on seed and seedlings with a luminometer, a scintillation counter, and with a charge coupled device (CCD) camera. A limitation encountered was that the strain could not be directly detected if field soil were used, possibly due to microbial competition, which reduced metabolic activity (and light production) of the test strain. The authors suggested that addition of a strong promoter to the *lux* operon should greatly increase light production, improving direct detection in field soil. The problem of the high energy demand upon bacterial light production was circumvented by de Weger et al. (1992), who utilized a system in which *lux* operon expression was dependent on a promoter that could be induced upon exposure of the bacterium to naphthalene or salicylate. The addition of these inducers to roots immediately prior to detection by autophotography allowed sensitive detection in a sand–vermiculite system. Use of this system with field soil was not reported. Beauchamp et al. (1991) examined the effect of host range on root colonization by rhizobacterial strains using introduced *lux* genes to visualize sites of root colonization. Root sections from seedlings grown in a plexiglass rhizobox were embedded in a selective agar medium and, after 3 days of further bacterial growth, sites of root colonization by the bioluminescent bacteria were observed. The Lux[+] population size was also evaluated by dilution plating followed by detection of light production.

In some instances, a particular characteristic of a rhizobacterial strain can be used to allow measurement of population size and determine spatial distribution in the rhizosphere. Inbar and Chet (1991a) used an image analysis technique to determine chitinolytic activity associated with roots as a result of colonization by the chitinase-producing biocontrol strain *Aeromonas caviae*. They found a logarithmic correlation between root chitinolytic activity (as determined by the image analysis) and the number of chitinolytic colony forming units along the roots. This method also provided qualitative data concerning the distribution of *Aeronomas caviae* along the roots with various inoculation methods.

A variety of immunological methods have been described for the detection of soil bacteria from soil samples (e.g., Tsuchiya, 1991; Levanony and Bashan, 1991; van Vuurde, 1991). These techniques have high sensitivity and specificity, making them useful in measurement of bacterial persistence in the environment, and in some instances can be adapted for *in situ* study of root colonization sites and bacterial cell densities (van Vuurde, 1991). Note also that use of immunological techniques avoids the controversy associated with marker genes concerning whether host fitness has been compromised in any way.

The DNA probe technology represents another choice for monitoring microbes introduced into the environment. Its sensitivity is particularly useful for following the persistence and spread in the environment of the introduced strain. Barry et al. (1990) described a general method for generating DNA probes that can discriminate among even closely related bacterial species. By designing universal polymerase chain reaction primers that anneal to conserved regions on either side of variable regions within 16S rRNA genes, they were able to easily amplify species specific probes from a variety of bacterial species. It was demonstrated that a probe specific for one *Aeromonas* species would not hybridize with DNA from three other *Aeromonas* species. Of course, in instances where the particular soil environment contains organisms from the same species as that of the introduced organism, other approaches might be required for the generation of a probe capable of intraspecies discrimination (Gill et al., 1991). The DNA amplification strategies, such as the polymerase chain reaction (PCR) (Mullis and Faloona, 1987), can also be diagnostic for the presence of bacteria in environmental samples. Again, the goal is to generate primers that will only amplify a diagnostic DNA fragment from the specific strain being monitored. Polymerase chain reaction amplification strategies detecting one cell of *Pseudomonas cepacia* per gram of river sediment (Steffan and Atlas, 1988) and one cell of *E. coli* per 100 ml of water (Bej et al., 1990) have been reported.

3.2. Monitoring Expression of Genes Involved in Biocontrol

The ability to measure gene expression is critical to developing an understanding of parameters within which a biocontrol agent will function optimally. For example, if antibiotic biosynthesis by a biocontrol strain makes a major contribution to the biocontrol efficacy of the strain, it is necessary to determine which factors regulate this synthesis to make predictions about which soil types or environmental conditions will be compatible with strain performance (e.g., Gutterson, 1990). In the following examples, gene expression measurements provided information about *in vitro* and *in situ* activity of biocontrol-related genes, allowed the identification of genes responding to components of the rhizosphere, allowed studies of nutrient availability in various soil types and rhizospheres, and allowed the identification of putative positive regulatory genes involved in siderophore biosynthesis by a PGPR strain.

The use of promoterless reporter genes introduced, either by transposition or by *in vitro* recombinant DNA methodologies, into the coding regions of genes involved in biocontrol is increasing. A transcriptional fusion was generated between the *afuE* locus of *P. fluorescens* Hv37a, which is involved in synthesis of the antibiotic oomycin A, and a promoterless *lux* gene cassette (Gutterson, 1990). This construction was introduced into the *P. fluorescens* chromosome, and it was possible to monitor bioluminescence to gain insight into environmental factors affecting oomycin A production, as well as the timing of antibiotic production. For example, it was possible to determine that in a single soil type (Hesperia Fine Sandy Loam) the expression of *afuE* was significantly reduced at 24°C relative to the level measured at 20°C. This type of analysis, then, allows the prediction that the biocontrol strain Hv37a might not function as effectively under warmer weather conditions, and points toward attempting to diminish variability in biocontrol under field conditions by altering regulation of antibiotic biosynthesis genes (Gutterson, 1990). It was noted that the *lux* system was useful at early times after bacterial inoculation, but that at later times the energy demands of bioluminescence caused the relationship between *lux* transcription and light production to break down. Pierson and Thomashow (1991) fused a phenazine antibiotic biosynthetic locus from *Pseudomonas aureofaciens* strain 30-84 to a β-galactosidase reporter gene. Influence of various liquid culture growth conditions upon β-galactosidase production were assessed. It was determined that iron stimulated transcription of the phenazine biosynthetic locus as much as eightfold, while the carbon sources glucose and gluconate exerted a catabolite repression effect. Loper and Lindow (1991) cloned an iron-regulated promoter from a biosynthesis gene for the siderophore pyoverdine upstream of a promoterless icenucleation gene. This construction allowed them to monitor *in situ* expression of the pyoverdine gene in the rhizosphere. Since the level of *inaZ* expression varied 100,000-fold depending on whether a medium was iron-deplete or iron-replete, it was also possible to use the strain harboring the gene fusion as a biological probe for determining the concentration of available iron in the rhizosphere. Measurements of ice-nucleation activity track *in situ* gene expression with great sensitivity based upon a quantitative relationship between the amount of InaZ protein in the bacterial outer membrane and ice-nucleation activity (Loper and Lindow, 1991; Lindow, 1991). The *inaZ* system may be most appropriate for *in situ* work with respect to the reporter gene approach, since detection of ice-nucleation activity does not require high levels of bacterial metabolic activity as is the case with the *lux* system (Lindow, 1991).

The ability to utilize reporter genes to generate biological sensors of rhizosphere nutrient availability was also demonstrated by de Weger et al. (1991). Using a Tn5 derivative, which contained a promoterless *lacZ* gene, they mutagenized *P. putida* strain WCS358 and screened the mutant collection for members in which β-galactosidase production was stimulated under phosphate-limited conditions.

One such mutant, when applied either to roots or sand and reisolated for β-galactosidase activity determinations, showed background levels of expression in phosphate-rich systems and high β-galactosidase levels in phosphate-deficient systems.

Leong et al. (1991) took advantage of information concerning how a gene involved in siderophore biosynthesis was regulated to devise a strategy using a *xylE* reporter gene, which encodes an enzyme capable of converting the substrate catechol to an assayable yellow product, to identify clones encoding potential positive regulators. A fusion between one of the genes involved in pseudobactin biosynthesis by *P. putida* WCS358 and *xylE* only expressed *xylE* when strain WCS358 was grown under iron limitation. However, when the construct was introduced into a different host, *Pseudomonas* A225, no response to iron limitation was observed. Reasoning that strain A225 lacked a positive regulator capable of activating the siderophore promoter, they mobilized a cosmid clone bank into the A225 derivative containing the reporter gene construct. Two cosmid clones were identified that allowed *xylE* expression only under iron limitation, and the potential activator genes are being further analyzed.

Reporter genes also have been delivered at random by transposition into the genomes of biocontrol strains in screening efforts aimed at identifying genes responding positively to the presence of rhizosphere components (Lam et al., 1990; Banks et al., 1991; see Section 3.3).

In summary, several gene expression monitoring tools have been developed or adapted for use with rhizosphere bacteria. These tools have been used in a variety of creative approaches, and their use is expanding rapidly. Perhaps one of their most important contributions will be to explain the observed inconsistencies in field performance of certain biological control strains by defining conditions under which the expression of genes involved in biocontrol efficacy is either induced or repressed. Conceivably, the increased predictability that such tools can provide will lead to greater agricultural use of biological control strains.

3.3. Genetic Approaches

Mutant analysis has been used to establish the roles of a number of biological activities in disease control. The approach is particular useful in determining the relative contributions of different genetic traits in complicated processes. By altering one trait at a time, its effect on the whole process can be observed. One must keep in mind, however, that the interpretations are based entirely on the assumption that the trait in question is the only alteration in the mutant. Certain chemical mutagens, such as nitrosoguanidine, are notorious for the induction of multiple mutations. The use of transposons has largely circumvented the problem of unsuspected multiple mutations, but even single insertions can result in pleiotropic mutants, as illustrated by some of the Phz⁻ mutants isolated from strain 2-79 (see Section 2.2). It is prudent to examine a number of independently derived mutants.

Besides deciphering the significance of known traits, genetic analysis can also be used to uncover previously unknown traits involved in complicated processes such as root colonization and disease suppression. One starts with a random collection of transposon mutants, and directly examine the phenotype of each mutant with respect to the process of interest (e.g., the ability to control disease). Mutants that exhibit altered phenotype are presumed to have insertions in genes that are involved in the process. The approach is laborious but can potentially reward one with the discovery of genetic traits that may not be suspected otherwise. Lam et al. (1990) illustrated this approach with two examples. In the first example, mutants defective in competitive colonization were sought. A collection of insertion mutants of *P. fluorescens* strain 2-79 was generated using a Tn5 derivative containing a constitutively expressed *lacZ* gene. Since strain 2-79 was naturally *Lac⁻*, the insertion mutants could be distinguished from the parent on X-gal plates (on which Lac⁺ colonies turn blue and Lac⁻ colonies stay white). Each mutant was examined for its ability to colonize wheat seedling roots in the presence of the wild-type parent. If an insertion mutant was as competitive as the wild-type parent, then the mutant/wild-type ratio would be the same at the beginning (input) and at the end of the experiment (output). Mutants that were unable to compete with the wild-type parent were obtained (output ratio <1:20 after normalization of input ratio to 1). Recently, Lugtenberg et al. (1992) initiated an effort to identify colonization-defective mutants of *P. fluorescens* strain WCS365 by such an approach. So far, 13 mutants have been identified. Some of the mutants were auxotrophic, some were nonmotile, and others were nonmotile as well as fluorescence negative. The mutants were not different from the wild-type parent with respect to the ability to colonize plant roots in the absence of the wild-type parent, growth rates in different lab media, outer membrane protein and lipopolysaccharide patterns, and attachment to plant roots. In the second example, bacterial genes inducible by plant root exudate were sought in strain 2-79 (Lam et al., 1990). Such genes were expected to be involved in bacteria–plant interactions. A Tn5 derivative containing a promoterless *lacZ* gene was used to generate a mutant collection. Expression of the *lacZ* gene depended on the presence of a promoter outside the transposon. Mutants that showed differential *lacZ* gene expression in the presence and absence of plant root exudate were identified. In a similar approach, Banks et al. (1991) introduced Tn*phoA*, a Tn5 derivative capable of generating translational fusions between an interrupted gene and an alkaline phosphatase reporter, into *P. putida* strain GR12-2 as a step toward identifying seed exudate-responsive genes encoding non-cytoplasmic proteins.

4. CONCLUDING REMARKS

The use of *A. radiobacter* strain K84 to control crown gall is one of the most successful examples of biological control. Besides what we discussed earlier (Section 2.2), two additional features contribute to the success of this biocontrol system. First, there is a single specific target pathogen. Second, the pathogen is a bacterium closely related to the control agent. The responses of the pathogen and the control agent to environmental changes are expected to be closely matched; conditions that favor the development of the pathogen will also favor proliferation of the control agent. Unfortunately, most economically significant plant pathogens are fungi, and some conditions that favor fungal development may be unfavorable for the biocontrol bacteria. In the field, crop losses are frequently due to the presence of more than one pathogen. As a result, bacterial strains that have been shown to provide good protection against specific fungal pathogens in the greenhouse often fail to perform consistently in the field.

For most practical applications, we must extend the performance range, both in terms of environmental conditions and pathogen spectrum, of any potential biocontrol product. One approach may be to use a mixture of strains (see, e.g., Cook et al., 1988), each of which presumably performs optimally under somewhat different conditions. This approach has not been given much attention but is well worth investigating. One question of concern is whether a stable population structure can be maintained within a mixture such that the component strains are not lost as a result of competition. Another approach may be to genetically improve our current candidate strains. Before we can do that rationally, we need better understanding of the environmental effects on the population as well as the genetic expression of critical biological activities of the bacterial strains. Current method developments discussed in Section 3 can be expected to contribute significantly to research in these areas.

As we seek biocontrol strains with broad disease control spectrum, we may find that the strains that pass the increasingly demanding tests possess a large arsenal of weapons. The *P. fluorescens* strain CHA0, for example, produces a variety of antagonistic compounds, each of which may contribute differently to the suppression of different pathogens (Défago et al., 1990; Haas et al., 1991). Even in the suppression of a single pathogen, multiple control mechanisms are the rule rather than exception. Most antibiotic negative mutants, for example, still retain significant disease control activity (Thomashow and Weller, 1988; Voisard et al., 1989; Howie and Suslow, 1991; Keel et al., 1992).

Meanwhile, attempts to genetically modify biocontrol strains have begun, with very interesting results. A mutant of Hv37a that produced oomycin A constitutively was shown to provide improved control of *Pythium* infections (Howie and Suslow, as cited in Gutterson, 1990). The production of pyoluteorin (Plt) and 2,4-diacetylphloroglucinol (Phl) were increased when the cosmid pME3090 from strain CHA0 was reintroduced into the CHA0 background. The resulting strain, CHA0/pME3090, provided improved disease suppression on cucumber but showed a toxic effect on cress and sweet corn. Both Plt and Phl were shown to have phytotoxic effects, but cucumber was less sensitive to Phl than the other plants (Maurhofer et al., 1992). These results indicate that more antibiotics is not always better. The overall effect depends on the host–pathogen system, and has to be determined empirically. Several DNA regions have been isolated which, when introduced into unrelated recipient bacteria, conferred the ability to produce antibiotics that were undetectable in the recipient previously (Hill et al., 1990; Thomashow, 1991; Vincent et al., 1991). Improvement in disease suppression was observed in one case where the recipient was not normally biocontrol effective (Hill et al., 1990). No improvement was observed in another case where the recipient strains already had good disease suppressive activity (Vincent et al., 1991).

In the past decade, an interdisciplinary approach has contributed much to our current understanding of the biological activities involved in disease control by selected bacteria. Now an even broader coalition of disciplines will be needed as detailed investigations extend into the natural rhizosphere. Attempts to improve biocontrol strains by genetic modification are beginning.

And the pace quickens.

ACKNOWLEDGMENTS

We thank all our colleagues who generously provided us with reprints and unpublished manuscripts.

REFERENCES

Ahl P, Voisard C, Défago G (1986): Iron bound-siderophores, cyanic acid and antibiotics involved in suppression of *Thielaviopsis basicola* by a *Pseudomonas fluorescens* strain. J Phytopathol, 116:121–134.

Albert F, Anderson AJ (1987): The effect of *Pseudomonas putida* colonization on root surface peroxidase. Plant Physiol 85:537–541.

Anderson AJ, Guerra D (1985): Responses of bean to root colonization with *Pseudomonas putida* in a hydroponic system. Phytopathology 75:992–995.

Anderson, AJ, Habibzadegah-Tari P, Tepper CS (1988): Molecular studies on the role of a root surface agglutinin in adherence and colonization by *Pseudomonas putida*. Appl Environmental Microbiol 54:375–380.

Baker KF (1987): Evolving concepts of biological control of plant pathogens. Annu Rev Phytopathol 25:67–85.

Bakker PAHM, Bakker AW, Marugg JD, Weisbeek PJ, Schippers B (1987): Bioassay for studying the role of siderophores in potato growth stimulation by *Pseudomonas spp.* in short potato rotations. Soil Biol Biochem, 19:443.

Bakker PAHM, Lamers JG, Bakker AW, Marugg JD, Weisbeek PJ, Schippers B (1986): The role of siderophores in potato tuber yield increase by *Pseudomonas putida* in a short rotation of potato. Neth J Plant Pathol 92:249–256.

Banks E, Bayliss C, Brown G, Cass L, Kerpal R, Lifshitz R, Wood JM (1991): Identification of *Pseudomonas putida* genes induced by canola seed exudate. In Keel C, Koller B, Défago G (eds): Plant Growth-Promoting Rhizobacteria— Progress and Prospects I. OBC/WPRS Bulletin, 14/8: 374–375.

Barry T, Powell R, Gannon F (1990): A general method to generate DNA probes for microorganisms. Bio/Technol, 8:233–236.

Beauchamp CJ, Kloepper JW, Lemke PA (1991): Root colonization of various host plants by *Tn5* bioluminescent pseudomonads. In Keel C, Koller B, Défago G (eds): Plant Growth-Promoting Rhizobacteria—Progress and Prospects. IOBC/WPRS Bulletin 14/8:243–247.

Becker JO, Cook RJ (1988): Role of siderophores in suppression of *Pythium* species and production of increased-growth response of wheat by fluorescent pseudomonads. Phytopathology, 78:778–782.

Bej AK, Steffan RJ, DiCesare J, Haff L, Atlas RM (1990): Detection of coliform bacteria in water by polymerase chain reaction and gene probes. Appl Environ Microbiol, 56:307–314.

Bennett RA, Lynch JM (1981): Colonization potential of rhizosphere bacteria. Curr Microbiol, 6:137–138.

Bol JF, Linthorst HJM, Cornelissen BJC (1990): Plant pathogenesis-related proteins induced by virus infection. Annu Rev Phytopathol 28:113–138.

Boller T (1985): Induction of hydrolases as a defense reaction against pathogens. In Key JL, Kosuge T (eds): Cellular and Molecular Biology of Plant Stress. New York: Alan R. Liss, pp 247–262.

Bowen GD (1991): Microbial dynamics in the rhizosphere: Possible strategies in managing rhizosphere populations. In Keister DL, Cregan PB (eds): The Rhizosphere and Plant Growth. Dordrecht: Kluwer Academic Publishers, pp 25–32.

Brand I, Lugtenberg BJJ, Glandorf DCM, Bakker PAHM, Schippers B, de Weger LA (1991): Isolation and characterization of a superior potato root-colonizing Pseudomonas strain. In Keel C, Koller B, Défago G (eds): Plant Growth-Promoting Rhizobacteria—Progress and Prospects. IOBC/WPRS Bulletin, 14/8:350–354.

Brisbane PG, Janik JJ, Tate ME, Warren RFO (1987): Revised structure for the phenazine antibiotic from Pseudomonas fluorescens 2-79 (NRRL 15132). Antimicrob Agents Chemother 31:1967–1971.

Buell CR, Anderson AJ (1991): Genetic analysis of agglutination in Pseudomonas putida. In Keel C, Koller B, Défago G (eds): Plant Growth-Promoting Rhizobacteria—Progress and Prospects. IOBC/WPRS Bulletin, 14/8:318–322.

Bull CT, Weller DM, Thomashaw LS (1991): Relationship between root colonization and suppression of *Gaeumannomyces graminis* var. *tritici* by *Pseudomonas fluorescens* strain 2-79. Phytopathology, 81:954–959.

Carr JP, Klessig DF (1989): The pathogenesis-related proteins of plants. In Setlow JK (ed): Genetic Engineering, Principles and Methods, Vol 11. New York: Plenum, pp 65–109.

Catska V, Afifi AF, Vancura V (1975): The effect of volatile and gaseous metabolites of swelling seeds on germination of fungal spores. Folia Microbiol, 20:152–156.

Chao WL, Li RK, Chang WT (1988): Effect of root agglutinin on microbial activities in the rhizosphere. Appl Environ Microbiol, 54:1838–1841.

Cook RJ, Baker KF (1983): The Nature and Practice of Biological Control of Plant Pathogens. American Phytopathological Society, St. Paul, Minnesota.

Cook RJ, Weller DM, Bassett EN (1988): Effect of bacterial seed treatments on growth and yield of recropped wheat in western Washington, 1987. Biol Cult Tests Control Plant Diseases, 3:53.

Cooksey DA, Moore LW (1982): High frequency spontaneous mutations to Agrocin 84 resistance in *Agrobacterium tumefaciens* and *A. rhizogenes*. Physiol Plant Pthol, 20:129–135.

Défago G, Berling CH, Burger U, Haas D, Kahr G, Keel C, Voisard C, Wirthner P, Wuthrich B (1990): Suppression of back root rot of tobacco and other root diseases by strains of *Pseudomonas fluorescens*: potential applications and mechanisms. In Hornby D (ed): Biological Control of Soil-Borne Plant Pathogens. Oxon: CAB International, pp 93–108.

de Weger LA, Bakker PAHM, Schippers B, van Loosdrecht MCM, Lugtenberg BJJ (1989): *Pseudomonas* spp. with mutational changes in the O-antigenic side chain of their lipopolysaccharide are affected in their ability to colonize potato roots. In Lugtenberg BJJ (ed): Signal Molecules in Plants and Plant–Microbe Interactions. NATO ASI Series, Vol H36. Heidelberg: Springer-Verlag, pp 197–202.

de Weger LA, Dekkers LC, Lugtenberg BJJ (1991): Use of reporter bacteria for studing the availability of phosphate in the rhizosphere and in soil. In Keel C, Koller B, Défago G (eds): Plant Growth-Promoting Rhizobacteria—Progress and Prospects. IOBC/WPRS Bulletin 14/8:223–226.

de Weger LA, Dunbar P, Mahafee WF, Lugtenberg BJJ, Sayler GS (1992): Use of bioluminescence reporters to detect *Pseudomonas* bacteria in the rhizosphere. In press.

de Weger LA, Jann B, Jann K, Lugtenberg B (1987a): Lipopolysaccharides of *Pseudomonas* spp. that stimulate plant growth: composition and use for strain identification. J Bacteriol 169:1441–1446.

de Weger LA, Lugtenberg BJJ (1990): Plantgrowth-stimulating rhizobacteria. In Heslot H, Davies J, Florent J, Bobichon L, Durand G, Penasse L (eds): Proceedings 6th International Symposium on Genetics of Industrial Microorganisms. Strasbourg: Société Francaise de Microbiologie, pp 827–838.

de Weger LA, van der Vlugt CIM, Wijfjes AHM, Bakker PAHM, Schippers B, Lugtenberg B (1987b): Flagella of a plant-growth-stimulating *Pseudomonas fluorescens* strain are required for colonization of potato roots. J Bacteriol, 169:2769–2773.

Drahos DJ, Hemming BC, McPherson S (1986): Tracking recombinant organisms in the environment: β-galactosidase as a selectable non-antibiotic marker for fluorescent pseudomonads. Bio/Technol, 4:439–444.

Dunsmuir P, Howie W, Newbigin E, Suslow T (1991): Resistance to *Rhizoctonia solani* in transgenic tobacco expressing the *Serratia marcescens* chitinase gene. In Hallick RB (ed): Program and Abstracts, The International Society for Plant Molecular Biology, Third International Congress: Molecular Biology of Plant Growth and Development. Tucson: p 1313.

Elad Y, Chet I (1987): Possible role of competition for nutrients in biocontrol of *Pythium* damping-off by bacteria. Phytopathology, 77:190–195.

Ellis JG, Kerr A (1978): Developing biological control agents for soil borne pathogens. In Ridé M (ed): Proceedings of the Fourth International Conference on Plant pathogenic Bacteria. Angers, France: pp 245–250.

Fiske MJ, Tobey-Fincher KL, Fuchs RL (1990): Cloning of two genes from *Bacillus circulans* WL-12 which encode 1,3-β-glucanase activity. J Gen Microbiol, 136:2377–2383.

Flaishman M, Eyal Z, Voisard C, Haas D (1990): Suppression of *Septoria tritici* by phenazine- or siderophore-deficient mutants of *Pseudomonas*. Curr Microbiol, 20:121–124.

Gill S, Mercier J-P, Lemieux C, Gagne S, Brown G, Dion P (1991): General approach to the design of strain-specific probes for plant growth-promoting rhizobacteria. In Keel C, Koller B, Défago G (eds): Plant Growth-Promoting Rhizobacteria—Progress and Prospects. IOBC/WPRS Bulletin, 14/8:408–411.

Gorecki RJ, Harman GE, Matick LR (1985): The volatile exudates from germinating pea seeds of different viability and vigor. Can J Bot 63:1035–1039.

Gurusiddaiah S, Weller DM, Sarkar A, Cook RJ (1986): Characterization of an antibiotic produced by a strain of Pseudomonas fluorescens inhibitory to *Gaeumannomyces graminis* var. *tritic* and *Pythium* spp. Antimicrob Agents Chemother 29:488–495.

Gutterson N (1990): Microbial fungicides: recent approaches to elucidating mechanisms. Crit Rev Biotechnol, 10:69–91.

Haas D, Keel C, Laville J, Maurhofer M, Oberhansli T, Schnider U, Voisard C, Wuthrich B, Défago
    G (1991): Secondary metabolites of *Pseudomonas fluorescens* strain CHAO involved in the sup-
    pression of root diseases. In Hennecke H, Verma DPS (eds): Advances in Molecular Genetics of
    Plant–Microbe Interactions, Vol. 1. Dordrecht: Kluwer Academic Publishers, pp 450–456.
Hadar Y,Harman GE, Taylor AG, Norton JM (1983): Effects of pregermination of pea and cucumber
    seeds and of seed treatment with *Enterobacter cloacae* on rots caused by *Pythium* spp. Phytopa-
    thology, 73:1322–1325.
Hamdan H, Weller DM, Thomashow LS (1991): Relative importance of fluorescent siderophores and
    other factors in biological control of *Gaeumannomyces graminis* var. *tritici* by *Pseudomonas
    fluorescens* 2-79 and M4-80R. Appl Environ Microbiol, 57:3270–3277.
Handelsman J, Parke J (1989): Mechanisms in biocontrol of soilborne plant pathogens. In Kosuge T,
    Nester EW (eds): Plant–Microbe Interactions: Molecular and Genetic Perspectives, Vol. 3. New
    York: McGraw-Hill, pp 27–61.
Harman GE, Mattick LR, Nash GT, Nedrow BL (1980): Stimulation of fungal sore germination and
    inhibition of sporulation in fungal vegetative thalli by fatty acids and their volatile peroxidation
    products. Can J Bot, 58:1541–1547.
Harman GE, Nedrow B, Nash GT (1978): Stimulation of fungal spore germination by volatiles from
    aged seeds. Can J Bot., 56:2124–2127.
Harrison L, Kenfield D, Bunkers G, Strobel G (1988): Pseudomycin, a broadrange antimycotic from
    Pseudomonas syringae. Phytopathology, 78:1528 (Abstract).
Hill SD, Stein JI, Ligon JM (1990): Isolation and characterization of genes from Pseudomonas
    fluorescens that encode the biosynthesis of an antibiotic effective against Rhizoctonia solani. In
    Abstracts, 5th International Symposium on the Molecular Genetics of Plant–Microbe Interactions,
    Interlaken, Switzerland, p 35.
Homma Y, Chikuo Y, Ogoshi A (1991): Mode of suppression of sugar beet damping-off caused by
    Rhizoctonia solani by seed bacterization with *Pseudomonas cepacia*. In Keel C, Koller B, Défago
    G (eds): Plant Growth-Promoting Rhizobacteria—Progress and Prospects. IOBC/WPRS Bulletin,
    14/8:115–118.
Homma Y, Suzui T (1989): Role of antibiotic production in suppression of radish damping-off by
    seed bacterization with *Pseudomonas cepacia*. Annu Phytopathol Soc Jpn, 55:643–652.
Howell CR, Beier RC, Stipanovic RD (1988): Production of ammonia by *Enterobacter cloacae* and
    its possible role in the biological control of Pythium preemergence damping-off by the bacterium.
    Phytopathology, 78:1075–1078.
Howell CR, Stipanovic RD (1979): Control of *Rizoctonia solani* on cotton seedlings with *Pseudomo-
    nas fluorescens* and with an antibiotic produced by the bacterium. Phytopathology, 69:480–482.
Howell CR, Stipanovic RD (1980): Suppression of *Pythium ultimum*-induced damping-off of cotton
    seedlings by *Pseudomonas fluorescens* and its antibiotic, pyoluteorin. Phytopathology, 70:712–715.
Howie WJ, Cook RJ, Weller DM (1987): Effects of soil matric potential and cell motility on wheat
    root colonization by fluorescent pseudomonads suppressive to take-all. Phytopathology, 77:286–292.
Howie WJ, Correll M, Gutterson N, Suslow T (1988): Indirect evidence for oomycin A expression in
    situ: Effect of soil temperature, moisture, and texture. Phytopathology, (Abstract) 78:1558.
Howie WJ, Suslow T (1986): Effect of antifungal compound biosynthesis on cotton root colonization
    and *Pythium* suppression by a strain of *Pseudomonas fluorescens* and its antifungal minus isogenic
    mutant. Phytopathology, (Abstract) 76:1069.
Howie WJ, Suslow T (1991): Role of antibiotic biosynthesis in the inhibition of *Pythium ultimum* in
    the cotton spermosphere and rhizosphere by *Pseudomonas fluorescens*. Mol Plant Microbe Int,
    4:393–399.
Inbar J, Chet I (1991a): Detection of chitinolytic activity in the rhizosphere using image analysis.
    Soil Biol Biochem, 23:239–242.
Inbar J, Chet I (1991b): Evidence that chitinase produced by *Aeromonas caviae* is involved in the
    biological control of soil-borne plant pathogens by this bacterium. Soil Biol Biochem 23:239–242.

Iswandi A, Bossier P, Vandenabele J, Verstraete W (1987): Influence of the inoculation density of the rhizopseudomonad strain 7NSK2 on the growth and the composition of the root microbial community of maize (*Zea mays*) and barley (*Hordeum vulgare*). Biol Fertil Soils, 4:119–123.

Jones JDG, Dean C, Gidoni D, Gilbert D, Bond-Nutter D, Lee R, Bedbrook J, Dunsmuir P (1988): Expression of bacterial chitinase protein in tobacco leaves using two photosynthetic gene promoters. Mol Gen Genet, 212:536–542.

Jones JDG, Grady KL, Suslow TV, Bedbrook JR (1986): Isolation and characterization of genes encoding two chitinase enzymes from *Serratia marcescens*. EMBO J, 5:467–473.

Keel C, Defago G (1991): The fluorescent siderophore of Pseudomonas fluorescens strain CHAO has no effect on the suppression of root diseases of wheat. In Keel C, Koller B, Défago G (eds): Plant Growth-Promoting Rhizobacteria—Progress and Prospects. IOBC/WPRS Bulletin, 14/8:136–141.

Keel C, Voisard C, Berling CH, Kahr G, Défago G (1989): Iron sufficiency, a pre-requisite for the suppression of tobacco black root rot by *Pseudomonas fluorescens* strain CHAO under gnotobiotic conditions. Phytopathology, 79:584–589.

Keel C, Wirthner P, Oberhansli TH, Voisard C, Burger U, Haas D, and Défago G (1990): Pseudomonads as antagonists of plant pathogens in the rhizosphere: Role of the antibiotic 2,4-diacetylphloroglucinol in the suppression of black root rot of tobacco. Symbiosis 9:327–341.

Keel C, Schnider U, Maurhofer M, Voisard C, Laville J, Burger U, Wirthner P, Haas D, Défago G (1992): Suppression of root diseases by *Pseudomonas f luorescens* CHAO: importance of the bacterial secondary metabolite 2,4-diacetylphloroglucinol. Mol Plant Microbe Int, 5:4–13.

Keen NT, Yoshikawa M (1983): β-1,3-Endoglucanase from soybean releases elicitor-active carbohydrates from fungus cell walls. Plant Physiol, 71:460–465.

Kempe J, Sequeira L (1983): Biological control of bacterial wilt of potatoes: attempts to induce resistance by treating tubers with bacteria. Plant Disease, 67:499–503.

Kempf H-J, Schroth MN (1991): Detection of herbicolin A in crown and root tissues of wheat seedlings after inoculation with *Erwinia herbicola*. In Keel C, Koller B, Défago G (eds): Plant Growth-Promoting Rhizobacteria—Progress and Prospects. IOBC/WPRS Bulletin, 14/8:197.

Kempf H-J, Wolf G (1989): *Erwinia herbicola* as a biocontrol agent of *Fusarium culmorum* and *Puccinia recondita* f. sp. *tritici* on wheat. Phytopathology, 79:990–994.

Kloepper JW, Leong J, Tientze M, Schroth MN (1980): *Pseudomonas* siderophores: a mechanism explaining disease-suppressive soils. Curr Microbiol, 4:317–320.

Kloepper JW, Schroth MN (1981): Relationship of *in vitro* antibiosis of plant growth-promoting rhizobacteria to plant growth and the displacement of root microflora. Phytopathology, 71:1020–1024.

Kobayashi Y, Tanaka H, Ogasawara (1974): Purification and properties of F-1a, a β-1,3-glucanase which is highly lytic toward cell walls of *Piricularia oryzae* P2. Agric Biol Chem 38:973–978.

Kraus J, Loper JE (1991): Biocontrol of Pythium damping-off of cucumber by *Pseudomonas fluorescens* Pf-5: Mechanistic studies. In Keel C, Koller B, Défago G (eds): Plant Growth-Promoting Rhizobacteria—Progress and Prospects. IOBC/WPRS Bulletin, 14/8:172–176.

Kuc J (1987): Plant immunization and its applicability for disease control. In Chet I (ed): Innovative Approaches to Plant Disease Control. New York: Wiley, pp 255–274.

Lam BS, Strobel GA, Harrison LA, Lam ST (1987): Transposon mutagenesis and tagging of fluorescent *Pseudomonas*: Antimycotic production is necessary for control of Dutch elm disease. Proc Natl Acad Sci USA 84:6447–6451.

Lam ST, Ellis DM, Ligon JM (1990): Genetic approaches for studying rhizosphere colonization. Plant Soil, 129:11–18.

Lam ST, Torkewitz NR, Nautiyal CS, Dion P (1991): Impact of the ability to utilize a single substrate on colonization competitiveness. Phytopathology, (Abstract) 81:1177.

Leben SD, Wadi JA, Easton GD (1987): Effects of *Pseudomonas fluorescens* on potato plant growth and control of Verticillium dahliae. Phytopathology, 77:1592–1595.

Leong J, Bitter W, Koster M, Marugg JD, Venturi V, Weisbeek PJ (1991): Molecular analysis of iron transport in palnt growth-promoting *Pseudomonas putida* WCS358. In Keel C, Koller B, Défago G (eds): Plant Growth-Promoting Rhizobacteria—Progress and Prospects. IOBC/WPRS Bulletin, 14/8:127–135.

Levanony H, Bashan Y (1991): Enumeration and identification of rhizosphere bacteria by advanced immuno techniques. In Keel C, Koller B, Défago G (eds): Plant Growth-Promoting Rhizobacteria—Progress and Prospects. IOBC/WPRS Bulletin, 14/8:231–237.

Lim H, Kim Y, Kim S (1991): *Pseudomonas stutzeri* YPL-1 genetic transformation and antifungal mechanism against *Fusarium solani,* an agent of plant root rot. Appl Environ Microbiol 57:510–516.

Lindow SE (1991): Tests of specificity of competition among *Pseudomonas syringae* strains on plants using recombinant ice-strains and use of ice nucleation genes as probes of *in situ* transcriptional activity. In Hennecke H, Verma DPS (eds): Advances in Molecular Genetics of Plant–Microbe Interactions, Vol. 1. Dordrecht: Kluwer Academic Publishers, pp 457–464.

Lockwood JL (1977): Fungistasis in soils. Biol Rev, 52:1–43.

Loper JE (1988): Role of fluorescent siderophore production in biological control of Pythium ultimum by a *Pseudomonas fluorescens* strain. Phytopathology, 78:166–172.

Loper JE, Buyer JS (1991): Siderophores in microbial interactions on plant surfaces. Mol Plant–Microbe Inter, 4:5–13.

Loper JE, Ishimaru CA (1991): Factors influencing siderophore-mediated biocontrol activity of rhizosphere *Pseudomonas* spp. In Keister DL, Cregan PB (eds): The Rhizosphere and Plant Growth. Dordrecht: Kluwer Academic Publishers, pp 253–261.

Loper JE, Lindow SE (1991): A biological sensor for available iron in the rhizosphere. In Keel C, Koller B, Défago G (eds): Plant Growth-Promoting Rhizobacteria—Progress and Prospects. IOBC/WPRS Bulletin, 14/8:177–181.

Lugtenberg BJJ, Brand I, de Weger LA (1992): Plant root colonization by Pseudomonas. In Proceedings of the 3rd International Symposium on Pseudomonas Biology and Biotechnology, Trieste, Italy: in press.

Lugtenberg BJJ, de Weger LA, Bennett JW (1991): Microbial stimulation of plant growth and protection from disease. Curr Opinion Biotechnol, 6, in press.

Mahaffee WF, Backman PA, Shaw JJ (1991): Visualization of root colonization by rhizobacteria using a luciferase marker. In Keel C, Koller B, Défago G (eds): Plant Growth-Promoting Rhizobacteria—Progress and Prospects. IOBC/WPRS Bulletin, 14/8:248–251.

Maurhafer M, Keel C, Schnider U, Voisard C, Haas D, Défago G (1992): Influence of enhanced antibiotic production in *Pseudomonas fluorescens* strain CHA0 on its disease suppressive capacity. Phytopathology 82:190–195.

Mitchell R, Alexander M (1961): The mycolytic phenomenon and biological control of *Fusarium* in soil. Nature (London), 190:109–110.

Mitchell R, Alexander M (1963): Lysis of soil fungi by bacteria. Can J Microbiol 9:169–177.

Moore LW (1988): Use of Agrobacteriuym radiobacter in agricultural ecosystems. Microbiol Sci, 5:92–95.

Morris J, O'Sullivan DJ, Koster M, Leong J, Weisbeek PJ, O'Gara F (1992): Characterization of fluorescent siderophore-mediated iron uptake in *Pseudomonas* sp. strain M114: evidence for the existence of an additional ferric siderophore receptor. Appl Environ Microbiol, 58:630–635.

Mullis KB, Faloona FA (1987): Specific synthesis of DNA *in vitro* via a polymerase-catalyzed chain reaction. In Wu R (ed): Methods in Enzymology, Vol. 155. San Diego: Academic Press, pp 335–350.

Nelson EB (1987): Rapid germination of sporangia of Pythium species in response to volatiles from germinating seeds. Phytopathology, 77:1108–1112.

Nelson EB (1988): Biological control of *Pythium* seed rot and pre-emergence damping-off of cotton with *Enterobacter cloacae* and *Erwinia Herbicola* applied as seed treatments. Plant Dis., 72:140–142.

Nelson EB, Chao WL, Norton JM, Nash GT, Harman GE (1986): Attachment of *Enterobacter* cloacae to hypae of *Pythium ultimum*: Possible role in the biological control of *Pythium* pre-emergence damping-off. Phytopathology, 76:327–335.

Nelson EB, Craft CM (1991): Metabolism of *Pythium* sporangium germination stimulants by *Enterobacter* cloacae and other soil bacteria. Phytopathology, (Abstract). 81:1178.

Norton JM, Harman GE (1985): Responses of soil microorganisms to volatile exudates from germinating pea seeds. Can J Bot, 63:1040–1045.

Ordentlich A, Elad Y, Chet I (1988): The role of *Serratia marcescens* in biocontrol of *Sclerotium rolfsii*. Phytopathology, 78:84–88.

Parke JL (1991): Root colonization by indigenous and introduced microorganisms. In Keister DL, Cregan PB (eds): The Rhizosphere and Plant Growth. Dordrecht: Kluwer Academic Publishers, pp 33–42.

Paulitz TC (1991): Effect of *Pseudomonas putida* on the stimulation of *Pythium ultimum* by seed volatiles of pea and soybean. Phytopathology, 81:1282–1287.

Paulitz TC, Loper JE (1991): Lack of a role for fluorescent siderophore production in the biological control of *Pythium* damping-off of cucumber by a strain of *Pseudomonas putida*. Phytopathology, 81:930–935.

Pierson LS III, Thomashow LS (1991): Analysis of phenazine antibiotic production by *Pseudomonas aureofaciens* strain 30-84. In Keel C, Koller B, Défago G (eds): Plant Growth-Promoting Rhizobacteria—Progress and Prospects. IOBC/WPRS Bulletin, 14/8:119–121.

Roberts WK, Selitrennikoff CP (1988): Plant and bacterial chitinases differ in antifungal activity. J Gen Microbiol, 134:169–176.

Ryals J, Ward E, Metraux J-P (1991): Systemic acquired resistance: an inducible defense mechanism in plants. In Wray JL (ed): The Biochemistry and Molecular Biology of Inducible Enzymes and Proteins in Higher Plants. Cambridge: Cambridge University Press, in press.

Ryder MH, Jones DA (1990): Biological control of crown gall. In Hornby D (ed): Biological Control of Soil-Borne Plant Pathogens. Oxon: CAB International, pp 45–63.

Scher FM, Kloepper JW, Singleton CA (1985): Chemotaxis of fluorescent *Pseudomonas* spp. to soybean seed exudates *in vitro* and in soil. Can J Microbiol, 31:570–574.

Scher FM, Kloepper JW, Singleton CA, Zaleska I, Laliberte M (1988): Colonization of soybean roots by *Pseudomonas* and *Serratia* species: relationship to bacterial motility, chemotaxis, and generation time. Phytopathology 78:1055–1059.

Schippers B, Bakker AW, Bakker PAHM (1987): Interactions of deleterious and beneficial rhizosphere microorganisms and the effect of cropping practices. Annu Rev Phytopathol, 25:339–358.

Sequeira L (1983): Mechanisms of induced resistance in plants. Annu Rev Microbiol, 37:51–79.

Shapira R, Ordentlich A, Chet I, Oppenheim AB (1989): Control of plant diseases by chitinase expressed from cloned DNA in *Escherichia coli*. Phytopathology, 79:1246–1249.

Shaw JJ, Kado CI (1986): Development of a *Vibrio* bioluminescence gene-set to monitor phytopathogenic bacteria during the ongoing disease process in a non-disruptive manner. Bio/Technol, 4:560–564.

Sietsma JA, Wessels JGH (1979): Evidence for covalent linkages between chitin and β-glucan in a fungal wall. J Gen Microbiol, 114:99–108.

Skujins JJ, Potgieter HJ, Alexander M (1965): Dissolution of fungal cell walls by a streptomycete chitinase and β-(1,3)-glucanase. Arch Biochem Biophys, 111:358–364.

Sneh B (1981): Use of rhizosphere chitinolytic bacteria for biological control of *Fusarium oxysporum* f. sp. *dianthi* in carnation. Phytopath Z, 100:251–256.

Steffan RJ, Atlas RM (1988): DNA amplification to enhance detection of genetically engineered bacteria in environmental samples. Appl Environ Microbiol, 54:2185–2191.

Stutz EW, Défago G, Kern H (1986): Naturally occurring fluorescent pseudomonads involved in suppression of black root rot of tobacco. Phytopathology, 76:181–185.

Sundheim L, Poplawsky AR, Ellingboe AH (1988): Molecular cloning of two chitinase genes from *Serratia marcescens* and their expression in *Pseudomonas* species. Physiol Mol Plant Pathol, 33:483–491.

Suslow TV (1982): Role of root-colonizing bacteria in plant growth. In Mount MS, Lacy GH (eds): Phytopathogenic Prokaryotes, Vol. 1. London: Academic Press, pp 187–223.

Suslow TV, Matsubara D, Jones J, Lee R, Dunsmuir P (1988): Effect of expression of bacterial chitinase on tobacco susceptibility to leaf brown spot *Alternaria longipes*. Phytopathology, 78:1556.

Tari PH, Anderson AJ (1988): Fusarium wilt suppression and agglutinability of *Pseudomonas putida*. Appl Environ Microbiol, 54:2037–2041.

Thomashow LS (1991): Molecular basis of antibiosis mediated by rhizosphere pseudomonads. In Keel C, Koller B, Défago G (eds): Plant Growth-Promoting Rhizobacteria—Progress and Prospects. IOBC/WPRS Bulletin, 14/8:109–114.

Thomashow LS, Pierson LS III (1991): Genetic aspects of phenazine antibiotic production by fluorescent pseudomonads that suppress take-all disease of wheat. In Hennecke H, Verma DPS (eds): Advances in Molecular Genetics of Plant–Microbe Interactions, Vol. 1. Dordrecht: Kluwer Academic Publishers, pp 443–449.

Thomashow LS, Weller DM (1988): Role of a phenazine antibiotic from *Pseudomonas fluorescens* in biological control of *Gaeumannomyces graminis* var. *tritici*. J Bacteriol, 170:3499–3508.

Thomashow LS, Weller DM, Bonsall RF, Pierson LS (1990): Production of the antibiotic phenazine-1-carboxylic acid by fluorescent *Pseudomonas* species in the rhizosphere of wheat. Appl Environ Microbiol, 56:908–912.

Torkewitz NR, Lam ST (1991): The effect of the population level of biocontrol bacteria on disease control efficacy. Phytopathology, (Abstract) 81:1179.

Tsuchiya K (1991): The use of monoclonal antibodies as probes for the detection and characterization of *Pseudomonas cepacia*. In Keel C, Koller B, Défago G (eds): Plant Growth-Promoting Rhizobacteria—Progress and Prospects. IOBC/WPRS Bulletin, 14/8:256–259.

van Peer RG, Niemann GJ, Schippers B (1991): Induced resistance and phytoalexin accumulation in biological control of *Fusarium* wilt of carnation by *Pseudomonas* sp. strain WCS417r. Phytopathology, 81:728–734.

Van Vuurde JWL (1991): Immunofluorescence colony-staining (IFC) and immunofluorescence cell-staining as tools for the study of rhizosphere bacteria. In Keel C, Koller B, Défago G (eds): Plant Growth-Promoting Rhizobacteria—Progress and Prospects. IOBC/WPRS Bulletin, 14/8:215–222.

Vancura V, Stotzky G (1976): Gaseous and volatile exudates from germinating seeds and seedlings. Can J Bot, 54:518–532.

Vincent MN, Harrison LA, Brackin JM, Kovacevich PA, Mukerji P, Weller DM, Pierson EA (1991): Genetic analysis of the antifungal activity of a soilborne *Pseudomonas aureofaciens* strain. Appl Environ Microbiol, 57:2928–2934.

Voisard C, Keel C, Haas D, Défago G (1989): Cyanide production by *Pseudomonas fluorescens* helps suppress black root rot of tobacco under gnotobiotic conditions. EMBO J, 8:351–358.

Ward ER, Uknes SJ, Williams SC, Dincher SS, Wiederhold DL, Alexander DC, Ahl-Goy P, Metraux J-P, Ryals JA (1991): Coordinate gene activity in response to agents that induce systemic acquired resistance. Plant Cell, 3:1085–1094.

Wei G, Kloepper JW, Tuzun S (1991): Induction of systemic resistance of cucumber to *Colletotrichum orbiculare* by select strains of plant growth-promoting rhizobacteria. Phytopathology, 81:1508–1512.

Weller DM (1988): Biological control of soilborne plant pathogens in the rhizosphere with bacteria. Annu Rev Phytopathol, 26:379–407.

Weller DM, Cook RJ (1983): Suppression of take-all of wheat by seed treatments with fluorescent pseudomonads. Phytopathology, 73:463–469.

Weller DM, Howie WJ, Cook RJ (1988): Relationship between *in vitro* inhibition of *Gaeumannomyces graminis* var. *tritici* and suppression of take-all of wheat by fluorescent pseudomonads. Phytopathology, 78:1094–1100.

Xu GW, Gross DC (1986): Selection of fluorescent pseudomonads antagonistic to *Erwinia carotovora* and suppressive of potato seed piece decay. Phytopathology, 76:414–422.

≡17

DIAGNOSTIC TECHNIQUES FOR PLANT PATHOGENS

S. A. MILLER

The Ohio State University, Department of Plant Pathology, Wooster, Ohio 44691

T. R. JOAQUIM

Agri-Diagnostics Associates, Cinnaminson, New Jersey 01809

1. INTRODUCTION

Immunoassay and nucleic acid hybridization based technology, two distinct aspects of modern plant disease diagnostics, are the subjects of this chapter. Diagnostic tests based on these techniques are among the first products of agricultural biotechnology to have an impact on crop management. During the last decade, immunoassays, particularly the enzyme-linked immunosorbent assay (ELISA) and immunofluorescence assay (IFA), have become widely accepted as valid tools in the diagnosis of plant disease and detection of pathogens. Immunoassays have now been developed for nearly all classes of plant pathogens, including viruses, mycoplasma-like organisms (MLOs), bacteria, fungi, and nematodes. The ELISA kits suitable for use in research laboratories and clinical settings are commercially available for detection of a broad range of plant pathogenic viruses, bacteria, and fungi through a number of companies. Simplified ELISA kits that can be completed in less than 10 min also have been introduced recently into the market place, and are being used by growers, crop consultants, county agents, and others involved in crop disease management (Miller et al., 1990).

Nucleic acid hybridization technology has been introduced more recently than immunoassays, but holds considerable promise for application to pathogen detec-

Biotechnology in Plant Disease Control, pages 321–339
© 1993 Wiley-Liss, Inc.

tion and disease diagnosis. This technology has been extended to all classes of plant pathogenic microorganisms, and nucleic acid hybridization based products and services are beginning to become available commercially. The recently developed polymerase chain reaction (PCR) technology and novel nonradioactive probe labeling techniques are also being applied.

This chapter reviews the basic principles and applications of immunoassays and nucleic acid hybridization techniques relevant to this area of plant science. It is intended not as an exhaustive technical review of specific techniques but rather an introduction to some modern methods of plant disease diagnosis and pathogen detection.

2. IMMUNOASSAYS

2.1. Polyclonal and Monoclonal Antibodies as Reagents

Polyclonal antibodies are effective serological reagents that have been utilized in research and commercial applications in human and veterinary medicine, food quality assurance and plant disease diagnosis for many years. Details of the mammalian immune response and antibody production are presented in a number of recent books and will not be dealt with here (see Harlow and Lane, 1988; Hampton et al., 1990). Briefly, antibodies are immunoglobulin proteins produced in higher vertebrates in response to the presence of foreign or nonself-substances. Molecules that induce production of antibodies are called immunogens, while any substance capable of reacting with an antibody is called an antigen. An epitope is the part of the antigen that reacts with the antibody. Lymphocytes called B cells are induced by the presence of an immunogen to produce antibodies, which are secreted into the bloodstream and are also present in a modified form on the surface of these cells and act as antigen receptors. Many B lymphocytes are produced during an immune response, and each cell secretes only one type of antibody into the bloodstream. Polyclonal antibodies are thus heterogeneous with respect to specificity, affinity for antigens, and other characteristics.

Antisera are usually easy and inexpensive to produce in small and large experimental animals, including mice, rats, rabbits, sheep, goats, and chickens. Antibodies can be purified by a number of techniques, but commonly, they are precipitated in the presence of ammonium sulfate, then further purified by ion exchange chromatography, Protein A affinity chromatography, or specific affinity chromatography. In some instances absorption with nontarget antigens can improve the specificity of the antiserum. Detailed procedures for antibody production and purification have been published (Weir et al., 1986; Harlow and Lane,

1988; Hampton et al., 1990). Polyclonal antibodies often, but not always, have appropriate specificity for use in practical immunoassays for detection of viruses, mycoplasma-like organisms, and bacteria. However, for some pathogens in these groups and for more physiologically complex microorganisms, such as fungi and nematodes, extensive cross-reactivity within and among taxa often occurs and the usefulness of such antisera for many practical applications is limited. In these cases purification of specific components of the target pathogen for production of a more specific antiserum and/or development of monoclonal antibodies (MAbs) may provide the needed specificity.

Monoclonal antibodies are homogeneous populations of immunoglobulins produced by hybrid cells called hybridomas. The practice of MAb production has been described in great detail elsewhere (Weir et al., 1986; Goding, 1987; Harlow and Lane, 1988), and will be mentioned only briefly. Hybridomas are produced by fusing specialized myeloma cells with antibody-producing spleen cells of a laboratory animal (usually a mouse) that has been immunized previously with the target immunogen. Hybridomas are grown on a selective medium and cloned, and antibodies secreted by individual clones are harvested and tested for specificity, affinity for the antigen, suitability for selected immunoassays, and other characteristics. Because hybridomas are immortal, a constant, uniform supply of antibodies can be available, a distinct advantage over polyclonal antisera, which may vary significantly from batch to batch. Monoclonal antibodies have been produced in recent years to nearly all classes of plant pathogenic microorganisms (see reviews by Miller and Martin, 1988, Chu et al., 1989; Dewey et al., 1991).

2.2. Enzyme-Linked Immunosorbent Assay

Enzyme-linked immunosorbent assays are characterized by the specific interaction of antigen and antibody that is visualized through the action of an antibody–enzyme conjugate upon its substrate. The interactions occur on a solid phase support, allowing the separation of bound and unbound materials by washing. Many different types of ELISA exist, and some of the more commonly used variations are described below. Factors affecting the type of ELISA used for a particular application include: the type of application (field or laboratory), the sample milieu (soil, water, or plant tissue), the level of sensitivity required, the need for rapid results, the need for quantitation, the sophistication of the user and, where MAbs are used, the suitability of the MAbs for the particular format. More extensive descriptions of immunoassays are available (Chu et al., 1989; Rittenburg, 1990).

The double antibody sandwich ELISA (Voller et al., 1976; Clark and Adams, 1977) has become widely used for the detection of plant pathogens in a variety of samples. It is particularly useful for detecting pathogens in complex mixtures such as plant sap and soil. The target antigen or antigens are sandwiched between two layers of target-specific antibodies: ''capture antibodies'' immobilized on a solid phase and ''tag antibodies'' that have been conjugated to an enzyme. Antigen–antibody binding is visualized by the use of an appropriate substrate for the antibody-conjugated enzyme. Alkaline phosphatase is the enzyme most commonly used for conjugation with the tag antibody, although other enzymes, especially horseradish peroxidase, are also used (Polak and Kristek, 1988). Single species (e.g., rabbit:rabbit) or mixed species (e.g., rabbit:mouse) antibodies may be used in this type of immunoassay as the capture and tag antibodies. However, some MAbs may perform better as capture antibodies than as tag antibodies, or vice versa.

Indirect sandwich ELISAs may also be used to detect plant pathogens (Bar-Joseph and Malkinson, 1980). In this type of assay, the target antigen is sandwiched between target-specific antibodies raised in different animal species, one of which is immobilized on a solid phase. A third antibody, conjugated with an enzyme and specific to the immunoglobulin of the second antibody, is then used as a ''tag'' reagent (e.g., solid phase:rabbit target-specific capture antibody:target antigen:mouse target-specific antibody:antimouse enzyme-conjugated antibody). Antiimmunoglobulin–enzyme conjugates are available commercially and, thus, it is not necessary to prepare conjugates for target-specific antibodies. This method, however, is somewhat more time consuming than the direct method, and requires production of antibody in two different animal species.

The time and expense required to produce enzyme–antibody conjugates or anti-sera in multiple species may not be justified in instances where only a few samples are tested. In such cases, an alternative may be to use $F(ab')_2$ fragments from target-specific antibodies as the capture reagent (Barbara and Clark, 1982). Target antigens are sandwiched between these fragments and intact specific second antibodies. The reaction is visualized by the addition of an enzyme–antibody conjugate that reacts specifically with the Fc portion of the second antibody, followed by addition of the appropriate substrate. Another alternative is to use a form of indirect ELISA in which the target antigen is immobilized on a solid support and binds target-specific antibody added in solution. The antigen–antibody complex is detected by the addition of an immunoglobulin-specific antibody–enzyme conjugate. This method has been shown to be more generally reactive to strains of a particular virus than sandwich ELISA (Bar-Joseph and Salomon, 1980). However, due to competition between the target antigen and nontarget proteins in a sample for binding sites on the solid support, this assay would not provide a quantitative estimate of target antigen in a crude extract (Lommel et al., 1982). It can be a cost-effective, useful tool in pathogen diagnosis, however.

In the last several years a number of different ELISA formats have been developed that fulfill specific needs for plant pathogen detection. Most of the early assays utilized 96-well plastic microtiter plates. This format is still extremely useful, particularly where large numbers of samples are to be run. Manufacturers of microtiter plates have introduced helpful innovations, including configuring the plates in 8-well removable strips and treating the plates to improve protein binding and reduce well-to-well variability. Improvements have also been made in assay sensitivity and "user-friendliness." Microtiter plate or "multiwell" ELISAs for plant pathogen detection are available commercially that can be completed in 60 min or less (Miller et al., 1990).

An ELISA format that has been used extensively in recent years is the immunoblot. The technique is similar to multiwell ELISA, except that the assay is carried out on a membrane and a precipitating substrate is required. Sandwich ELISA is preferred in most instances, as chlorophyll and other interfering components of plant sap can be removed easily by washing. Capture antibodies are bound to membranes composed of nitrocellulose (Banttari and Goodwin, 1985; Leach et al., 1987; Dietzgen and Francki, 1990), cellulose (Sherwood, 1987; Haber and Knapen, 1989), or other materials. The membrane can be attached to a support, such as a dipstick (Dewey et al., 1989), and the assay then carried out by transferring the dipstick from solution to solution. Plant tissue can also be directly blotted onto antibody-bound nitrocellulose membranes to allow determination of pathogen distribution in the tissue (Hsu and Lawson, 1991; Gwinn et al., 1991).

The sensitivity of ELISA can be improved by using beads that increase the surface area available for antibody binding (Marshall and Bush, 1987). Fins built into the inside of plastic test tubes serve the same purpose. "Flow-through" assays have also been developed that utilize absorbent materials to force solutions through the reaction area, resulting in improved reaction kinetics and increased sensitivity (Valkirs and Barton, 1985). Flow-through immunoassays are available commercially for the detection of a number of plant pathogenic fungi. These assays are completed in 10 min or less and can achieve sensitivities as low as 25–250 ng protein/ml of mycelial extract (Miller et al., 1990; Mittermeier et al., 1990).

2.3. Immunofluorescence and Immunogold Labeling

Immunofluorescence (IF) has many applications in research and clinical settings. There are some disadvantages of IFAs, in that they can be difficult to standardize and interpret, and specialized equipment is needed to view and/or quantify the antibody-binding reaction. However, in many laboratories IF is used routinely, and much of the work has been directed towards detection of phytopathogenic bacteria. Immunofluorescence is considered to be more sensitive than ELISA and other serological methods for this application (Franken and van Vuurde, 1990).

The fluorochromes most widely used in IF are fluorescein, which emits a yellow-green color, and rhodamine, which emits a red color. Both can be conjugated directly to the detecting antibodies (direct IF), or to secondary antibodies (anti-immunoglobulin antibodies), or other molecules, such as avidin or protein A (indirect IF). Fluorochrome conjugates are available from a number of commercial sources and thus indirect IF can be readily applied to systems where specific primary antisera are available. Direct IF requires fewer steps, however, and is more appropriate for large numbers of samples. General methods (Johnson and Holborow, 1986; Harlow and Lane, 1988), as well as step-by-step protocols for IF tests for bacteria and viruses, have been published recently (De Boer, 1990; Gingery, 1990).

Colloidal gold labeling, while first developed for electron microscopy, has been used recently to visualize antibody–antigen binding in light microscopic and visual systems (Hsu, 1984; Cassab and Varner, 1987). The gold precipitate is pink to red in color, depending on the intensity of the antigen–antibody interaction. The sensitivity of the system can be increased by using a silver enhancement method, which results in the development of a black-brown color. Good quality colloidal gold conjugates are available from a number of commercial sources for indirect immunoassays, although they can also be easily prepared in the laboratory (Langenburg, 1990). In most applications for plant pathogen detection the silver enhancement step is necessary to achieve an adequate level of sensitivity; this adds an extra step that can limit the use of this labeling method in "rapid" or "field-usable" applications. However, for clinical and research applications, gold labeling can be quite effective.

3. APPLICATIONS OF IMMUNOASSAY IN DETECTION AND DIAGNOSIS

The availability of MAb technology and rapid, sensitive immunoassay formats has opened the door to development of sophisticated, "high tech" detection–diagnostic tools. Applications of these technologies in basic research, that is, studies of pathogen taxonomy, disease physiology, host–parasite interactions, pathogen ecology, and epidemiology, are potentially immense. For many plant pathogens, particularly fungi, immunoassays are information-gathering devices that allow researchers to develop profiles of population movements that could not be obtained by standard techniques. Pathogen populations in soil, water, or crop systems can be assessed and correlated with crop damage due to disease, allowing development of treatment thresholds. Once the basic parameters have been determined by researchers, the farmer can carry out a relatively simple assay and use the results to decide on a course of treatment. In addition to the specificity and ease of use of modern immunoassays, such as field ELISAs, these assays can also exhibit high sensitivity, permitting very early detection and quantification of plant pathogens. Detection of 1% or less infection of plant tissues has been demonstrated for several crop systems (MacDonald et al., 1990; Benson, 1991; Ricker et al., 1991).

3.1. Detection of Pathogens in Plant Tissue

Proper identification of the causal agent of disease is the first step in devising an effective management strategy. Currently, the greatest use of immunoassays is for the diagnosis of pathogens in diseased plant tissue. Immunoassays, particularly ELISA and IF, are being used routinely by plant disease diagnostic clinics for the diagnosis of viruses and bacteria. These immunoassays are also utilized in stock plant indexing systems and in seed testing and certification programs (van Vuurde et al., 1983; Raju and Olson, 1985; Lawson and Horst, 1987).

Immunofluorescence has been particularly useful in seed testing, although contamination of the sample with large numbers of nontarget microorganisms may interfere with the assay unless highly specific antibodies are used. The use of semiselective media can be a partial remedy but is not always effective. In addition, the inability of IF to distinguish between live and dead cells can be a disadvantage in some applications, for example, testing seed lots for bacterial infestation. Workers in The Netherlands have developed a number of innovative techniques to address these issues; in one of these, target bacteria can be removed selectively from a sample containing large numbers of saprophytic bacteria by a technique called immunoisolation, resulting in a cleaner preparation and an effective increase in sensitivity when combined with IF (van Vuurde, 1987). Standard cultural methods have been combined with IF in another procedure that permits detection/identification of viable cells, called immunofluorescence colony staining (IFC) (van Vuurde, 1987, 1990; Franken and van Vuurde, 1990). In IFC, bacterial colonies embedded in agar medium are stained directly with fluorescein-conjugated antisera. After washing to remove unbound antisera, fluorescing colonies can be viewed using the low power objective of an appropriately equipped microscope. Target colonies are easily differentiated from nontarget bacterial colonies growing in the same medium.

The ELISA may also prove to be a useful tool in plant certification programs. Latent infection of tomato by *Clavibacter michiganensis* subsp. *michiganensis* was detected in stems of symptomless tomato transplants by using a commercially available ELISA kit. The ELISA was shown to be as effective as culture plating in detecting the bacteria in greenhouse and field-grown plants, and was considered to be the most practical method for rapid testing of large numbers of plants (Gitaitis et al., 1991). Immunoassays for diagnosis of diseases caused by fungi are also being used in the clinical setting (Pscheidt et al., 1990).

Although the inability of immunoassays to distinguish between viable and dead cells may be a disadvantage in seed testing, it can be an advantage to the diagnostician who often receives samples for analysis that are in a condition too poor to allow isolation and culture of the fungal or bacterial pathogen by classical methods. Of the diagnostic techniques currently available, ELISA (multiwell format) is probably the most useful for diagnostic clinics; the procedure is easy to learn, commercial kits are available for many pathogens, and equipment costs can be minimized. Interpretation of ELISA is usually straightforward, although appropriate controls must be included to avoid false positive or false negative results.

While multiwell ELISA, IF, and related tests are gaining acceptance in diagnostic clinics, they are impractical for most commercial applications. Only the most sophisticated growers have facilities and staff available to carry out these types of tests. Therefore, on-site, user-friendly tests are needed for the majority of growers who must make a rapid diagnosis in the greenhouse or field. Such rapid tests are being developed by a number of companies in the United States and Europe. All of the assays currently available commercially are designed to detect fungal plant pathogens and include tests for species of *Phytophthora, Pythium, Rhizoctonia,* and *Sclerotinia* (Agri-Diagnostics Associates, Cinnaminson NJ). Products that will become available in the near future include rapid immunoassays for *Pseudocercosporella herpotrichoides,* the causal agent of cereal eyespot, for other cereal pathogens, and for *Botrytis* spp. (Saunders D, Kenyon L, E.I. DuPont de Nemours Corporation, Wilmington DE, personal communication) and for *Septoria tritici* and *S. nodorum,* foliar pathogens of wheat (Mittermeier et al., 1990). Rapid tests can be used for both disease diagnosis and quantification of pathogen biomass, depending on the crop system being tested. For example, in the ornamental crop nursery industry, early detection and diagnosis of *Phytophthora* spp. is critical for production of high quality plants. Benson (1991) demonstrated that rapid assays were effective in detecting *Phytophtora cinnamomi* in the roots of diseased azaleas prior to the development of above-ground symptoms. Immunoassay results correlated well with standard culture plate methods for *Phytophthora* detection, as had been shown previously using a multiwell ELISA in nursery crops (MacDonald et al., 1990).

A more advanced application of immunoassays is the quantification of pathogen levels in a crop and correlation of those levels to crop damage. This type of information will allow growers to manage fungicide applications more accurately, which may reduce crop inputs as well as address environmental concerns. This finding is particularly important in countries where farmers are under pressure from government agencies and environmental groups to reduce the use of pesticides. A number of researchers have monitored disease development in crops through a growing season by using ELISA (Smith et al., 1990; Benson, 1991; Miller et al., 1992), but more research is needed to develop firm treatment threshold values for a wide range of crops.

3.2. Detection of Plant Pathogens in Soil and Water

Detection and quantification of plant pathogens in soil by any method is complicated by a number of factors, including (1) low pathogen concentrations, (2) uneven distribution of pathogens, (3) the presence of large numbers of nontarget microorganisms in the soil, and (4) the presence of soil components, such as organic acids and clay particles that may interfere with an assay. Therefore, in addition to a specific and sensitive assay, a means of propagule concentration and extraction, a representative sampling protocol, and a means of blocking and/or excluding interfering factors are needed.

Phytophthora spp. have been detected and quantified in soil samples by using a multiwell ELISA coupled with a relatively simple sample concentration/extraction protocol (Miller et al., 1989a and b). Soil samples are air-dried, then mixed thoroughly with water, and allowed to stand for 15 min, during which time organic material floats to the surface, where it is captured, then extracted, filtered, and tested by ELISA. In a study of the use of this system to detect *P. sojae* (= *P. megasperma* f. sp. *glycinea*) oospores in Midwestern U.S. soil samples, the results of ELISA correlated well with those of a standard baiting protocol. The same multiwell ELISA for *Phytophthora* spp. detection was assessed as a method of estimating levels of *Phytophthora citrophthora* and *P. parasitica* in citrus soils in Florida and California. The assay was not effective in determining propagule densities. However, ELISA results for citrus fibrous roots were significantly correlated with propagule densities determined by dilution plating, to percentage root rot and to isolation frequency (Timmer et al., 1991). Since treatment thresholds, based on dilution plating data, have already been suggested in Florida and California for these diseases (Menge, 1986; Sandler et al., 1987), ELISA-based threshold values could be determined by extrapolation.

Recycling of irrigation water has become part of crop management in areas where restrictions on water use and irrigation run-off are in place. The reuse of irrigation water increases the risk of disease to crops, particularly diseases caused by pythiaceous fungi. Ali-Shtayeh et al. (1991) developed a method to detect *Phytophthora* and *Pythium* species in recycled irrigation water in California ornamental crop nurseries by using commercially available ELISA kits. Large volumes of water were filtered through 0.45-μm filters to concentrate propagules of the pathogens. The filters were boiled to disrupt the propagules and extracts were tested by ELISA (multiwell format). As few as 17–23 zoospores per liter of water were detected in a sample. The regular use of such kits (or simplified versions) in nursery operations to check the pathogen population levels in recycled water could serve to reduce disease incidence by signaling the nursery manager to initiate procedures to prevent introduction of *Phytophthora* or *Pythium* species into the nursery.

4. NUCLEIC ACID PROBES

Nucleic acid hybridization technology has been used routinely and effectively in disciplines as diverse as food microbiology (Curiale et al., 1990; Evans and Towner, 1990), forensics (Landegren et al., 1988), and human diagnostics (Caskey, 1987). In recent years, nucleic acid based techniques have emerged as viable and practical alternatives to antibody-based technology for a number of applications. One important advantage of nucleic acid based technology over immunoassays is that all of the genetic material of the target pathogen is available for construction of a probe. The antigen profile of a microorganism is known to be affected by numerous factors, including life stage (Barak et al., 1985) and growth substrate (Carter and Lynch, 1991). On the other hand, nucleic acid based assays are not yet competitive with immunoassays for routine applications in diagnostic laboratories or in field settings. Other comparative advantages and disadvantages of these technologies have been reviewed recently (Chu et al., 1989).

The following is a brief review of the general principles of nucleic acid hybridization and the most widely used nucleic acid based techniques, such as dot blot hybridization, restriction fragment length polymorphism (RFLP) analysis, and the recently developed PCR technology.

4.1. Nucleic Acid Hybridization Principles

The DNA duplex consists of two complementary polynucleotide strands where adenine (A) of one strand always pairs with thymine (T) of the other strand and similarly, cytosine (C) always pairs with guanine (G). Ribonucleic acid (RNA) differs from DNA in regard to nucleotide base composition in that uracyl (U) replaces thymine. The specificity of base pairing is dictated by steric factors, such as the size of the nucleotide bases and position of the amino ($-NH_2$) and carboxyl ($C = O$) groups for proper hydrogen binding. The double helical duplex is held together by hydrogen bonds occurring between the complimentary nucleotide base pairs (A-T and G-C). Denaturation (i.e., disruption of the hydrogen bonds) of the double-stranded DNA can be achieved by exposure to high temperatures or alkaline pH. The dissociation of the DNA duplex as the temperature increases is referred to as melting. The temperature at which one-half of the double helical structure is dissociated is referred to as the melting temperature (T_m). The T_m of the duplex DNA is a function of several factors including nucleotide base composition, length of the DNA duplex, and ionic strength. Dissociated strands of DNA can be immobilized on a solid phase support, such as latex, magnetic beads, microtiter plates, or more commonly nitrocellulose or nylon-based membranes, and then hybridized with single-strand, labeled nucleic acid (usually DNA) probes. The probes will hybridize only with the denatured strands of the nucleic acid that are complementary. Hybridization is confined to those regions where the DNA probe and the specimen's DNA share homologous stretches of nucleotides. The reassociation of the two denatured strands is therefore highly specific and can occur either between DNA–DNA, DNA–RNA, or RNA–RNA strands. The RNA probes, however, are less stable than DNA probes due to degradation by RNase. The formation of double-stranded complexes between the nucleic acids of a specimen and the labeled DNA probes constitutes the basic foundation of many nucleic acid based detection techniques.

4.2. Nucleic Acid Hybridization Based Detection Techniques

4.2.1. Dot-Blot Hybridization Assay

Of the several filter hybridization techniques that are employed in plant disease diagnostics, dot-blot, or spot hybridization is often the simplest and most effective for the detection of plant pathogens (Burrows, 1988; Miller and Martin, 1988; Chu et al., 1989). Typically, samples of denatured nucleic acid from healthy and infected plants are directly spotted onto a prewetted, solid matrix such as a nitrocellulose or nylon-based membrane. Nucleic acids are then firmly immobilized on the solid support by baking the membrane in an oven for 2 h at 80°C. To block the remaining free DNA binding sites, the membrane is incubated several minutes in a sealed plastic bag with prehybridization solution (hybridization solution minus the probe). This solution consists of a mixture of denatured nonhomologous DNA (herring or salmon sperm DNA, or calf-thymus DNA) and protein, usually bovine serum albumin (BSA) or nonfat dry milk. After removing the prehybridization solution, the hybridization solution containing the denatured specific DNA probe is added to the plastic bag containing the sample membrane, and incubated for several hours to allow hybridization between the probe and the target nucleic acid. The membrane is washed several times to remove the unbound DNA probe following hybridization. Hybridization between the target nucleic acid immobilized on the membrane and the DNA probe is detected either by autoradiography on X-ray film, if the probe was labeled radioactively, or by a colorimetric reaction, if an enzyme-labeled probe was used. This method is easy to carry out in appropriately equipped laboratories, does not require isolation of target nucleic acid, and can be very sensitive. Many studies have shown that the sensitivity of dot-blot hybridization may equal or exceed that of ELISA (Sela et al., 1984; Rosner et al., 1986; Varveri et al., 1988; Polston et al., 1989).

4.2.2. Restriction Fragment Length Polymorphism Analysis

A typical RFLP analysis encompasses the first isolation and purification of genomic or organelle DNA, followed by digestion of the purified DNA with restriction endonucleases. These enzymes cleave the DNA at specific recognition sequences of four to six nucleotides in length. Commercially available and commonly used restriction endonucleases are *Bam*H1, *Eco*R1, *Hind*III, and *Pst*1. Digested DNA fragments are separated by electrophoresis in agarose gels. Differences in RFLP profiles produced by the same restriction enzyme are due to deletions, insertions, or base substitutions. The DNA bands can be visualized by either staining with ethidium bromide or by transfer of the DNA to nitrocellulose or nylon membranes followed by hybridization with a labeled DNA probe in a procedure known as Southern blotting. Hybridization of DNA probes with DNA fragments can then be detected by autoradiography on X-ray film where radioactive probes are used, or by the use of colorimetric or chemiluminescent substrate systems where the probes are labeled nonradioactively.

4.2.3. Polymerase Chain Reaction

A recently developed *in vitro* procedure already playing a dramatic role in clinical medicine and diagnosis of human genetic diseases (Landegren et al., 1988), and with important applications in plant disease diagnosis, is PCR (Mullis et al., 1986). With PCR, exponential amplification of DNA sequences is achieved by repetitive cycles of DNA synthesis. Each PCR cycle consists of three reaction steps. The PCR begins with thermal denaturation of the duplex DNA containing the sequences of interest. Following denaturation, primers (synthetic, single-stranded DNA oligomers ~20 bp in length) are annealed, one on each strand, to complementary DNA sequences flanking the target DNA. Finally, beginning at the primer, the primed-template DNA is extended in the 3′ direction with DNA polymerase and deoxynucleoside triphosphates to yield one copy of the target DNA per strand. Assuming that doubling of the target DNA occurs with each cycle (at 100% efficiency), 1 millionfold amplification of the initial DNA can be achieved in 20 cycles in approximately 1 h. Simple, easy-to-operate automated DNA thermal cycling devices for PCR amplification are available commercially (e.g., Perkin-Elmer Corporation, Norwalk, CT). The amplified PCR products can then be either sequenced or identified by traditional DNA detection systems, such as dot-blot hybridization assay, or digested with restriction enzymes to determine RFLPs.

The DNA probes labeled with nonradioactive markers are often less sensitive than their ^{32}P labeled counterparts (Zwadyk et al., 1986). With the advent of PCR technology, where 1 millionfold amplification of the target DNA can be easily achieved, the higher sensitivity imparted by radioisotope-labeled DNA probes may not be necessary in many instances. The PCR can thus lead to the widespread use of safer, less expensive, more stable detection systems, such as those employing colorimetric, fluorescent, or chemiluminescent substrates.

5. APPLICATIONS OF NUCLEIC ACID HYBRIDIZATION TECHNOLOGY

The importance of nucleic acid hybridization techniques as diagnostic tools for plant diseases incited by viroids (Salazar et al., 1983), viruses (Ronco et al., 1989; Weideman and Koenig, 1990; Zilberstein et al., 1989), mycoplasma-like organisms (Lee and Davis, 1988; Davis et al., 1990), and bacteria (Thompson et al., 1988; Gilbertson et al., 1989; Cuppels et al., 1990) has been well documented. In most of the studies, dot-blot hybridizations with specific DNA or complementary DNA (cDNA) probes were used successfully to detect target pathogens in plant sap. Cuppels et al. (1990) and Zilberstein et al. (1989) developed nonradioactively labeled DNA probes to detect *Pseudomonas syringae* pv. *tomato* and tomato yellow leaf curl virus, respectively. In both cases, the probes were labeled with a commercially available nonradioactive reporter system.

The application of nucleic acid hybridization techniques to the detection of plant pathogenic fungi (Goodwin et al., 1909 a and b; Hamer, 1991; Shesser et al. 1991) and nematodes (Besal et al., 1988; Powers and Sandall, 1988; Harris et al., 1990) has also accelerated rapidly in recent years. In addition, this technology has been extended to studies of population biology (Hulbert and Michelmore, 1988; MacDonald and Martinez 1990), phylogeny, and taxonomy (Forster et al., 1989; Martin, 1990) of fungal plant pathogens.

Several laboratory-oriented, nucleic acid based diagnostic products or services are now available commercially to detect viroids, viruses, and mycoplasma-like organisms (e.g., AGDIA, Inc., Elkhart, IN, USA; and Bresatec Ltd., Adelaide, Australia). With the continued development of safe, stable, sensitive labels for nucleic acid probes, the use of this technology can be expected to increase in both diagnostic labs and field settings.

The RFLP analysis has proven to be a powerful tool in assessing genetic variation in populations of plant pathogens (Powers and Sandall, 1988; Mogen et al., 1990; MacDonald and Martinez, 1990) and differentiating aggressive from nonaggressive strains in populations of fungi (Koch et al., 1991) and bacteria (Graham et al., 1990). The application of RFLP analysis to the study of taxonomic and phylogenetic relationships of plant pathogenic fungi (Smith and Anderson, 1989; Martin, 1990; Koch et al., 1991), bacteria (Hartung and Civerolo, 1989; Graham et al., 1990), and nematodes (Radice et al., 1988; Harris et al., 1990) at the species and subspecies level has also been well documented. However, RFLP analysis is not particularly useful as a routine diagnostic tool for plant diseases due to its complexity and subsequent time and cost constraints.

Polymerase chain reaction is the most recent addition to an arsenal of biotechnology-based methodologies that can be applied to plant disease diagnostics and plant pathogen detection, as well as to basic studies of genetic relationships between taxa (Harris et al., 1990; Chase et al., 1991; Cubeta et al., 1991; Kohn et al., 1991; Meyer, 1991). Its use in the monitoring of microorganisms of environmental interest (Steffan and Atlas, 1991) also bodes well for related studies of plant pathogen detection and monitoring. Diagnostic applications of PCR have been shown for viroids (Hadidi and Yang, 1990; Levi et al., 1991), MLOs (Schaff et al., 1990), bacteria (Prossen et al., 1991), fungi (Rollo et al., 1990; Leung et al., 1991; Nazar et al., 1991; Schesser et al., 1991), and nematodes (Harris et al., 1990). Schesser et al. (1991) successfully employed PCR to detect a low-copy *Gaeumannomyces graminis*-specific DNA fragment in infected wheat seedlings that could not be detected by direct hybridization of the probe with DNA extracted from *G. graminis*-infected tissue. In a different study, PCR was used to detect and differentiate *Verticillium dahliae* and *V. albo-atrum* in plant tissue (Nazar et al., 1991). The two fungi were differentiated on the basis of five nucleotide differences within two internal transcribed spacers (ITS1 and ITS2) of ribosomal DNA. Wiglesworth et al. (1991) used PCR to amplify the ITS2 region of several fungi belonging to the Order Peronosporales, including *Pythium ultimum, Phytophthora infestans, P. parasitica* f. sp. *nicotianae, Peronospora tabacina, P. sacchari,* and *P. maydis*. The DNA sequencing of the ITS2 region showed that it could be used to distinguish the examined fungi at the genus and species, but not subspecies, levels.

6. CONCLUSIONS

The technologies described in this chapter represent significant improvements in the art and science of plant disease diagnostics and pathogen detection. Both immunoassays and nucleic acid based tests have the potential to change the way disease management decisions are made, by providing accurate diagnostic information in a timely way. In the long run, these technologies will also provide the tools to establish a better understanding of the behavior of pathogen populations in crops, soil, and water. This information will eventually reach the grower in the form of practical disease treatment thresholds.

Immunoassays were the first to be developed and are certainly the first to be used "on-site" in commercial applications. They are being integrated into research programs, diagnostic clinics, certification and quality testing programs, and into decision making at the grower level. The use of monoclonal antibodies and improved assay formats has resulted in increased specificity and sensitivity in assays that address a broad range of pathogens.

The widespread acceptance and routine use of nucleic acid based techniques in plant disease diagnostics will depend in large part on future technological advances that will lead to simpler, safer, less laborious, more user-friendly, and more cost-effective methodology. Amplification of DNA by PCR and the labeling of nucleic acid probes with nonradioactive reporter molecules such as digoxigenin, biotin, and photobiotin are some examples of the latest developments towards that end.

ACKNOWLEDGMENTS

The authors thank J. Labriola and R. Frederick for critically reading the manuscript.

REFERENCES

Ali-Shtayeh MS, MacDonald JD, Kabashima J (1991): A method for using commercial ELISA tests to detect zoospores of *Phytophthora* and *Pythium* species in irrigation water. Plant Dis 75:305–311.

Banttari EE, Goodwin PH (1985): Detection of potato viruses X, S, and Y by enzyme-linked immunosorbent assay on nitrocellulose membranes (dot-ELISA). Plant Dis 69:202–205.

Barak R, Maoz A, Chet I (1985): Antigenic differences among several *Trichoderma* isolates. Can J Microbiol 31:810–816.

Barbara DJ, Clark MF (1982): A simple indirect ELISA F(ab′)$_2$ fragments of immunoglobulin. J Gen Virol 58:315–322.

Bar-Joseph M, Malkinson M (1980): Hen egg yolk as a source of antiviral antibodies in the enzyme-linked immunosorbent assay (ELISA): a comparison of two plant viruses. J Virol Methods 1:179–183.

Bar-Joseph M, Salomon R (1980): Heterologous reactivity of tobacco mosaic virus strains in enzyme-linked immunosorbent assays. J Gen Virol 47:509–512.

Benson DM (1991): Detection of *Phytophthora cinnamomi* in azalea with commercial serological assay kits. Plant Dis 75:478–482.

Besal EA, Powers TO, Radice AD, Sandall LJ (1988): A DNA hybridization probe for detection of soybean cyst nematode. Phytopathology 78:1136–1139.

Burrows PR (1988): The use of nucleic acid probes to identify plant parasitic nematodes. Brighton Crop Protection Conference—Pests and Diseases, Vol. 2. Surrey, England: BCPC Registered Office, pp 811–820.

Carter J, Lynch JM (1991): Substrate-dependent variation in the protein profile and antigens of *Trichoderma harzianum*. Enzyme Microb Technol 13:537–543.

Caskey CT (1987): Disease diagnosis by recombinant DNA methods. Science 236:1223–1228.

Cassab GI, Varner JE (1987): Immunocytolocalization of extensin in developing soybean seed coats by immunogold-silver staining and by tissue printing on nitrocellulose paper. J Cell Biol 105: 2581–2588.

Chase TE, Otrosina WJ, Spieth PT, Cobb FW (1991): Use of PCR to distinguish biological species within *Heterobasidion annosum* complex. (Abstract) Phytopathology 81:1190.

Chu PWG, Waterhouse PM, Martin RR, Gerlach WL (1989): New approaches to the detection of microbial plant pathogens. Biotechnol Genet Eng Rev 7:45–111.

Clark MF, Adams AN (1977): Characteristics of the microplate method of enzyme-linked immunosorbent assay (ELISA) for the detection of plant viruses. J Gen Virol 34:475–483.

Cubeta MA, Echandi E, Abernethy T, Vigalys R (1991): Characterization of anastomosis groups of binucleate *Rhizoctonia* species using restriction analysis of an amplified ribosomal RNA gene. Phytopathology 81:1395–1400.

Cuppels DA, Moore RA, Morris VL (1990): Construction and use of a nonradioactive DNA hybridization probe for detection of *Pseudomonas syringae* pv. *tomato* in tomato plants. Appl Environ Microbiol 56:1743–1749.

Curiale MS, Mciver D, Weathersby S, Planer C (1990): Detection of Salmonellae and other *Enterobacteriaceae* by commercial deoxyribonucleic acid hybridization and enzyme immunoassay kits. J Food Protection 53:1037–1046.

Davis RE, Lee I-M, Douglas SM, Dally EL (1990): Molecular cloning and detection of chromosomal and extrachromosomal DNA of the mycoplasma-like organism associated with little leaf disease in Periwinkle (Catharanthus roseus). Phytopathology 80:789–793.

De Boer SH (1990): Immunofluorescence for bacteria. In Hampton R, Ball E, De Boer S (eds): Serological Methods for Detection and Identification of Viral and Bacterial Plant Pathogens—A Laboratory Manual. St. Paul: APS Press, pp 295–298.

Dewey FM, Evans D, Coleman J, Priestly R, Hull R, Horsley D, Hawes C (1991): Antibodies in plant science. Acta Bot Neerl Physiol Pharmacol Microbiol 40:1–27.

Dewey FM, MacDonald MM, Phillips SI (1989): Development of monoclonal-antibody-ELISA, -DOT-BLOT and -DIP-STICK immunoassays for *Humicola lanuginosa* in rice. J Gen Microb 135:361–374.

Dietzgen RG, Francki RIB (1990): Reducing agents interfere with the detection of lettuce necrotic yellows virus in infected plants by immunoblotting with monoclonal antibodies. J Virol Methods 28:199–206.

Evans C, Towner KJ (1990): A note on the use of microwaves in an improved non-radioactive DNA hybridization procedure for detection of bacteria in foodstuffs. Lett Appl Microbiol 10:233–235.

Forster H, Kinscher TG, Leong SA, Maxwell DP (1989): Restriction length polymorphisms of the mitochondrial DNA of *Phytophthora megasperma* isolated from soybean, alfalfa, and fruit trees. Can J Bot 67:529–537.

Franken AAJM, van Vuurder JWL (1990): Problems and new approaches in the use of serology for seedborne bacteria. Seed Sci Technol 18:415–426.

Gilbertson RL, Maxwell DP, Hagedorn DJ, Leong SA (1989): Development and application of a plasmid DNA probe for detection of bacteria causing common bacterial blight of bean. Phytopathology 79:518–525.

Gingery RE (1990): Fluorescent antibody test, viruses. In Hampton R, Ball E, De Boer S (eds): Serological Methods for Detection and Identification of Viral and Bacterial Plant Pathogens—A Laboratory Manual. St. Paul: APS Press, pp 287–293.

Gitaitis R, Beaver RW, Voloudakis AE (1991): Detection of *Clavibacter michiganensis* subsp. *michiganensis* in symptomless tomato transplants. Plant Dis 75:834–838.

Goding JW (1987): Monoclonal Antibodies: Principles and Practice. 2nd ed. London: Academic Press.

Goodwin PH, English JT, Neher DA, Duniway JM, Kirkpatrick BC (1990a): Detection of *Phytophthora parasitica* from soil and host tissue with a species-specific DNA probe. Phytopathology 80:277–281.

Goodwin PH, Kirkpatrick BC, Duniway JM (1990b): Identification of *Phytophthora citrophthora* with cloned DNA probes. Appl Environ Microbiol 56:669–674.

Graham JH, Hartung JS, Stall RE, Chase AR (1990): Pathological, restriction-fragment length polymorphism, and fatty acid profile relationships between *Xanthomonas campestris* from citrus and noncitrus hosts. Phytopathology 80:829–836.

Gwinn KD, Collins-Shepard MH, Reddick BB (1991): Tissue print-immunoblot, an accurate method of detection of *Acremonium coenophialum* in tall fescue. Phytopathology 81:747–748.

Haber S, Knapen H (1989): Filter paper sero-assay (FiPSA): A rapid, sensitive technique for serodiagnosis of plant viruses. Can J Plant Pathol 11:109–113.

Hadidi A, Yang X (1990): Polymerase chain reaction: Molecular technology to enhance detection and diagnosis of pome fruit viroids. Phytopathology 80:982–983.

Hamer JE (1991): Molecular probes for Rice Blast Disease. Science 252:632–633.

Hampton R, Ball E, De Boer S (eds) (1990): Serological Methods for Detection and Identification of Viral and Bacterial Plant Pathogens—A Laboratory Manual. St. Paul: APS Press, 389 pp.

Harlow E, Lane D (1988): Antibodies—A Laboratory Manual. Cold Spring Harbor: Cold Spring Harbor Laboratory, 726 pp.

Harris TS, S andall LJ, Powers TO (1990): Identification of single *Meloidogyne* juveniles by polymerase chain reaction amplification of mitochondrial DNA. J Nematol 22:518–524.

Hartung JS, Civerolo EL (1989): Restriction fragment polymorphisms distinguish *Xanthomonas campestris* strains isolated from Florida citrus nursuries from *X. c.* pv. *citri*. Phytopathology 79:793–799.

Hsu HT, Lawson RH (1991): Direct tissue blotting for detection of tomato spotted wilt virus in Impatiens. Plant Dis 75:292–295.

Hsu Y-H (1984): Immunogold for detection of antigen on nitrocellulose paper. Anal Biochem 142:221–225.

Hulbert SH, Michelmore RW (1988): DNA restriction fragment polymorphism and somatic variation in the lettuce downy mildew fungus, *Bremia lactucae*. Mol Plant–Microbe Inter 1:17–24.

Johnson GD, Holborow EJ (1986): Preparation and use of fluorochrome conjugates. In Weir DM (ed): Handbook of Experimental Immunology. Volume 1: Immunochemistry. 4th ed. Oxford: Blackwell Scientific Publications, pp 28.1–28.21.

Koch E, Song K, Osborn TC, Williams PH (1991): Relationship between pathogenicity and phylogeny based on restriction fragment length polymorphism in *Leptosphaeria maculans*. Mol Plant–Microbe Inter 4:341–349.

Kohn LM, Stasovski E, Carbone I, Royer J, Anderson JB (1991): Mycelial incompatibility and molecular markers identity genetic variability in field populations of *Sclerotinia sclerotiorum*. Phytopathology 81:480–485.

Landegren U, Kaiser R, Caskey CT, Hood L (1988): DNA diagnostics—Molecular techniques and automation. Science 242:229–237.

Langenburg WG (1990): Colloidal gold, its generation and use in immunoelectron microscopy. In Hampton R, Ball E, De Boer S (eds): Serological Methods for Detection and Identification of Viral and Bacterial Plant Pathogens—A Laboratory Manual. St. Paul: APS Press, pp 321–329.

Lawson RH, Horst RK (1987): Detection technology builds clean stock. Greenhouse Manager, March 1987:175–182.

Leach JE, Ramundo BA, Pearson DL, Claflin LE (1987): Dot-immunobinding assay for detecting *Xanthomonas campestris* pv. *holcicola* in sorghum. Plant Dis 71:30–33.

Lee I-M, Davis RE (1988): Detection and investigation of genetic relatedness among aster yellows and other mycoplasmalike organisms by using cloned DNA and RNA probes. Mol Plant–Microbe Inter 1:303–310.

Leung H, Loomis P, Shi Y (1991): Differentiation of the wheat bunt fungi by random amplification of polymorphic DNA. (Abstract) Phytopathology 81:1190.

Levy L, Hadidi A, Garnsey SM (1991): Multiplex reverse transcription/polymerase chain reaction for the detection of mixed citrus viroids in a single reaction. (Abstract) Phytopathology 81:1212.

Lommel SA, McCain AH, Morris TJ (1982): Evaluation of enzyme-linked immunosorbent assay for the detection of plant viruses. Phytopathology 72:1018–1022.

MacDonald BA, Martinez JP (1990): DNA restriction fragment polymorphisms among *Mycosphaerella graminicola* (anamorph *Septoria tritici*) isolates collected from a single wheat field. Phytopathology 80:1368–1373.

MacDonald JD, Stites J, Kabashima J (1990): Comparison of serological and culture plate methods for detecting species of *Phytophthora*, *Pythium* and *Rhizoctonia* in ornamental plants. Plant Dis 74:655–659.

Marshall DL, Bush GA (1987): Latex particle immunoassay. Am Biol Lab May/June 1987:48–53.

Martin FN (1990): Taxonomic classification of asexual isolates of *Pythium ultimum* based on cultural characteristics and mitochondrial DNA restriction patterns. Exp Mycol 14:46–56.

Menge JA (1986): Use of new systemic fungicides on citrus. Citrograph 71:245–252.

Meyer RJ (1991): Molecular systematics of *Trichoderma* species by restriction analysis of PCR-amplified ribosomal DNA fragments. (Abstract) Phytopathology 81:1240.

Miller SA, Korjagin V, Miller SM, Petersen FP, Klopmeyer MJ, Lankow RK, Grothaus GD (1989a): Detection and quantitation of *Phytophthora megasperma* f. sp. *glycinea* in field soil by immunoassay. (Abstract) Phytopathology 79:1139.

Miller SA, Martin RR (1988): Molecular diagnosis of plant disease. Annu Rev Phytopathol. 26:409–432.

Miller SA, Rittenburg JH, Petersen FP, Grothaus GD (1990): Development of modern diagnostic tests and benefits to the farmer. In Schots A (ed): Monoclonal Antibodies in Agriculture. Wageningen: Pudoc, pp 15–20.

Miller SA, Rittenburg JH, Petersen FP, Grothaus GD (1992): From the research bench to the market place: Development of commercial diagnostic kits. In Duncan J, Torrance L (eds):

Techniques for Rapid Detection of Plant Pathogens. Oxford: Blackwell Scientific Publications Ltd, pp. 208–221.

Miller SM, Petersen FP, Miller SA, Rittenburg JH, Wood SC, Grothaus GD (1989b): Development of a direct immunoassay to detect *Phytophthora megasperma* f. sp. *glycinea* in soil. Phytopathology 70:1139.

Mittermeier L, Dercks W, West SJE, Miller SA (1990): Field results with a diagnostic system for the identification of *Septoria nodorum* and *Septoria tritici*. Proceedings 1990 British Crop Protection Conference, Vol 2, pp 757–762.

Mogen BD, Olson HR, Sparks RB, Gudmestad NC, Orleson AE (1990): Genetic variation in strains of *Clavibacter michiganense* subsp. *sepedonicum*: polymorphisms in restriction fragments containing a highly repeated sequence. Phytopathology 80:90–96.

Mullis K, Faloona F, Scharf S, Saiki R, Horn G, Erlich H (1986): Specific enzymatic amplification of DNA in vitro: the polymerase chain reaction. Cold Spring Harbor Symp Quant Biol 51:263–273.

Nazar RN, Hu X, Schmidt J, Culham D, Robb J (1991): Potential use of PCR-amplified ribosomal intergenic sequences in the detection and differentiation of Verticillium wilt pathogens. Physiol Mol Plant Pathol 39:1–11.

Polak J, Kristek J (1988): Use of horse-radish peroxidase labelled antibodies in ELISA for plant virus detection. J Phytopathol 122:200–207.

Polston JE, Dodds JA, Perring TM (1989): Nucleic acid probes for detection and strain discrimination of cucurbit geminiviruses. Phytopathology 79:1123–1127.

Powers TO, Sandall LJ (1988): Estimation of genetic divergence in *Meloidogyne* mitochondrial DNA. J Nematol 20:505–511.

Prossen D, Hatziloukas E, Panopoulos NJ, Schaad NW (1991): Direct detection of the halo blight pathogen *Pseudomonas syringae* pv. *phaseolicola* in bean seeds by DNA amplification. (Abstract). Phytopathology 81:1159.

Pscheidt JW, Burket J, Fischer S (1990): Use of *Phytophthora* specific immunoassay kits in a plant disease clinic. (Abstract) Phytopathology 80:962.

Radice AD, Powers TO, Sandall LJ, Riggs RD (1988): Comparison of mitochondrial DNA from sibling species *Heterodera glycines* and *H. schachtii*. J Nematol 20:443–450.

Raju BC, Olson CJ (1985): Indexing systems for producing clean stock for disease control in commercial floriculture. Plant Dis 69:189–192.

Ricker RW, Marois JJ, Dlott JW, Bostock RM, Morrison JC (1991): Immunodetection and quantification of *Botrytis cinerea* on harvested wine grapes. Phytopathology 81:404–411.

Rittenburg JH (1990): Fundamentals of immunoassay. In Rittenburg JH (ed): Development and Application of Immunoassay for Food Analysis. London: Elsevier Applied Science, pp 29–57.

Rollo F, Salvi R, Torchia P (1990): Highly sensitive and fast detection of *Phoma tracheiphila* by polymerase chain reaction. Appl Microbiol Biotechnol 32:572–576.

Ronco AE, Bo ED, Ghiringhelli PD, Medrano C, Romanowski V, Sarachu, AN, Grau, O. (1989): Cloned cDNA probes for the detection of tomato spotted wilt virus. Phytopathology 79:1309–1313.

Rosner A, Lee RF, Bar-Joseph M (1986): Differential hybridization with cloned cDNA sequences for detecting a specific isolate of citrus tristeza virus. Phytopathology 76:820–824.

Salazar LF, Owens RA, Smith DR, Diener TO (1983): Detection of potato spindle tuber viroid by nucleic acid spot hybridization: evaluation with tuber sprouts and true potato seed. Am Potato J 60:587–597.

Sandler HA, Timmer LW, Graham JH (1987): Effect of systemic fungicides on *Phytophthora* populations, feeder root densities, and yield of citrus. (Abstract) Phytopathology 77:1691.

Schaff DA, Lee I-M, Davis RE (1990): Sensitive detection of mycoplasmalike organisms (MLOs) by polymerase chain reaction (PCR). (Abstract) Phytopathology 80:959.

Schesser K, Luder A, Henson JM (1991): Use of polymerase chain reaction to detect the take-all fungus, *Gaeumannomyces graminis*, in infected wheat plants. Appl Environ Microbiol 57:553–556.

Sela I, Reichman M, Weissbach A (1984): Comparison of dot molecular hybridization and enzyme-linked immunosorbent assay for detecting tobacco mosaic virus in plant tissues and protoplasts. Phytopathology 74:385–389.

Sherwood JL (1987): Comparison of a filter paper immunobinding assay, western blotting and an enzyme-linked immunosorbent assay for the detection of wheat streak mosaic virus. J Phytopathol 118:68–75.

Smith CM, Saunders DW, Allison DA, Johnson LEB, Labit B, Kendall SJ, Hollomon DW (1990): Immunodiagnostic assay for cereal eyespot: novel technology for disease detection. Proceedings 1990 British Crop Protection Conference, Vol 2. Surrey, England: BCPC Registered Office, pp 763–770.

Smith ML, Anderson JB (1989): Restriction fragment polymorphisms in mitochondrial DNAs of *Armillaria*: identification of North American biological species. Mycolog Res 93:247–256.

Steffan RJ, Atlas RM (1991): Polymerase chain reaction: applications in environmental microbiology. Annu Rev Microbiol 45:137–161.

Timmer LW, Menge JA, Zitco SE, Pond E, Miller SA (1991): Comparison of ELISA with other assays for *Phytophthora* spp. in citrus orchards in Florida and California (Abstract). Phytopathology 81:1213.

Thompson E, Leary JV, Chun WWC (1989): Specific detection of *Clavibacter michiganense* subsp. *michiganense* by a homologous DNA probe. Phytopathology 79:311–314.

Valkirs GE, Barton R (1985): Immunoconcentration—a new format for solid phase immunoassay. Clin Chem 31:1427–1431.

van Vuurde JWL (1987): New approach in detecting phytopathogenic bacteria by combined immuno-isolation and immunodetection assays. EPPO Bull 17:139–148.

van Vuurde JWL (1990): Immunofluorescence colony staining. In Hampton R, Ball E, De Boer S (eds): Serological Methods for Detection and Identification of Viral and Bacterial Plant Pathogens—A Laboratory Manual. St. Paul: APS Press, pp 299–305.

van Vuurde JWL, van den Bovenkamp GW, Birnbaum Y (1983): Immunofluorescence microscopy and enzyme-linked immunosorbent assay as potential routine tests for the detection of *Pseudomonas syringae phaseolicola* and *Xanthomonas campestris* pv. *phaseoli* in bean seed. Seed Sci Technol 11:547–559.

Varveri C, Candresse T, Cugusi M, Ravelonandro M, Dunez J (1988): Use of a ^{32}P-labeled transcribed RNA probe for dot hybridization detection of plum pox virus. Phytopathology 78:1280–1283.

Voller A, Bartlett A, Bidwell DE, Clark MF, Adams AN (1976): The detection of viruses by enzyme-linked immunosorbent assay (ELISA). J Gen Virol 33:165–167.

Weideman HL, Koenig R (1990): Differentiation of isolates of potato virus S which infect Chenopodium quinoa systematically by means of quantitative cDNA hybridization tests. J Plant Dis Prot 97:323–327.

Weir DM, Herzenberg LA, Blackwell C, Herzenberg LA (1986): Handbook of Experimental Immunology, Vols 1 and 4. 4th ed. St. Louis: Mosby.

Wiglesworth MD, Schardl CL, Nesmith WC, Siegel MR (1991): Differentiation of Perosnosporales and isolates of *Peronospora tabacina* by direct sequencing of an internal transcribed spacer (ITS2). (Abstract) Phytopathology 81:1237.

Zilberstein A, Navot N, Ovadia S, Reinhartz A, Herzberg M, Czosnek H (1989): Technique—A J Methods Cell Mol Biol 1:118–124.

Zwadyk P, Cooksey RC, Thornsberry C (1986): Commercial detection methods for biotinylated gene probes: Comparison with ^{32}P-labelled DNA probes. Curr Microbiol 14:95–100.

≡18

APPLICATION OF DNA FINGERPRINTING FOR DETECTING GENETICAL VARIATION AMONG ISOLATES OF THE WHEAT PATHOGEN *MYCOSPHAERELLA GRAMINICOLA*

A. ZILBERSTEIN
S. PNINI-COHEN
Z. EYAL
S. SHUSTER
Department of Botany, The George S. Wise Faculty of Life Sciences, Tel Aviv University, Tel Aviv 69978, Israel

J. HILLEL
Department of Genetics, Faculty of Agriculture, The Hebrew University of Jerusalem, Rehovot 76100, Israel

U. LAVI
Department of Horticulture Genetics and Breeding, Agricultural Research Organization, The Volcani Center, Bet Dagan 50250, Israel

1. INTRODUCTION

Evaluation of genetic variability that exists among and within populations of certain phytopathogenic fungi depends mainly on pathogenic characteristics result-

Biotechnology in Plant Disease Control, pages 341–353
© *1993 Wiley-Liss, Inc.*

ing from assessing host response. In many of these fungi the sexual state cannot be obtained and manipulated in vitro and, therefore, genetic information concerning the pathogen itself is lacking.

Recent advances in recombinant DNA technologies enabled us to verify genetic relationships among organisms at the species, strain, cultivar, and individual level (Caetano-Anolles et al., 1991; Epplen et al., 1991; Jeffreys et al., 1991b; Hillel et al., 1991). These approaches offer tools to identify specific differences that exist among and within populations of plant pathogens.

2. HOST X PATHOGEN INTERACTIONS

Physiologic specialization in several fungal pathogens of wheat (*Triticum aestivum* L. and *Triticum durum* Desf.) is based on the statistical analysis of the cultivar X isolate interaction term (Eyal et al., 1987; Hetzler et al., 1991). The lack of distinct qualitative host response resorted to analyses of quantitative differences in symptoms. There are inherent difficulties in adapting quantitative assessments in establishing qualitative categorization of specificity in the host (genes for resistance) and the pathogen (genes for virulence). The interaction term, which is usually rather small, may be strongly affected by the choice of host genotypes, isolates of the pathogen, and by the environmental conditions. The expression of specificity is not always clear and the establishment of response classes in the host (susceptible, intermediate, and resistant) is a source for inconsistency. The recognition of specificity in the pathogen and the differential response of host genotypes is of great importance in screening resistant germplasm and its incorporation into breeding programs. Needless to say, the genetic factors associated with specificity in the pathogen have not been identified and characterized.

This lack of specificity is examplified in a number of necrophytic pathogens inciting foliar diseases of wheat. In the necrophytic foliar pathogen of wheat *Mycosphaerella graminicola* (Fuckel) Schroeter (anamorph: *Septoria tritici* Roberge. in Desmaz.) the causal agent of *Septoria tritici* blotch, the cultivar X isolate interactions served as the basis for recognizing specificity. In several studies on *S. tritici* no significant interaction terms were found between wheat cultivars X isolates. The variation among isolates was attributed to differences in aggressiveness, namely, no clear specificity (Marshall, 1985). The response of the cultivars was not specific but rather a continuous gradation. Thus, the discrepancies in conclusions regarding specificity for the same pathogen may be derived from the choice of isolates, the differentiating capacity of the test cultivars, differences in methodology, and environmental conditions. The differences in conclusions may have considerable implications on screening for resistance and on the adaption of breeding strategies (incorporation, fixation, deployment, etc.) in developing resistant wheat cultivars.

At the present time no specific toxin(s) was identified in *S. tritici* that can be associated with specificity (Eyal et al., 1987). A self-inhibiting compound (methyl 3-indole carboxylate) was suggested to be associated with aggressiveness in *S. tritici* (Zelikovitch et al., 1990). It was found that isolates of *S. tritici* that differed in their virulence on certain wheat genotypes did not fully express their expected virulence when mixtures of isolates were used to inoculate wheat genotypes. Questions related to the possible speciation of *S. tritici* on different *Triticum* species, stability over time of an isolate in terms of its restored virulence following continuous culturing, and the possible change in virulence upon passage through plant hosts may further complicate the issue of specificity.

3. DEOXYRIBONUCLEIC ACID PROBING METHODS

It is likely that some of the difficulties associated with establishing the fungal isolate as a well-defined biological unit may be clarified by employing DNA typing technologies. The development of molecular techniques has aided in resolving variations between related eukaryotic genomes and even within different individuals. Two linked approaches of analyzing DNA polymorphism are currently being developed by using defined DNA markers (Jeffreys et al., 1985a; Lansman et al., 1981; Tanksley et al., 1989). These approaches employ cloned DNA fragments homologous to a single or several loci (restriction fragment length polymorphism, RFLP), or to highly reiterated sequences that are scattered in the eukaryotic genome (variable number of tandem repeats, VNTR). Both types of DNA markers are screened directly at the DNA level and behave as codominant markers.

The RFLP genomic mapping detects sequence variations resulting from base pair changes that form or cancel a restriction site, causing a visible change in the length of a certain DNA fragment, when probed with a homologous sequence (usually radioisotope labeled). Such an RFLP marker usually recognizes two or several loci when corresponding to a single copy gene or to a gene family, respectively. Some of the RFLP probes are well defined and their corresponding phenotypes are known. Using these single locus markers, RFLP maps are being constructed for some eukaryotic organisms (Tanksley et al., 1989). Building of such maps requires numerous defined single locus markers, as well as suitable information related to the physical linkage existing between the markers.

Jeffreys and co-workers (Jeffreys et al., 1985a and b; Jeffreys et al., 1990; Jeffreys et al., 1991a and b) and others (Daly et al., 1991; Epplen et al., 1991; Hillel et al., 1990; Weising et al., 1991b) developed more versatile DNA markers. These markers are able to detect multiallelic variations that correspond to high heterozygocity. These *hypervariable* probes consist of tandem repeats of short sequences, also termed as ''minisatellites,'' which are spread throughout the eukaryotic genome. The polymorphism observed by employing these probes results from allelic differences in the number of repeats (Jeffreys et al., 1985a). These variations in length are easily detected in Southern blots, by using restriction enzymes that cut in the marginal DNA sequences, leaving the full-size tandem repeat regions as intact fragments. The resulting minisatellite length variation, whose simplified display is a ladder of fragments on a Southern blot, provides a reliable and stable tool to differentiate even between closely related individuals. Among several VNTR (Nakamura et al., 1987) markers two probes designated 33.6 and 33.15 (Jeffreys et al., 1985a), were isolated from a human genomic library constructed in a lambda vector. Sequencing data show that the "core" of the repeated "units" is partially homologous and common to many other VNTRs dispersed in the human genome and in certain related viruses (Nakamura et al., 1987). However, despite the homology found between the cores of various hypervariable probes, each of them may detect an independent set of fragments, so-called "fingerprints." The estimation of genetic variability by the utilization of 33.6 and 33.15 heterologous probes had been broadened to include a wide range of animals, birds, plants, and fungi (Burke et al., 1991; Daley et al., 1991; Epplen et al., 1991; Hillel et al., 1990; Lavi et al., 1991; Weising et al., 1991b).

Single locus minisatellite probes originally identified by certain VNTRs have been developed to specifically analyze variability between individuals in human and other mammals. These probes are used to find out correlations with phenotypic characteristics of inherited diseases, to supply evidence in forensic cases, and in paternity evaluations. When the amount of DNA is limited, it is possible to amplify a single minisatellite locus by the polymerase chain reaction (PCR) using specific primers homologous and complementary to the unique sequences present in the margins of the array of repetitive sequences (Boerwinkle et al., 1989; Jeffreys et al., 1990). This innovative technique allows a fine mapping of changes occurring within a certain VNTR locus, such as point mutations, differences in nucleotide number, deletions, and duplications (Jeffreys et al., 1990; Jeffreys et al., 1991a and b). This type of analysis of interallelic variability [so-called MVRs (Minisatellite Variant Repeats)] is highly informative (Boerwinkle et al., 1989). Using this strategy, differences between individuals are screenable using specific PCR primers that can differentiate between two types of variant repeats (Jeffreys et al., 1991b), or by performing further restriction enzyme analyses of the amplified products (Jeffreys et al., 1990; Jeffreys et al., 1991b).

Relying on the sequence of the "core" monomers, the use of corresponding oligonucleotides, as well as those comprising tandem repeats of GT, CA, and TCC as probes for DNA fingerprinting, has been developed. These short sequences are also termed "microsatellites" (Nakamura et al., 1987; Weising et al., 1991b). Epplen et al. (Epplen et al., 1991) summarized the strategy of using such simple repetitive oligonucleotides for DNA fingerprinting of many fungi, plants, and animals, to discriminate between individuals or isolates. They suggested that although different oligonucleotide probes have been informative for different organisms, about 20 combinations of repeats are required to cover efficient probing or all the eukaryotic genomes.

Recently with the advent of PCR methods, DNA amplification fingerprinting (DAF) has been developed. Single, short oligonucleotide primers (5–15b) of arbitrary sequence, also called RAPD after random amplified polymorphic DNA (Caetano-Anolles et al., 1991a; Welsh et al., 1991; Williams et al., 1990) are used to amplify short segments of a certain genome and to generate a range of distinct extension products. This polymorphism is currently used to distinguish between species, strains, and individuals among prokaryotic and eukaryotic organisms (Caetano-Anolles et al., 1991b). This approach offers a relatively rapid technique for verifying genetic variability without using Southern blotting and labeled probes (Caetano-Anolles et al., 1991b; Williams et al., 1990).

4. TYPING GENETIC DIFFERENCES IN FUNGAL PLANT PATHOGENS

The application of different RFLP and fingerprinting methods to verify genetic relatedness has recently been expanded to phytopathogenic fungi. A DNA amplification fingerprinting approach was used by Goodwin and Annis (1991) to distinguish between isolates of *Leptosphaeria maculans* (Desm.) Ces. et de Not the causal agent of blackleg disease of crucifers. Fifteen Canadian isolates of *L. maculans* (7 virulent and 8 avirulent) were examined for genetic relatedness using RAPD oligonucleotide probes. Six different decamers were tested as primers to amplify *L. maculans* DNA. Several primers such as R3 (5'-TTCCTCTAGG), R5 (5'-CGACATAATC), and R6 (5'-AGTTCTGTTC) did not reveal any amplification products. Polymorphic amplifications of DNA fragments from the isolates were observed with primers R2 (5'-AGTACAGGTC), R4 (5'-TCCTACGCAC), and R28 (5'-ATGGATCCGC). According to these amplified products three groups of isolates were distinguished: group 1 contained all isolates of the virulent pathotype; group 2 contained isolates of the avirulent pathotype from western Canada, and group 3 contained avirulent isolates from Ontario. The recently introduced virulent isolates, which included three isolates from Ontario, showed relatively little divergence. Geographic divergence was recorded for the indigenous populations of the avirulent pathotype. The authors suggested that in addition to pathotype identification, RAPD can be utilized in the management of the blackleg disease in rapeseed lots, since only minute amounts of fungal DNA are required to identify the presence of a certain pathotype. According to the arbitrary primed PCR approach, oligonucleotide primers with a G + C content ranging from 60 to 70% are more efficient in generating reliable RAPD patterns especially in eukaryotes (Caetano-Anolles et al., 1991a and b). Among primers with such a G + C content, the specific sequence of the decameric oligonucleotides dictates its adaptability as RAPD primers for a certain genome. Thus many oligonucleotides have to be screened to obtain reliable DAF patterns that reflect genetic relatedness in a certian fungal population. Synthetic oligonucleotides (microsatellites) were used for DNA fingerprinting of six different pathovars of *Ascochyta rabiei* (Pass.) a pathogen of chickpea (*Cicer arietinum* L.) (Weising et al., 1991a). The use of certain restriction enzyme–synthetic probe combinations allowed the discrimination between the fungal isolates. It was concluded that the use of simple repeats of $(GATA)_4$, $(GTG)_5$, and $(CA)_8$ sequences may be expanded to a more general identification of fungal races and pathotypes.

A 4-kb genomic fragment of *Erysiphe graminis* f. sp. *hordei* was used to detect RFLPs. This fragment contains several repeated motifs (Brown et al., 1990). This probe (E9) hybridizes to various fragments that are dispersed in the genome of the fungal pathogen, rather than being organized in tandem array. This multiloci RFLP probe was used to elucidate the composition and structure of the *E. graminis* f. sp. *hordei* isolates in the population. The probe identifies genetic variability among isolates that share similar virulence phenotypes previously differentiated on host genotypes. Yet, 28 polymorphic *E. graminis* f. sp. *hordei* isolates were differentiated when only 3 low-copy RFLP markers were used (Christiansen et al., 1990).

A distance of 6.5 cM between the RFLP locus G538 and the avirulence locus Avr6 of *Bremia lactucae* was reported by Hulbert et al. (1988). In *B. lactucae* the assumed genome size is about 2000 cM, corresponding to 5×10^7 bp in the haploid stage, thus a genetic distance of 6.5 cM represents about 150 kb. A middle repetitive DNA sequence isolated from genomic DNA of *Magnaporthe grisea* (anamorph: *Pyicularia grisea,* formerly *P. oryzae*), the causal agent of the rice blast fungus, was named MGR for *M. grisea* repeat (Valent and Chumley, 1991). The MGR sequences in a single strain of the fungal pathogen are polymorphic, with respect to restriction sites and sequence arrangement. The probe MGR586 identifies a highly polymorphic series of EcoRI restriction fragments that is unique for every rice pathogen tested so far. The MGR fingerprints are extremely useful for identifying strains, confirming parent–mutant relationships, mapping the genome, and studying the origin and evolution of populations of *M. grisea* in the field. The MGR fingerprints segregate as stable Mendelian markers in genetic crosses making it a valuable physical marker for genetic mapping.

The human hypervariable minisatellite probes 33.6 and 33.15 (Hulbert et al., 1988; Jeffreys et al., 1985a) were used successfully to distinguish between subgroups of *Colletotrichum gloeosporioides* (Penz.) Penz. ans Sacc. in Australia (Braithwaite and Manners, 1989).

5. *SEPTORIA TRITICI* BLOTCH OF WHEAT

McDonald and Martinez (1990a and b, 1991) developed a set of anonymous DNA probes from *S. tritici* plasmid library that identify RFLPs distributed through-out the nuclear DNA of the fungus. Most of these probes hybridized to a single copy sequences present on different chromosomes. Approximately 26% of the genomic clones hybridized to many different genomic fragments in Southern blots and probably contain highly repetitive sequences. When analyzing various *S. tritici* populations most RFLPs detected appear to result from a change in a single locus identified by restriction site mutations or sequence deletions. High level of genetic variation in *S. tritici* was found within microgeographical scale (leaf, field) (McDonald and Martinez, 1990b).

About 14 new bands were found in the California population that were not identified in the original screening of 6 isolates from Texas, Montana, and Israel. The single locus RFLP probes resolved differences between individuals in cer-tain populations that were not visualized by the repetitive probes. This genetic variation among isolates of the pathogen was probably a result of the heterogene-ity of ascospores of the sexual state (incoming or existing) serving as primary inoculum followed thereafter by propagation via the asexual state (pycnidio-spores). Since many alleles detected by most of the single-site RFLP probes were common to *S. tritici* isolates from Oregon and California populations, it was suggested that natural selection is operating to increase the frequency of specific alleles at single-site loci or that substantial gene flow occurs among these populations (McDonald and Martinez, 1991). The latter suggestion that gene flow affects genotypic frequency has to be carefully studied because of the large geographic distances from where the isolates originated (California, Oregon, Montana, Texas, and Israel). The authors suggested that the probes containing repetitive sequences can be successfully used to differentiate among different clones in a population of *S. tritici* (McDonald and Martinez, 1991), but they emphasized that a fine-mapping verification should also apply specific RFLP probes. However, when referring to pathogenicity, not all the minute differences that are detected by this strategy can be related to the pathogenic divergence exist-ing between clones or isolates.

The resolving power of two VNTR markers was utilized to differentiate between *S. tritici* isolates that vary in their virulence on wheat differentials. Isolates from Ethiopia (IPO88020), Israel (ISR398, ISR7901, ISR8036, and ISR88025), The Netherlands (IPO98011), Portugal (POR), Uruguay (URU), and USA (MTN) were used in this comparative study.

The hybridization pattern obtained by the human probe 33.6 (Jeffreys et al., 1985a) and three different restriction enzyme digestions of total *S. tritici* DNA is presented in Figure 18.1. Most of the isolates exhibited a completely different pattern of fingerprints. The hybridization pattern of the isolates MTN (Montana),

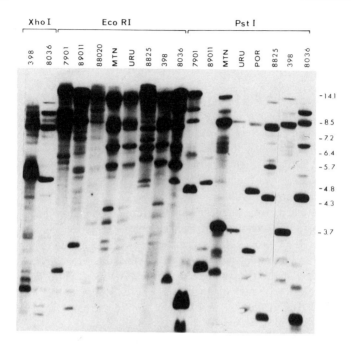

Fig. 18.1. DNA fingerprints of geographically divergent Septoria tritici isolates revealed by the human minisatellite probe 33.6. Samples of 10 µg total DNA isolated from conidia of the following *S. tritici* isolates : Ethiopia (IPO) 88020; Israel (ISR) 398, 7901, 8036, and 88025; The Netherlands (IPO) 89011; Portugal (POR); Uruguay (URU); and USA (Montana - MTN) were digested with XhoI, EcoRI, or PstI as indicated above the lanes. Following separation on a 0.8% agarose gel and blotting onto a nylon filter, the blot was hybridized with the human probe 33.6 in 0.263 M Na$_2$HPO$_4$, 7% sodium dodecyl sulfate (SDS), 1 mM ethylenediaminetetraacetic acid (EDTA), 1% bovine serum albumin (BSA) at 65°C for 24 h and then washed once with 0.263 M Na$_2$HPO$_4$, twice with 2 standard saline citrate (SSC), 0.1% SDS and twice with SSC, 0.1% SDS, at 65°C. The sizes in kb of the lambda HindIII fragments are indicated on the right.

URU (Uruguay) and ISR398 (Israel) reflected only small band differences. Hybridization with a completely different VNTR probe denoted as 22.3 (Haberfeld et al., 1991) that originated from a cattle revealed a more complex pattern of fingerprints (Fig. 18.2). The combination of probe 33.6 and the restriction enzymes PstI and EcoRI digests revealed an overall lower banding (Fig. 18.1), and lower band sharing among the eight tested *S. tritici* isolates (Table 18.1). In avocado, about 55% band sharing was reported for both probes (Lavi et al., 1991).

Three Israeli *S. tritici* isolates were examined with respect to their pathogenicity pattern on seedlings of three spring bear wheat accessions—Bobwhite"S"

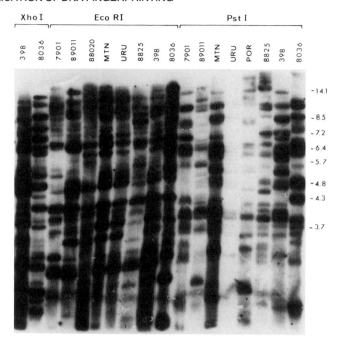

Fig. 18.2. The DNA fingerprints of geographically divergent *S. tritici* isolates revealed by the cattle minisatellite probe 22.3. The blot presented in Figure 18.1 was reprobed with probe 22.3 using identical stringency conditions. The sizes in kb of the lambda HindIII fragments are indicated on the right.

(AUR/KAL-BB/WOP"S", CM33203-K-9M-2Y-1M-1Y-2M-0Y-1PTZ-0Y, CIMMYT, Mexico), Seri 82 (KVZ/BUHO"S"/KAL/BB = VEE No. 3"S", CM33037-F-15M-500Y-0M-87B-0Y, CIMMYT, Mexico), and Shafir (SON64/ TZPP//NAI60/3/FA, Hazera Seed Co., Israel) (Table 18.2).

Isolates ISR398 and ISR8036 varied in their virulence on Seri 82, which possesses the Bezostaya 1-Kavkaz genes for resistance to *S. tritici*. The resistant Bobwhite"S" expressed a moderate level of pycnidia following inoculation with ISR88025. The significant (p = 0.01) cultivar X isolate interaction explained 16.1% of the total variance.

The genetic relatedness among the three *S. tritici* isolates was assessed by comparing the hybridization patterns obtained from probes 33.6 and 22.3 and the two restriction enzymes digests. A pairwise comparison of the similarity indexes of the three Israeli *S. tritici* isolates (Fig. 18.1 and 18.2) shows that probe 33.6 is a more powerful tool for illustrating the variations that exist among the three isolates (Table 18.3).

TABLE 18.1. Genetic Variability Among Eight *S. tritici* Isolates From Diverse Geographic Origin Assessed by Probes 33.6 and 22.3[a]

	PstI[c]		EcoRI[c]	
Criterion[b]	22.3	33.6	22.3	33.6
Average No. N + SE	19.0 ± 0.9	12.3 ± 1.6	18.6 ± 2.4	15.8 ± 2.1
No. S + SE	10.9 ± 0.5	9.4 ± 0.7	9.4 ± 1.2	10.0 ± 1.2
Range in S	7 – 15	2 – 17	0 – 16	0 – 18
BS (%) + SE	43.1 ± 1.7	20.2 ± 2.8	54.5 ± 5.8	43.6 ± 6.6

[a]The number of scoreable bands per individual (N), number of specific bands (S), and level of band sharing (BS) between eight *Septoria tritici* isolates for probes 22.3 and 33.6 and the restriction enzymes PstI and EcoRI estimated from Figures 18.1 and 18.2 (Hillel et al., 1990).
[b]Standard error (SE).
[c]Restriction enzymes.

TABLE 18.2. Assessment of Pycnidia Coverage of *Septoria tritici* Isolates on Wheat Seedlings

	S. tritici Isolates		
Wheat Cultivar	ISR8036	ISR398	ISR88025
Bobwhite "S"	2.4 ± 4.2[a]	0.1 ± 0.2	14.5 ± 10.3
Shafir	81.1 ± 6.5	75.9 ± 13.0	59.2 ± 6.7
Seri 82	78.3 ± 13.2	4.8 ± 7.2	7.5 ± 7.5

[a]Standard deviation.

TABLE 18.3. Estimation of Genetic Similarity by Band Sharing Comparison (%) of Three *S. tritici* isolates for Probes 22.3 and 33.6 Based on Data Presented in Figures 18.1 and 18.2

Restriction Enzymes and Probe	*S. tritici* Isolate Pairs		
	ISR8036/ISR398	ISR8036/ISR88025	ISR88025/ISR398
EcoRI/33.6	50[a]	46	34
EcoRI/22.3	49	45	54
PstI/33.6	19	32	26
PstI/22.3	32	48	41

[a]Genetic similarity index (Lansman et al., 1981) expressed in percent, where

$$S_{xy} = \frac{[2N_{xy}]}{(N_x + N_y)]} \; 100;$$

N_x and N_y = the number of fragments in isolates a and b, respectively; and N_{xy} = the number of fragments shared by the isolates.

The best display of variability between isolates was observed when the enzyme Pstl was used for digesting the total DNA samples. Because there are fewer bands visualized with probe 33.6 and they have a low band sharing value, a unique fingerprint pattern specific to each isolate is obtained (Fig. 18.1). This can be utilized to explore the genetic variation that exists within and among populations. The specific hybridization patterns revealed by isolates ISR398 and ISR8036 can be further used in studying population dynamics that otherwise may require the use of the quantitative assessments of host response on specific differentiating wheat cultivars.

6. FUTURE STRATEGIES

Further characterization of such repetitive *S. tritici* sequences detected by the VNTR probes, such as the 33.6 probe, could contribute to further development of more sensitive single locus probes (MVR-like) capable of discriminating between individuals in populations that are vegetatively propagated and remain haploid. Such an approach can be used to resolve the genetic differences between *S. tritici* isolates that otherwise lack distinct qualitative differentiating parameters. In this respect the detection of interallelic divergence by amplifying a specific allele, as proposed by Jeffreys et al. (1985b), seems to be a promising approach in verifying events occurring in a certain locus. These premeiotic changes that occur in a single locus are probably due to somatic mutations and are easily screened following PCR amplification (Jeffreys et al., 1990; Jeffreys et al., 1991b). This approach allows the direct estimation of minisatellite mutation rates that reflect differences among individuals without using numerous RFLP or RAPD probes. Such a strategy can be applied to other plant pathogenic fungi that lack specific markers and their sole indication for specificity is derived from assessing the interaction between the host and the pathogen.

REFERENCES

Boerwinkle E, Xiong W, Fourest E, Chan L (1989): Rapid typing of tandemly repeated hypervariable loci by the polymerase chain reaction: Application to the apolipoprotein B hypervariable region. Proc Natl Acad Sci USA 86:212–216.

Braithwaite KS, Manners JM (1989): Human hypervariable minisatellite probes detect DNA polymorphisms in the fungus *Colletotrichum gloeosporioides*. Curr Genet 16:473–475.

Brown JKM, O'Dell M, Simpson CG, Wolfe MS (1990): The use of DNA polymorphisms to test hypotheses about a population of *Erysiphe graminis* f.sp. *hordei*. Plant Pathol 39:391–401.

Burke T, Hanotte O, Bruford MW, Cairns E (1991): Multilocus and single locus minisatellite analysis in population biological studies. In Burke T, Dolf G, Jeffreys AJ, Wolff R (eds): DNA fingerprinting approaches and applications. Switzerland: Birkhauser Verlag, pp 154–168.

Caetano-Anolles G, Bassam BJ, Gresshoff PM (1991a): DNA amplification fingerprinting: A strategy for genome analysis. Plant Mol Biol Reporter 9:294–307.

Caetano-Anolles G, Bassam BJ, Gresshoff PM (1991b): DNA amplification fingerprinting using very short arbitrary oligonucleotide primers. Bio/Technol 9:553–557.

Christiansen SK, Giese H (1990): Genetic analysis of the obligate parasitic barley powdery mildew fungus based on RFLP and virulence loci. Theor Appl Genet 79:705–712.

Daly A, Kellam P, Berry ST, Chojecki AJS, Barnes SR (1991): Isolation and characterization of plant sequences homologous to human hypervariable minisatellites. In Burke T, Dolf G, Jeffreys AJ, Wolff R (eds): DNA fingerprinting approaches and applications. Switzerland: Birkhauser Verlag, pp 330–341.

Epplen JT, Ammer H, Epplen C, Kammerbauer C, Mitreiter R, Roewer L, Schwaiger W, Steimle V, Zischler H, Albert E, Andreas A, Beyermann B, Meyer W, Buitkamp J, Nanda I, Schmid M, Nurnberg P, Pena SDJ, Poche H, Sprecher W, Schartl M, Weising K, Yassouridis A (1991): Oligonucleotide fingerprinting using simple repeat motifs: A convenient ubiquitously applicable method to detect hypervariability for multiple purposes. In Burke T, Dolf G, Jeffreys AJ, Wolff R (eds): DNA fingerprinting approaches and applications. Switzerland: Birkhauser Verlag, pp 50–69.

Eyal Z, Scharen AL, Prescott JM, van Ginkel M (1987): The *Septoria* diseases of wheat: Concepts and methods of disease management. Mexico, CIMMYT: D.F. 52 pp.

Goodwin PH, Annis SL (1991): Rapid identification of genetic variation and pathotype of *Leptosphaeria maculans* by random amplified polymorphic DNA assay. Appl Environ Microbiol 57:2482–2486.

Haberfeld A, Cahaner A, Yoffe O, Plotzky Y, Hillel J (1991): DNA fingerprinting of farm animals generated by micro and minisatellite DNA probes. Anim Genet 22:299–305.

Hetzler J, Eyal Z, Mehta YR, Campos LA, Fehrmann H, Kushnir U, Zekaria-Cohen J, Cohen L (1991): Interactions between spot blotch (*Cochliobolus sativus*) and wheat cultivars. In Saunders DA (ed): Wheat for the non-traditional warm areas. Mexico: CIMMYT, pp 146–164.

Hillel J, Schaap T, Haberfeld A, Jeffreys AJ, Plotzky Y, Cahaner A, Lavi U (1990): DNA fingerprints applied to gene introgression in breeding programs. Genetics 124:783–789.

Hulbert SH, Ilott TW, Legg EJ, Lincoln SE, Lander ES, Michelmore RW (1988): Genetic analysis of the fungus *Bremia lactucae*, using Restriction Fragment Length Polymorphisms. Genetics 120:947–958.

Jeffreys AJ, McLeod A, Tamaki K, Neil DL, Monckton DG (1991a): Minisatellite repeat coding as a digital approach to DNA typing. Nature (London) 354:204–209.

Jeffreys AJ, Neumann R, Wilson V (1990): Repeat unit sequence variation in minisatellites: A novel source of DNA polymorphism for studying variation and mutation by single molecule analysis. Cell 60:473–485.

Jeffreys AJ, Royle NJ, Patel I, Armour JAL, MecLeod A, Collick A, Gray IC, Neumann R, Gibbs M, Crosier M, Hill M, Signer E, Monckton D (1991b): Principles and recent Advances in human DNA fingerprinting. In Burke T, Dolf G, Jeffreys AJ, Wolff R (eds): DNA fingerprinting approaches and applications. Switzerland: Birkhauser Verlag, pp 1–19.

Jeffreys AJ, Wilson V, Thein SL (1985a): Hypervariable minisatellites in human DNA. Nature (London) 314:67–73.

Jeffreys AJ, Wilson V, Thein SL (1985b): Individual-specific 'fingerprints' of human DNA. Nature (London) 316:76–79.

Lansman R, Shade RO, Shapira JF and Avise JC (1981): The use of restriction endonucleases to measure mitochondrial DNA sequence relatedness in natural populations. J Mol Evol 17:214–226.

Lavi U, Hillel J, Vainstein A, Lahav E, Sharon D (1991): Application of DNA fingerprints for identification and genetic analysis of avocado. J Am Soc Hort Sci 116:1078–1081.

Marshall D (1985): Geographic distribution and aggressiveness of *Septoria tritici* on wheat in the United States. Phytopathology 75:1319.

McDonald BA, Martinez JP (1990a): Restriction fragment length polymorphisms in *Septoria tritici* occur at a high frequency. Curr Genet 17:133–138.

McDonald BA, Martinez JP (1990b): DNA restriction fragment length polymorphisms among *Mycosphaerella graminicola* (anamorph *Septoria tritici*) isolates collected from a single wheat field. Phytopathology 80:1368–1373.

McDonald BA, Martinez JP (1991): DNA fingerprinting of the plant pathogenic fungus Mycosphaerella graminicola (anamorph *Septoria tritici*). Exp Mycol 15:146–159.

Nakamura Y, Leppert M, O'Connell P, Wolff R, Holm T, Culver M, Martin C, Fujimoto E, Hoff M, Kumlin E, White R (1987): Variable number of tandem repeat (VNTR) markers for human gene mapping. Science 235:1616–1622.

Tanksley SD, Young ND, Paterson AH, Bonierbale MW (1989): RFLP mapping in plant breeding: New tools for an old science. Bio/Technol 7:257–264.

Valent B, Chumley FG (1991): Molecular genetic analysis of the rice blast fungus, *Magnaporthe grisea*. Annu Rev Phytopathol 29:443–467.

Weising K, Kaemmer D, Epplen JT, Weigand F, Saxena M, Kahl G (1991a): DNA fingerprinting of *Ascochyta rabiei* with synthetic oligodeoxynucleotides. Curr Genet 19:483–489.

Weising K, Ramser J, Kaemmer D, Kahl G, Epplen JT (1991b): Oligonucleotide fingerprinting in plants and fungi. In Burke T, Dolf G, Jeffreys AJ, Wolff R (eds): DNA finger printing approaches and applications. Switzerland: Birkhauser Verlag, pp 312–329.

Welsh J, Peterson C, McClelland M (1991): Polymorphisms generated by arbitrarily primed PCR in the mouse : application to strain identification and genetic mapping. Nucleic Acids Res 19:303–306.

Williams JGK, Kuelik AR, Livak AR, Rafalski JA, Tingey SV (1990): Polymorphisms amplified by arbitrary primers are useful as genetic markers. Nucleic Acids Res 18:6531–6535.

Zelikovitch N, Eyal Z, Kashman Y (1990): Methyl 3-indole carboxylate in *Septoria tritici* cultures. Proceedings of the 4th International Mycological Congress, Germany: Regensburg, p 264.

INDEX

ABA. *See* Abscisic acid
Abscisic acid (ABA)
  in stress tolerance, 10
  in wound-induced gene activation,
    162-163, 165-167
Ac/Ds elements, in gene tagging, 77
Acetaldehyde, in fungal spore germination,
  306
ACR-toxin, in plant diseases, 118
Adenosine triphosphate (ATP)-binding
  protein, HlyB translocator as, 43
Adenylyl cyclase, inhibition of by fungal
  killer toxin, 262-263
Adhesion zones, in bacterial envelopes, 41
Aerolysin, secretion of, 44, 53
Aeromonas caviae, detection of, 307
Aeromonas hydrophila
  aerolysin in, 44, 53
  chitinase in, 214
  detection of, 308
Aeromonas salmonicida, outer membrane of
  involved in protein secretion, 41
African cassava mosaic geminivirus,
  defective interfering DNA of, 90
AF-toxin, in plant diseases, 118
afuE locus, of *Pseudomonas fluorescens*,
  309
Agricultural productivity, increases in,
  combined with improved
  environmental protection, 1-11, 16
Agrobacterium radiobacter strain 84, against
  crown gall, 277, 279
Agrobacterium rhizogenes, in mannopine
  utilization, 295
Agrobacterium tumefaciens
  colonization by, 287
  in functional cloning, 81
  neoplastic transformation induced by, in

    molecular plant breeding, 3
  pathogenicity of, 118
  Sec-dependent secretion pathway and,
    49-51
  in *ttr* gene transfer, 127-131
Agrocin 84, activity of, 277-281, 298, 312
AK-toxin, in plant diseases, 118
Alcaligenes spp., as delivery organisms, 286
Alfalfa
  avirulence gene elicitation of
    hypersensitive response in, 69
  coat protein-mediated resistance in, 92
  DNA segments introduced into, 4
  pin2 homologs in, 159
Alfalfa mosaic virus (AlMV)
  chitinases in, 142
  coat protein-mediated resistance to, 93,
    95-97
  salicylic acid against, 199
Alginate capsule techniques, in
  microencapsulation technology, 28
AlMV. *See* Alfalfa mosaic virus
Alternaria alternata, pathogenicity of, 118
Alternaria solani, pathogenicity of, 118
Alternaria spp., chitinases against, 303
AL-toxin, in plant diseases, 118
Alumina, in microencapsulation technology,
  28
amdS gene, in improvement of *Trichoderma*
  as biocontrol agent, 223, 225-227
α-Aminooxyacetic acid (AOA), in PAL
  inhibition, 194
α-Aminooxy-β-phenylpropionic acid
  (AOPP), in PAL inhibition, 194
N-(*O*-Aminophenyl)sulfinamoyl-tertiobutyl
  acetate (NH_2-PAS), in lignification,
  194
Ammonia, against damping-off diseases,

300-301
Amphibians, in disease resistance, 177
AM-toxin, in plant diseases, 118
Amylase, *Xanthomonas campestris* pv.
 campestris secretion of, 45
Anthers, mechanical removal of, 10
Anthranilate, in biocontrol, 299
Anthranilic acid, production of, 297
Antibiotics, in biocontrol, 295-298; *see also specific agents*
Antisense RNA, viral replication limited by, 6, 90
Anubis-2 peptide, in disease resistance, 183-185
AOA. *See* α-Aminooxyacetic acid
AOPP. *See* α-Aminooxy-β-phenylpropionic acid
Apples
 phenolics in surface wax of, 194
 phloridzin in, 195
Arabidopsis spp.
 avirulence gene elicitation of
 hypersensitive response in, 69
 chitinases in, 140
 chromosome walking and, 76
 gene tagging in, 77-78
 subtractive hybridization cloning in, 78
Arabidopsis thaliana, chitinase in, 214
Arum spp., salicylic acid in, 200
Ascochyta rabiei, typing genetic differences in isolates of, 346
Ascomycetes, killer phenomenon in, 257-258, 262
Asparagus, transgenic, introduction of DNA segments in, 4
Aspartate-metalloproteinase inhibitors, accumulation of upon wounding, 169
Aspergillus giganteus, killer phenomenon in, 258
Aspergillus nidulans, protoplast fusion in, 243
Aspergillus spp.
 compatibility groups in, 246-247
 progeny from protoplast fusion in
 Trichoderma spp. and, 246-248
ATP. *See* Adenosine triphosphate
ATPase, SecA, 42
AT-toxin, in plant diseases, 118
Avirulence genes, disease resistance gene recognition of, 68-69, 74, 80
Avocado, *Phytophthora cinnamonmi* as pathogen of, 74

Bacillus circulans, β-1,3-glucanase in, 303

Bacillus spp., as delivery organisms, 286
Bacillus subtilis, Sec-dependent secretion pathway and, 49
Bacillus thuringiensis
 against European corn borer, 284
 against lepidopterans, 282
 toxicity of to insects, 5
Bacteria
 crop plant resistance to, 7, 9
 killer phenomenon in, 257
 protein secretion by, 39-58
 see also individual genera and species
Bacterial scorch, of coffee plants, 120
Bacteriocins, activity of, 277, 280
Bacteroides fragilis, levanase gene in, 284
Bakanae disease, *Gibberella fujikuroi* in, 119
bar gene, in bialophos acetyltransferase, 121
Barley
 chitinases in, 140
 fungal killer toxins in, 269
 ribosome-inhibiting protein in, 7
Basidiomycetes, killer phenomenon in, 257-258, 262
Bayer junctions, in bacterial envelopes, 41
Bean
 chitinases in, 140, 143-145, 148-152
 phytoalexins in, 197
 systemic resistance to *Fusarium solani* induced in, 304
 Trichoderma spp. protoplast fusion progeny in biocontrol in, 252
Beet necrotic yellow vein virus (BNYVV), coat protein-mediated resistance to, 93
Beets, damping-off diseases in, 300
Belgian endive, nuclear male sterility and, 10
Beneficial genes, genetic linkage betewen, 3
Benomyl, progeny from protoplast fusion in *Trichoderma* spp. and, 246
Bialophos, production of, 120-121
Binary vector, construction of with *ttr* gene, 129-130
Bioconservation, associated with agricultural productivity, 3
Biocontrol. *See different aspects of biocontrol*
Biofungicide research team,
 interrelationships between functional tasks of, 23
Biotechnology, major trends in, 17-20
Biotechnology products, sales in worldwide, 15

Biotin, in subtraction hybridization cloning, 78

Birds-foot trefoil, transgenic, introduction of DNA segments in, 4

Blackleg disease, *Leptosphaeria maculans* in, 345

Black point, present losses in Australia due to, 22

Black root rot
2,4-diacetylphloroglucinol against, 297-298
hydrogen cyanide against, 301

Black rot, *Xanthomonas campestris* pv. *campestris* in, 45

BNYVV. *See* Beet necrotic yellow vein virus

Botrytis cinerea
cell wall-degrading enzyme levels after infection by, 143
proanthocyanidins against, 195
Trichoderma spp. protoplast fusion progeny against, 252

Botrytis spp., in plant tissue, 328

Bradyrhizobium spp, *nod* genes in, 192

Brassica napus cv. Westar, chimeric 35S-chitinase gene introduced into, 147

Bremia latucae, typing genetic differences in isolates of, 346

Broad host range vectors, development of, 285-286

Bs2 resistance gene, in pepper plants, 74

BT gene *Tn5* genomic insertion, in *Pseudomonas fluorescens*, 30

Bunt disease, losses in Australia due to, 22

Cabbage, transgenic, introduction of DNA segments in, 4

CaMV. *See* Cauliflower mosaic virus

Candida albicans, chromosomes in, 245

Candida malodendra, killer phenomenon in, 262

Candida spp., killer phenomenon in, 262

Canteloupe, coat protein-mediated resistance in, 92

Carnation mosaic virus, phenolics as signals for induced resistance to, 198

Carnations, *Fusarium* wilt in, 304

Carrot, transgenic, introduction of DNA segments in, 4

Cassava, genetic transformation in, 73

Catechol, as spore germination inhibitor, 195

Cathepsin D, wound-induced, 166-167

Cauliflower mosaic virus (CaMV)
coat protein-mediated resistance to, 91, 96, 99

dual 35S promoter, with viral leader sequence, 80

*Cbh*I promoter and terminator, in cellulase gene induction, 212-213

cDNA. *See* Complementary DNA

Cecropins, in disease resistance, 177-179, 183-184

Celery, transgenic, introduction of DNA segments in, 4

Cell envelope, in gram-negative bacteria, 40-44

Cellulase, *Erwinia* spp. secretion of, 56

Cellulose, cloned gene product degradation of, 287

Cell wall, plant, phenolics as components of, 192-194

Cell wall-degrading enzymes, in fungal disease resistance, 139-153; *see also specific enzymes*

Cephalosporium spp., progeny from protoplast fusion in *Trichoderma* spp. and, 246

Cereal cyst nematode, potential and present losses in Australia due to, 22

Charge coupled device camera, in bacterial detection, 307

Chicken egg-white lysozyme, peptides incubated in, effects of on disease resistance, 185-187

Chickpea
Ascochyta rabiei in, 346
phytoalexins in, 197

Chimeric genes
in male sterility, 10
in molecular plant breeding, 4-5, 7, 9

Chitinase[poly[1,4-(*N*-acetyl-β-D-glucosaminade)]glycanohydrolase], activation of during fungal attack, 140

Chitinases
activation of during fungal attack, 140-153
in biocontrol, 7, 213-214, 302-303
in *Escherichia coli*, 214-222
in hypersensitive response, 67
modification of expression of in plants, 145-147
plant disease control by, 7, 214-215, 230
in rhizobacteria, 221-223
in *Trichoderma* spp., 223, 225-228, 231

Chitosan, in wound-induced gene activation, 162

Chlorogenic acid, as preformed antimicrobial defense chemical, 195

Cholera toxin, secretion of, 44, 53, 55
Chromosome electrophoresis, of progeny
    from protoplast fusion in
    Trichoderma spp., 244-245, 250-251
Chromosome jumps, to ends of pulse-field
    gel electrophoresis bands, 76
Chromosome walking, in cloning disease
    resistance genes, 75-77
Cicer arietinum L., *Ascochyta rabiei* in, 346
Cinnamic acid hydroxylase, in lignification,
    193
Cinnamyl alcohol dehydrogenase, in lignin
    precursor synthesis and
    polymerization, 193-194
cis-zeatin, in plant diseases, 118
Cladosporium cucumerinum, scab caused by,
    71
Cladosporium fulvum
    cell wall-degrading enzyme levels after
        infection by, 142
    peptide carrying avirulence gene 9 in, 68
Clavibacter michiganense subsp.
        michiganense
    cecropin B against, 178
    ELISA detection of in plant tissue, 327
Clavibacter xyli, *Bacillus thuringiensis* toxin
        gene cloned into, 284
Cloning
    of disease resistance genes, 71, 73-83
    of tattoxin resistance genes, 121-122
Clover, transgenic, introduction of DNA
        segments in, 4
CMV. *See* Cucumber mosaic virus
Coat protein genes
    background information on, 89-90
    coat protein-mediated resistance, 90-91
        breadth of protection afforded by,
            97-98
        efficacy of under field conditions,
            99-101
        examples and phenotypes of, 92-95
        levels of coat protein vs. levels of CP-
            MR, 98-99
        molecular and cellular mechanisms
            of, 92, 95
        overcoming CP-MR by inoculation
            with viral RNA, 96-97
        reduces numbers of infection sites on
            inoculated leaves, 95-96
        technical aspects of developing,
            91-92
Cochliobolus heterostrophus
    phytoalexins against, 197

sexual compatibility in, 247
Coffee plants, bacterial scorch of, 120
Colicins
    killer phenomenon and, 257
    release of, 42
Colletotrichum gloeosporioides, typing
        genetic differences in isolates of, 346
Colletotrichum lagenarium
    chitinase levels after infection by, 141
    induced resistance to, 198
    salicylic acid against, 200, 203
Colletotrichum lindemuthianum,
        phytoalexins against, 197
Colletotrichum spp., systemic resistance
        induction against, 304-305
Colloidal gold labeling, in diagnosis of plant
        pathogens, 326
Complementary DNA (cDNA)
    bean chitinase, 142
    chromosome walking and, 76
    in functional cloning, 80, 81
    fungal killer toxin, 261, 266
    in subtraction hybridization cloning, 78
    libraries, differential hybridization of to
        plant mRNAs, 9
    pin2, 159
Complementary spore color mutants, in
        selection of progeny from protoplast
        fusion in *Trichoderma* spp., 240, 242
Complementation of traits, in *Trichoderma*
        spp., 240-241
Com proteins, in *Bacillus subtilis*, 49-50
Coniferyl alcohol, antimicrobial effects of,
        194
Cooperative agreements, between academia
        and industry, 18-19
Corn
    avirulence gene elicitation of
        hypersensitive response in, 69
    chromosome walking in, 75
    fungal killer toxins in, 269
    hydroxamic acids in, 196
    DNA segments introduced into, 4
    nuclear male sterility and, 10
    phytoalexins in, 197
    Trichoderma spp. protoplast fusion
        progeny in biocontrol in, 252-253
    Ustilago maydis in, 258
Coronatine, in plant diseases, 118-119
Corynebacterium spp., as delivery
        organisms, 286
Cotton
    avirulence gene elicitation of

hypersensitive response in, 69
chitinases in, 140
damping-off diseases in, 300
DNA segments introduced into, 4-5
Trichoderma spp. protoplast fusion
    progeny in biocontrol in, 252-253
volatile compounds in, 306
4-Coumarate-CoA ligase, in lignin precursor
    synthesis and polymerization, 193
Cowpea
    avirulence gene elicitation of
        hypersensitive response in, 69
    DNA segments introduced into, 4
Crown gall disease
    Agrobacterium tumefaciens in, 128
    agrocin 84 against, 277-281, 298, 312
Crucifers, blackleg disease in, 345
Cucumber
    biocontrol activity in, 212
    chitinases in, 140-141
    coat protein-mediated resistance in, 92
    damping-off diseases in, 300
    disease resistance genes in, 71
    DNA segments introduced into, 4
    induced resistance in, 198
    salicylic acid in, 200-202
    systemic resistance induction in, 304-305
Cucumber mosaic virus (CMV)
    coat protein-mediated resistance to,
        93-98, 100-101
    disease resistance genes against, 71

Damping-off diseases, *Enterobacter cloacae*
    against, 300-301
Debaromyces hansenii, killer phenomenon
    in, 262
Defensins, in disease resistance, 177
Deleterious genes, separation of from
    beneficial genes, 3
Dephosphorylation, of plant proteins, elicitor
    activity in, 70
Detoxification, of toxins, in disease
    resistance, 115-135
Dextran sulfate, in subtraction hybridization
    cloning, 79
Diacetaphenone derivatives, in take-all
    disease control, 21
2,4-Diacetylphloroglucinol
    against black root rot, 297-298, 304
    increased production of, 312-313
    against take-all disease, 297-298
Diagnostic techniques, for plant pathogens
    background information on, 321-322

immunoassays
    applications, 326-329
    enzyme-linked immunosorbent assay,
        323-325
    immunofluorescence, 325-326
    immunogold labeling, 325-326
    polyclonal and monoclonal antibodies
        as reagents, 322-323
    nucleic acid hybridization technology
        applications, 332-334
    nucleic acid probes, 329-330
    dot-blot hybridization assay, 331
    nucleic acid hybridization principles,
        330-331
    polymerase chain reaction, 332
    RFLP analysis, 331-332
Dianthus barbatus, carnation mosaic virus
    in, 198
2,6-Dichloroisonicotinic acid, in induced
    resistance, 204
Diet, improved, increased agricultural
    productivity role in, 16
Disease resistance genes
    advantages and problems associated with
        use of, 72
    background information on, 65-66
    characteristics of disease resistance and
        resistance genes, 66-67
    cloning of
        chromosome walking, 75-77
        directed by isolation of resistance
            gene proteins, 79
        functional cloning approaches, 79-82
        gene tagging, 77-78
        modification of recognitional
            specificity of cloned disease
            resistance genes, 82-83
        subtractive hybridization cloning,
            78-79
    current status of disease and pest control
        using, 70-72
    functioning of, 69-70
    improved disease control with cloned
        resistance genes, 71-74
    pathogen interaction with plants carrying
        defined disease resistance genes,
        67-69
DNA fingerprinting, of genetic variation in
    Mycosphaerella graminicola isolates,
    341-351
DNA probing methods, in DNA
    fingerprinting, 343
Dot-blot hybridization assay, in diagnosis of

plant pathogens, 331
Double-stranded RNA (dsRNA), killer
    phenomenon in yeasts encoded by,
    258, 260-261, 263-267
dsRNA. See Double-stranded RNA
Durability, of disease resistance genes, 74

Eggplant, pin2 homologs in, 159
Eldana saccharina
    Bacillus thuringiensis against, 282
    Pseudomonas fluorescens against,
        282-283
Elicitor—receptor model, of disease
    resistance, 68, 70, 73, 79, 82
Elicitors
    cell wall-degrading enzymes and,
        141-142
    plant recognition of, 67-70, 73, 79, 81-82
ELISA. *See* Enzyme-linked immunosorbent
    assay
Elite phenotype, in molecular plant breeding,
    3
Emericellopsis spp., progeny from protoplast
    fusion in *Trichoderma* spp. and, 246
En-1/Spm elements, gene tagging in, 77
Endochitinases
    activation of during fungal attack, 140
    in biocontrol, 213-214
    in induced resistance, 199
Endoglucanase
    Pseudomonas solanacearum secretion of,
        45, 52
    Xanthomonas campestris pv. *campestris*
        secretion of, 45
Endo-polygalacturonases, in wound-induced
    gene activation, 161-162
Enterobacter cloacae
    in biocontrol activity, 212
    competition for nutrients and, 306
    against damping-off diseases, 300-301
    elimination of signals and, 306
Enterobacter spp., as delivery organisms,
    286
Environmental protection, combined with
    increased agricultural productivity,
    1-11, 16
Enzyme-linked immunosorbent assay
    (ELISA), in diagnosis of plant
    pathogens, 323-325
Erwinia amylovora
    cell envelope-associated protein
        produced by, 58
Erwinia carotovora

extracellular virulence proteins of, 45-46
mutants of, with deficiencies in plant cell
    wall degrading enzymes, 39-40,
    42, 47-50, 52-53, 55-56
Sec-independent secretion of proteases
    by, 46
Erwinia carotovora subsp. *atroseptica*
    Anubis-2 peptide against, 184
    cecropin B against, 184
    Hecate-1 peptide against, 184
    magainin-2 against, 184
    mellitin against, 184
    in rhizosphere colonization, 292
    SB-37 peptide against, 182
    Shiva-11 peptide against, 184
Erwinia carotovora subsp. *carotovora*
    cecropin B against, 178
    SB-37 peptide against, 182
Erwinia chrysanthemi
    cecropin B against, 178
    extracellular virulence proteins of, 45-46
    mutants of, with deficiencies in plant cell
        wall degrading enzymes, 39-42,
        47-50, 53-57
    Sec-independent secretion of proteases
        by, 46
Erwinia spp., as delivery organisms, 286
Erysiphe graminis f.sp. *hordei*, typing
    genetic differences in isolates of, 346
Escherichia coli
    Anubis-2 peptide against, 184
    cecropin B against, 184
    cell envelope of, 40
    chitinase in, 214-222, 302-303
    Hecate-1 peptide against, 184
    hemolysin in, 40, 43
    lac operon genes in, 25
    magainin-2 against, 184
    mellitin against, 184
    protein secretion apparatus lacking in
        laboratory strains of, 40
    Shiva-11 peptide against, 184
Ethanol, in fungal spore germination, 306
Ethylene, genetic interference in metabolism
    of, 11
European corn borer, *Bacillus thuringiensis*
    toxin gene against, 284
Exo-polygalacturonases, *Pseudomonas*
    solanacearum secretion of, 45
Exo-poly-α-D-galacturonosidase, *Erwinia*
    spp. secretion of, 45, 56
Expression cloning. *See* Functional cloning
Extracellular virulence proteins, of plant

pathogenic bacteria, 44-46

Fatty acid derivatives, as induced
   antimicrobial defense chemicals, 196
Fiber, increased demand for, agricultural
   productivity role in satisfying, 16, 19
Fine-structure mapping, of clones, 76
Fire blight disease, harpin protein in, 58
Flavobacterium spp., as delivery organisms,
   286
Flavonoids, in *nod* gene induction, 192
Flax, transgenic, introduction of DNA
   segments in, 4
Floral plants, genetic transformation in, 73
"Flowing helium" transformation gun, in
   functional cloning, 80-81
p-Fluorophenylalanine, progeny from
   protoplast fusion in *Trichoderma* spp.
   and, 246
Fluorescein, in immunofluorescence assays,
   325
Fluorochrome conjugates, in
   immunofluorescence assays, 325-326
Fluorochromes, in immunofluorescence
   assays, 325
Fmoc chemistry strategy, in peptide
   synthesis, 183
Food, increased demand for, agricultural
   productivity role in satisfying, 16, 19
Foreign genes, in molecular plant breeding,
   4-5, 7
French bean, transgenic, introduction of
   DNA segments in, 4
Frost damage, protection against, 282
Functional cloning, of cloning disease
   resistance genes, 79-82
Fungi
   killer phenomenon in, 257-270
   resistance to diseases caused by, 7-9,
      139-153
   see also individual genera and species
Fusaric acid, in plant diseases, 118
Fusarium graminearum, *Trichoderma* spp.
   protoplast fusion progeny against,
   252
Fusarium oxysporum
   cell wall-degrading enzyme levels after
      infection by, 142
   chitinase in, 302
   endo-chitinases in, 214
   β-1,3-glucanase in, 302
   pathogenicity of, 118
Fusarium oxysporum f.sp. *radolens*,

genetically engineered *Pseudomonas*
   fluorescens against, 221
Fusarium solani f.sp. *phaseoli*
   cell wall-degrading enzyme levels after
      infection by, 142
   phytoalexins detoxified by, 197
   systemic resistance to, 304
Fusarium solani f.sp. *pisi*, cell wall-
   degrading enzyme levels after
   infection by, 142
Fusarium wilt, systemic resistance to, 304

Gaeumannomyces graminis, polymerase
   chain reaction analysis of, 334
Gaeumannomyces graminis var. *tritici*
   2,4-diacetylphloroglucinol against,
      297-298
   hydrogen cyanide against, 301
   phenazine against, 297
   Pseudomonas fluorescens against, 221,
      292-293, 297
   siderophores against, 299
   in take-all disease, 25-27
β-Galactosidase
   in monitoring biocontrol bacteria in
      rhizosphere, 307
   Pseudomonas inability to make, 27
Gallic acid, as spore germination inhibitor,
   195
Gas chromatography—fatty acid methyl
   ester (GC-FAME) analysis, of
   Pseudomonas corrugata, 34
GC-FAME. *See* Gas chromatography—fatty
   acid methyl ester analysis
Gene IV protein, of filamentous phages, 57
Gene-for-gene relationship, in disease
   resistance, 68
Gene splicing, invention of, 17
Gene tagging
   in cloning disease resistance genes,
      77-78
   inactivation of stress tolerance by, 9
Gene technology, in molecular plant
   breeding, 3-5
General elicitors, plant recognition of, 67-68
Genes. *See specific genes and gene types*
Genetic engineering. *See specific aspects of
   genetic engineering*
Genomic subtraction. *See* Subtraction
   hybridization cloning
Gibberella fujikuroi, pathogenicity of,
   118-119
Gibberellin, in plant diseases, 118-119

Gliocladum spp., in biocontrol, 237-238, 244

β-1,3-Glucanases
    activation of during fungal attack, 140-142
    in biocontrol, 7, 302-303
    in hypersensitive response, 67

Glyceollin, as induced antimicrobial defense chemical, 196-197

Glycine max, *Pseudomonas* identified from root washings of, 34

Gramineae, hydroxamic acids in, 195

Gram-negative bacteria
    envelope of
        envelope structure, 40-41
        hemolysin secretion by Sec-independent pathway, 43
        multiple pathways into and across, 41-43
        pullulanase secretion by Sec-dependent pathway, 43-44
    protein secretion by, 39-58
    protein translocation pathways of, 40-44
    see also individual genera and species

Grapes
    agrocin 84 in, 277-279
    proanthocyanidins in, 195

Gray mould disease, *Botrytis cinerea* in, 195

Growth factors, cellular, in molecular plant breeding, 3

gusA reporter gene, in coat protein-mediated resistance, 99

Hairy root disease, *Agrobacterium rhizogenes* in, 128

Halo blight, *Pseudomonas syringae* in, 119-120

Hanseniaspora uvarum, killer phenomenon in, 262

Hansenula spp., killer phenomenon in, 258
    Hansenula californica, killer phenomenon in, 262
    Hansenula canadensis, killer phenomenon in, 262
    Hansenula dimannae, killer phenomenon in, 262
    Hansenula marakii, killer phenomenon in, 262
    Hansenula saturnus, killer phenomenon in, 262

Haploidization agents, progeny from protoplast fusion in *Trichoderma* spp. and, 246

harpin protein, synthesis of, 58

HC-toxin, in plant diseases, 118

Hecate-1 peptide, in disease resistance, 183-184

Helminthosporium carbonarum, pathogenicity of, 118

Helminthosporium maydis, pathogenicity of, 118, 119

Helminthosporium oryzae, pathogenicity of, 119

Helminthosporium sacchari, pathogenicity of, 118

Helminthosporium sativum, pathogenicity of, 118

Helminthosporium victoriae, pathogenicity of, 117-119

Helminthosporol, in plant diseases, 118

Hemolysin, *Escherichia coli* secretion of, 40, 43

Herbicolin A, in biocontrol, 298

Hewlett–Packard 5898A Microbial Identification System, in analysis of *Pseudomonas* strains, 34

Hexafluoro-2-propanol, helicity in single-chain alpha-helical peptides induced by, 183, 185

HMT-toxin, in plant diseases, 118

Honey bee lymph gland protein, in bacterial inactivation, 9

hrp genes, protein secretion pathways and, 57-58

HS-toxin, in plant diseases, 118

HV-toxin, in plant diseases, 118-119

Hyalophora cecropia, lytic protein in humoral immune system of, 184

Hybrid seeds, production of, male sterile plants in, 10

Hydrogen cyanide, in disease suppression, 301, 304

Hydrolytic enzymes, in biocontrol, 302-303; *see also specific enzymes*

Hydroquinone, as preformed antimicrobial defense chemical, 195

Hydroxamic acids, as preformed antimicrobial defense chemicals, 195-196

N-(O-Hydroxyphenyl)sulfinamoyl-tertiobutyl acetate (OH-PAS), in lignification, 194

Hydroxyproline-rich glycoproteins, in hypersensitive reaction, 67

Hypersensitive reaction
    lignins in, 193
    phenolics in, 195
    in plant resistance to pathogens, 67,

69-70, 80-83
Hypervariable probes, in DNA
    fingerprinting, 343-344, 346, 348-349
Hyphal anastomosis, in *Trichoderma* spp.,
    240-241

Ice-nucleation active (INA) bacteria
    in frost damage, 282
    monitoring of, 309
IFA. *See* Immunofluorescence assay
Immunofluorescence assay (IFA), in
    diagnosis of plant pathogens, 325-326
Immunogold labeling, in diagnosis of plant
    pathogens, 325-326
INA bacteria. *See* Ice-nucleation active
    bacteria
Indolacetic acid, in plant diseases, 118
Induced resistance, phenolics as signals for,
    198-200
Industrial raw materials, increased demand
    for, agricultural productivity role in
    satisfying, 16
Insect tolerance, of transgenic crop plants,
    5-6
Integration vectors, development of,
    285-286
Isoflavonoids, as induced antimicrobial
    defense chemicals, 196
Isolectin liposomes, k1 toxin in, 263
Isozyme analysis, of progeny from
    protoplast fusion in *Trichoderma*
    spp., 243-244, 250

Japanese radish, lignins in, 193
Jasmonic acid, in wound-induced gene
    activation, 162-163, 165, 167

k1 killer toxin, in *Saccharomyces cerevisiae*,
    261-264, 266
k2 killer toxin, in *Saccharomyces cerevisiae*,
    264
k3 killer toxin, in *Saccharomyces cerevisiae*,
    264
K28 killer toxin, in *Saccharomyces
    cerevisiae*, 261-263
Kanamycin, bacterial resistance to, 25
Kievitone, detoxification of, 197
Killer phenomenon, in fungi, 257-270
King's medium B, nonfluorescence of
    Pseudomonas on, 33
Klebsiella oxytoca
    pectate lyase in, 56
    pullulanase in, 40, 57
Klebsiella spp., pullulanase in, 41-44, 48-51

Kluveromyces lactis, killer phenomenon in,
    260-263
Kluveromyces phaffi, killer phenomenon in,
    262
KP1 killer toxin, in *Ustilago maydis*,
    265-266
KP4 killer toxin, in *Ustilago maydis*,
    265-266
KP6 killer toxin, in *Ustilago maydis*,
    260-263, 265, 267-269

Lactone, expressing bacterial avirulence
    gene D, 68
Lactose, as *Pseudomonas* energy source, 27
Lactose permease, in monitoring biocontrol
    bacteria in rhizosphere, 307
*lac*ZY-microbial tracking system, for
    pseudomonads, in take-all disease
    biocontrol, 25-27, 29-33, 35
LcrD protein, in *Yersinia* spp., 58
Leaves, inoculated, coat protein-mediated
    resistance in reduction of number of
    infection sites on, 95-96
Lepidopterans
    Bacillus thuringiensis against, 282
    Pseudomonas fluorescens against,
        282-283
Leptosphaeria maculans, typing genetic
    differences in isolates of, 345-346
Lettuce
    DNA segments introduced into, 4
    nuclear male sterility and, 10
Leucine aminopeptidase, wound-induced,
    166, 168
Levanase gene, in sugarcane, 284
Lignins
    in hypersensitive response, 67
    as plant cell wall components, 192, 194
    systemic resistance induction and, 304
Linker-scanning mutagenesis, in assessment
    of wound-induced proteinase
    inhibitor expression, 164-165
Luminometer, in bacterial detection, 307
Luteone, as spore germination inhibitor, 194
lux gene cassette, promoterless, 309
lux operon, in monitoring biocontrol bacteria
    in rhizosphere, 307
Lycopersicon esculentum cv. Mountain
    Pride, *Pseudomonas corrugata* as
    minor pathogen of, 32
Lysozyme genes, in disease resistance,
    185-187
Lytic enzymes, activation of during fungal
    attack, 140-143

364

Lytic peptides, in disease resistance,
177-187

Maackiain, detoxification of, 197
Magainins, in disease resistance, 177,
183-184
Magnaporthe grisea, typing genetic
differences in isolates of, 346
Maize. *See* Corn
Mannopine utilization, in rhizosphere
colonization, 294-296
Medicarpin, detoxification of, 197
Melittins, in disease resistance, 177, 183-184
Melon
chitinases in, 140-141
coat protein-mediated resistance in, 92
lignins in, 193
Messenger RNA (mRNA)
cathepsin D inhibitor, 166-167
chitinase, 141
endochitinase, 213-214
leucine aminopeptidase, 166, 168
pathogenesis-related protein, 200-202
pin2, 160-163, 165
salicylic acid, 200-202
Shiva-1, 180
thiol proteinase inhibitor, 166-167
threonine dehydrase, 166, 168
Metalloproteinase inhibitors, accumulation
of upon wounding, 169
Methylcellulose, in microencapsulation
technology, 28
4-Methylumbelliferyl chitotriose, chitinase
in digestion of, 214
MGR586 probe, in typing genetic
differences in fungal isolates, 346
Microencapsulation technology, for
microbial pesticides, in take-all
disease biocontrol, 27-29
Microsatellite probes, in DNA
fingerprinting, 344-345
Minisatellite probes, in DNA fingerprinting,
343-344, 346, 348-349
Minisatellite variant repeats (MVRs), in
DNA fingerprinting, 344, 351
Mitochondrial matrix, eukaryotic, import of
nuclear-encoded proteins into, 41
Monoclonal antibodies, as reagents, in
diagnosis of plant pathogens, 322-323
mRNA. *See* Messenger RNA
Muskmelon, lignins in, 193
MVRs. *See* Minisatellite variant repeats
Mycosphaerella graminicola, DNA

fingerprinting of, 341-351

Nalidixic acid
in monitoring biocontrol bacteria in
rhizosphere, 306-307
Pseudomonas resistance to, 24-25, 33
Nectria spp., phytoalexins in, 197
Nematodes
crop plant resistance to, 7, 9
potential and present losses in Australia
due to, 22
Neomycin, bacterial resistance to, 25
Neurospora crassa, chromosomes in, 245
NH$_2$-PAS. *See* N-(O-Aminophenyl)
sulfinamoyl-tertiobutyl acetate
Nicotiana tabacum
SB-37 gene cloned into, 179
Shiva-1 gene cloned into, 179-182
nod genes, induction of, 192
Nonhost resistance, avirulence genes in, 69
Nopaline Ti plasmids, *Agrobacterium
tumefaciens* carrying, 277, 280
North Carolina Biotechnology Center,
survey of biotechnology funding and
research in academia conducted by,
18
Nuclear male sterility, production of hybrid
seeds and, 10
Nucleic acid hybridization
applications of, 332-334
detection techniques based on, 331-332
principles of, 330-331
Nucleic acid probes, in diagnosis of plant
pathogens, 329-332
Nutrients, competition of, 305-306

Oats
fungal killer toxins in, 269
halo blight in, 120
Helminthosporium victoriae-induced
symptoms in, 117
Pc-2 disease resistance gene in, 79
Octopine Ti plasmids, *Agrobacterium
tumefaciens* carrying, 280
OH-PAS. *See* N-(O-Hydroxyphenyl)
sulfinamoyl-tertiobutyl acetate
Oilseed rape, transgenic, introduction of
DNA segments in, 4
Olfactory perception, resistance genes for,
69
Oligosaccharides, in wound-induced gene
activation, 161-162
OMT. *See ortho*-diphenol-methyltransferase

Oncogenes, in modification of cellular
    growth factor activity, 3
Ondontoglossum ringspot virus (ORSV),
    coat protein-mediated resistance to,
    93
Oomycin A
    afuE locus in synthesis of, 309
    against *Pythium* spp., 298
Open reading frames (ORFs), fungal killer
    toxins encoded from, 261, 267
Ophiobolin, in plant diseases, 118-119
Ophiobolus miyabeanus, pathogenicity of,
    118
ORFs. *See* Open reading frames
ORSV. *See* Ondontoglossum ringspot virus
ortho-diphenol-methyltransferase (OMT), in
    lignin precursor synthesis and
    polymerization, 193
Oxalate, in induced resistance, 204

P1 killer toxin, in *Ustilago maydis*, 264-265
P4 killer toxin, in *Ustilago maydis*, 264-265
P6 killer toxin, in *Ustilago maydis*, 261, 264-
    267
PAL. *See* Phenylalanine ammonia lyase
Papaya, coat protein-mediated resistance in,
    92, 98
Papaya ringspot potyvirus, coat protein-
    mediated resistance to, 98
Parthenocissus quinquifolia, lysozyme in,
    187
Pathogenesis-related (PR) proteins
    accumulation of, 140-142
    in induced resistance, 199
    salicylic acid and, 200-202, 204
Pathogenicity
    disease protection and, 116-117
    role of toxins in, 117-119
Pc-2 disease resistance gene, elicitor
    interaction with, 79
PCR. *See* Polymerase chain reaction
PC-toxin, in plant diseases, 118
Pea
    damping-off diseases in, 300
    DNA segments introduced into, 4
    phytoalexins in, 197
    volatile compounds in, 306
Pea early browning virus (PEBV), coat
    protein-mediated resistance to, 93, 98
Pear
    DNA segments introduced into, 4
    fire blight disease in, 58
    hydroquinone in, 195

Peat and clay mixes, in microencapsulation
    technology, 28
PEBV. *See* Pea early browning virus
Pectate lyase
    in *Erwinia* spp., 45-46, 52-57
    in *Klebsiella oxytoca*, 56
    in *Xanthomonas campestris* pv.
        campestris, 45, 52
Pectin, cloned gene product degradation of,
    287
Pectin lyase
    in *Erwinia carotovora*, 42
    in *Erwinia* spp., 45
Pectin methylesterase, in *Erwinia* spp., 46,
    56
PeMV. *See* Pepper mottle virus
Penicillium spp., progeny from protoplast
    fusion in *Trichoderma* spp. and, 246,
    248
Pepper mild mottle virus (PPMV), coat
    protein-mediated resistance to, 93, 97
Pepper mottle virus (PeMV), coat protein-
    mediated resistance to, 94, 98-99
Peppers, disease resistance genes in, 74
Perception of risk, public, 17
Periconia circinata, pathogenicity of, 118
Peronospora maydis, polymerase chain
    reaction analysis of, 334
Peronospora parasitica, radish infected
    with, 193
Peronospora sacchari, polymerase chain
    reaction analysis of, 334
Peronospora tabacina
    induced resistance to, 199
    polymerase chain reaction analysis of,
        334
Peroxidase, in lignin precursor synthesis and
    polymerization, 193
Persea americana, cloning of disease
    resistance genes into, 74
Pesticides, integrated use of, 6
Petunia, transgenic, introduction of DNA
    segments in, 4
PFGE. *See* Pulse-field gel electrophoresis
Phage T4 lysozyme, in resistance to bacterial
    pathogens, 9
Pharmaceutical raw materials, increased
    demand for, agricultural productivity
    role in satisfying, 15-16, 35
Phaseollidine, detoxification of, 197
Phaseollin, detoxification of, 197
Phaseollinisoflavan, detoxification of, 197
Phaseolotoxin, in plant diseases, 118-119,

121
Phenazine
β-galactosidase reporter gene fusion to, 309
against take-all disease, 21, 26, 296-297
Phenolics
background information on, 191-192
endogenous signaling and, 200-202
as induced antimicrobial defense chemicals, 196-198
as plant cell wall components, 192-194
practical applications for, 202-204
as preformed antimicrobial defense chemicals, 194-196
as signals for induced resistance, 198-200
see also specific compounds
Phenotypes
coat protein-mediated resistance, 92
elite, 3
killer, 258-260
Sec, 42
Phenylalanine ammonia lyase (PAL), in lignin precursor synthesis and polymerization, 193-194
Phloretin, as preformed antimicrobial defense chemical, 195
Phloridzin, as preformed antimicrobial defense chemical, 195
Phloroglucinol derivatives, in take-all disease control, 21
Phosphate, in induced resistance, 204
Phosphorylation, of plant proteins, elicitor activity in, 70
Phycomyces blakesleeanus, endo-chitinases against, 214
Phyllostica maydis, pathogenicity of, 118
Phytoalexins
in hypersensitive response, 67
as induced antimicrobial defense chemicals, 196-198
Phytohormones, in plant diseases, 119
Phytophthora cinnamoni
cecropin B against, 179
in plant tissue, 328
resistance to, 74
Phytophthora citrophthora, in soil and water, 329
Phytophthora infestans
cell wall-degrading enzyme levels after infection by, 141-142
polymerase chain reaction analysis of, 334

in potato late blight disease, 73
Phytophthora megasperma
glyceollin against, 197
phytoalexins against, 196-197
Phytophthora parasitica
cell wall-degrading enzyme levels after infection by, 141
in soil and water, 329
Phytophthora parasitica f.sp. *nicotianae*, polymerase chain reaction analysis of, 334
Phytophthora sojae, in soil and water, 329
Phytophthora spp.
in plant tissue, 328
in soil and water, 329
Trichoderma spp. protoplast fusion progeny against, 252
Pichia acaciae, killer phenomenon in, 261-262
Pichia anomala, killer phenomenon in, 262
Pichia farinosa, killer phenomenon in, 262
Pichia kluyveri, killer phenomenon in, 262-263
Pilin assembly proteins, in *Pseudomonas aeruginosa*, 49-51
Pin2 protein, wound-induced, 159-165
Pisatin, detoxification of, 197
PKT240*tac*, restriction enzyme map of, 285
Plant breeding, molecular
background information on, 1-5
insect tolerance and, 5-6
nuclear male sterility and hybrid seed production, 10
prospects for, 10-11
resistance to bacteria and nematodes, 7, 9
stress tolerance and, 9-10
tolerance to fungal diseases and, 7-9
virus resistance and, 6
Plant tissue, detection of pathogens in, 327-328
PLRV. *See* Potato leaf roll virus
Plum pox virus (PPV), coat protein-mediated resistance to, 94
PM-toxin, in plant diseases, 118
Polyacetylenes, as induced antimicrobial defense chemicals, 196
Polyclonal antibodies, as reagents, in diagnosis of plant pathogens, 322-323
Polygalacturonase
in *Erwinia* spp., 45
in *Pseudomonas solanacearum*, 45, 52, 55
Polymerase chain reaction (PCR)

in diagnosis of plant pathogens, 308, 332
in DNA fingerprinting, 344-345, 351
inverted, in gene tagging, 77
linked markers identified via, 75
in subtraction hybridization cloning, 78
Polysaccharide lyases, in biocontrol, 213,
Poly(vinyl alcohol), *Pseudomonas*
  degradation of, 28
Poplar, transgenic, introduction of DNA
  segments in, 4
Pore formation, in plasma membrane, 177
Potato
  chitinase in, 214
  chlorogenic acid in, 195
  coat protein-mediated resistance in, 92
  DNA segments introduced into, 4-5
  gene tagging in, 77
  genetic transformation in, 73
  lignins in, 193
  proteinase inhibitors in response of to
    wounding, 157-171
  systemic resistance to *Pseudomonas
    solanacearum* induced in, 304
Potato droopy mutants, abscisic acid
  deficiency in, 165
Potato late blight disease, disease resistance
  genes against, 73
Potato leaf roll virus (PLRV)
  coat protein-mediated resistance to, 94
  expression of antisense RNAs against, 90
Potato virus X (PVX)
  coat protein-mediated resistance to,
    94-95, 97
  expression of replicase of, 90, 100
Potato virus Y (PVY)
  coat protein-mediated resistance to, 94
  expression of replicase of, 98-100
Potyvirus group, coat protein-mediated
  resistance to, 98-99
PPMV. *See* Pepper mild mottle virus
PPV. *See* Plum pox virus
Proanthocyanidins, as preformed
  antimicrobial defense chemicals, 195
Promoterless reporter genes, increasing use
  of, 309
Protease inhibitors, in hypersensitive
  response, 67
Proteases
  in *Erwinia* spp., 46
  in *Xanthomonas campestris* pv.
    campestris, 45
Protein
  cloned gene product degradation of, 287

pathogenic bacteria in secretion of, 39-58
Proteinase inhibitors, wound-induced
  accumulation of, 169
  background information on, 157-159
  chimeric promoter construction and,
    169-170
  genes induced in response to wounding
    and, 165-168
  manipulation of signal transduction
    pathway to generate hyperreactive
    plants, 170-171
  prospects for, 171
  proteinase inhibitor II gene family
    developmental and environmental
      factors influence pin2 gene
      expression, 160-163
    gene family, 159
    promoter elements, 163-165
    Shiva-1 gene and, 180-182
Protein kinase inhibitors, in inhibition of
  elicitor activity, 70
Protein kinases, in derepression of defense
  response genes, 70
Protein phosphatases, in derepression of
  defense response genes, 70
Protocatechuic acid, as spore germination
  inhibitor, 195
Protoplast fusion
  in *Kluveromyces lactis*, 260
  in *Trichoderma* spp., 237-253
PR proteins. *See* Pathogenesis-related
  proteins
Pseudobactin M114, in biocontrol, 300
Pseudobactin MT3A, in biocontrol, 300
Pseudocercosporella herpotrichoides, in
  plant tissue, 328
Pseudomonas aeruginosa
  23S rRNA gene in, 34
  Sec-dependent secretion pathway and,
    49-51
Pseudomonas aureofaciens
  phenazine locus in, 309
  against take-all disease, 25-27, 29-30, 33,
    35
Pseudomonas cepacia
  detection of, 308
  in rhizosphere colonization, 292
  against take-all disease, 34
Pseudomonas corrugata 2140 strain, against
  take-all disease, 29, 31-32, 34
Pseudomonas delafieldii, against take-all
  disease, 34
Pseudomonas F agar, nonfluorescence of

Pseudomonas on, 33
Pseudomonas fluorescens
  afuE locus of, 309
  chitinase in, 221, 302
  against lepidopterans, 282-284
  in monitoring biocontrol bacteria in
    rhizosphere, 307
  in mutant analysis, 311-312
  Pythium spp. controlled by, 298
  in rhizosphere colonization, 292-293,
    297
  siderophores and, 299
  systemic resistance induction and, 304
  against take-all disease, 30-32, 292-293,
    297
  volatile compounds and, 301
Pseudomonas gladioli, against take-all
    disease, 34
Pseudomonas marginalis, against take-all
    disease, 31
Pseudomonas methanolica, against take-all
    disease, 34
Pseudomonas putida
  chitinase in, 221-224
  competition for nutrients and, 306
  detection of, 307
  elimination of signals and, 306
  monitoring of, 309-310
  in rhizosphere colonization, 292
  siderophores and, 299-300
  systemic resistance induction and, 304
  tac promoter in, 284
  against take-all disease, 31
Pseudomonas solanacearum
  avirulence genes cloned from, 69
  cecropin B against, 178
  extracellular virulence proteins of, 45,
    52, 55
  mutants of, with deficiencies in plant cell
    wall degrading enzymes, 40-41
  SB-37 peptide against, 187
  Shiva-1 peptide against, 180-182, 187
  systemic resistance to, 304
Pseudomonas spp.
  Bacillus thuringiensis toxin gene cloned
    into, 284
  as delivery organisms, 286
  in rhizosphere colonization, 294
Pseudomonas stutzeri
  chitinase in, 302
  β-1,3-glucanase in, 302
Pseudomonas syringae
  cell wall-degrading enzyme levels after

    infection by, 141
  in frost damage, 282
  lactone expressing avirulence gene D in,
    68
  phytoalexins against, 198
  protein secretion pathways of, 57-58
Pseudomonas syringae pv. *atropurpurea*,
  pathogenicity of, 118-119
Pseudomonas syringae pv. *coronafaciens*
  pathogenicity of, 120
Pseudomonas syringae pv. *glycinea*,
  avirulence genes in, 69
Pseudomonas syringae pv. *phaseolica*,
  pathogenicity of, 118-119, 121
Pseudomonas syringae pv. *syringae*,
  pathogenicity of, 118
Pseudomonas syringae pv. *tabaci*
  cecropin B against, 178
  pathogenicity of, 118-123, 125, 131, 133
Pseudomonas syringae pv. *tomato*,
  avirulence genes cloned from, 69
Pseudomycin, in biocontrol, 298
Pullulanase, *Klebsiella* spp. secretion of,
  40-44, 48-51, 57
Pulse-field gel electrophoresis (PFGE), in
  establishment of long-range
  restriction maps, 76
pupB gene, in iron transport, 300
PVX. *See* Potato virus X
PVY. *See* Potato virus Y
Pyicularia grisea, typing genetic differences
  in isolates of, 346
Pyocyanine, in biocontrol, 298
Pyoluteorin
  in biocontrol, 298
  increased production of, 312-313
Pyoverdines, in biocontrol, 299, 309
Pyrrolnitrin, in biocontrol, 21, 298
Pythium aphanidermatum, cell wall-
  degrading enzymes in resistance to,
  147
Pythium aristosporum-based root disease,
  take-all fungus control and, 22, 26
Pythium nunn, in biocontrol activity, 212
Pythium spp.
  competition for nutrients and, 305-306
  in damping-off diseases, 300-301
  elimination of signals and, 305-306
  oomycin A against, 298
  in plant tissue, 328
  in soil and water, 329
Pythium ultimum
  polymerase chain reaction analysis of,

334
siderophores and, 299
Trichoderma spp. protoplast fusion
progeny against, 252

Radish, lignins in, 193
RAPD oligonucleotide probes, in typing
genetic differences in fungal plant
pathogens, 345-346, 351
Rapeseed
blackleg disease in, 345
nuclear male sterility and, 10
rbcs gene, coat protein-mediated resistance
to, 96
Recognitional specificity, of cloned disease
resistance genes, modification of,
82-83
Regulatory sequences, in molecular plant
breeding, 4-5
Reo virus type 3, segmented dsDNA
molecules from, 264
Reporter genes, promoterless, 309
Restorer genes, lethal effect of dominant
male sterility gene and, 10
Restriction fragment length polymorphism
(RFLP) analysis
in diagnosis of plant pathogens, 331-332
in DNA fingerprinting, 343, 347, 351
linked markers identified via, 75, 76
of loci involved in stress tolerance, 9, 10
of progeny from protoplast fusion in
Trichoderma spp., 245-246,
250-251
RFLP analysis. *See* Restriction fragment
length polymorphism analysis
Rhizobitoxine, in plant diseases, 118
Rhizobium japonici cus, pathogenicity of,
118
Rhizobium meliloti, chitinase in, 221-223
Rhizobium spp.
colonization by, 287
nod genes in, 192
Rhizoctonia solani
biocontrol of, 7-9
cecropin B against, 179
cell wall-degrading enzymes in resistance
to, 140, 145-152
chitinases against, 214-215, 223-224
in rhizosphere colonization, 293
take-all fungus control and, 22, 26
Trichoderma spp. protoplast fusion
progeny against, 252
Rhizoctonia spp., in plant tissue, 328

Rhodamine, in immunofluorescence assays,
325
Rhodotorula glutinis, killer phenomenon in,
262
Ribgrass mosaic virus (RMV), coat protein-
mediated resistance to, 93
Ribose biphosphate carboxylase, coat
protein-mediated resistance to, 96
Ribosomal RNA (rRNA)
Pseudomonas aeruginosa, 34
Pseudomonas corrugata 2140 strain, 29
Ribozome-inhibiting protein, in biocontrol
of plant pathogenic fungi, 7-9
Ribozymes, in control of plant viral diseases,
6
Rice
coat protein-mediated resistance in, 92
DNA segments introduced into, 4
Rice blast fungus, *Magnaporthe grisea* in,
346
Rice leaf spot disease, *Helminthosporium
oryze* in, 119
Rice stripe virus (RSV), coat protein-
mediated resistance to, 94
Rice tungro disease, viral resistance genes
in, 89
Rifampicin
in monitoring biocontrol bacteria in
rhizosphere, 306-307
Pseudomonas resistance to, 24-25, 27, 33
RIP. *See* Ribosome-inhibiting protein
Ripening and spoilage, control of, 11
RMV. *See* Ribgrass mosaic virus
RNA *N*-glycosidase, in modification of 28S
rRNA, 7
Rpg4 gene, in soybean, 68
rRNA. *See* Ribosomal RNA
RSV. *See* Rice stripe virus
Rust fungus, lignins in presence of, 193
Rye, hydroxamic acids in, 196
Ryegrass, halo blight in, 119

Saccharomyces cerevisiae, killer
phenomenon in, 257-258, 260-264,
266-267
Salicylic acid, endogenous, in disease
resistance, 200-204
Salmonella typhimurium, cell envelope of,
40
Sarcotoxins, in disease resistance, 177
Satellite RNA, viral replication limited by, 6,
90
SB-37 peptide, in disease resistance,

178-179, 182, 184, 187

Scab, disease resistance genes against, 71

Scintillation counter, in bacterial detection, 307

Sclerotinia homeocarpa, Trichoderma spp. protoplast fusion progeny against, 252

Sclerotinia spp., in plant tissue, 328

Sclerotium rolfsii
  biocontrol of, 7
  cell wall-degrading enzymes in resistance to, 145
  chitinases against, 214-215, 220, 223-224
  Trichoderma spp. protoplast fusion progeny against, 252

scrA gene, in sugarcane, 283-284

Sec-mediated general export pathway, across cell envelope, 41-44, 46, 49-57

Secretion pathway genes, mutagenesis and cloning of, 47-49

Self-pollination, male sterile plants in control of, 10

Septoria nodorum
  in plant tissue, 328
  potential and present losses in Australia due to, 22

Septoria tritici
  host—pathogen interactions and, 342-343
  in plant tissue, 328
  in wheat, 347-351

Serine-metalloproteinase inhibitors, accumulation of upon wounding, 169

Serratia marcescens
  in biocontrol of plant pathogenic fungi, 7
  chitinase in, 214, 216, 221, 302-303

Sexual crosses, progeny of, resistance genes with altered recognitional specificities recovered in, 70

Shipping costs, reduction of, 10

Shiva-1 peptide, in disease resistance, 179-182, 187

Shiva-11 peptide, in disease resistance, 183-184

SHMV. *See* Sunn hemp mosaic virus

Siderophores, in biocontrol, 298-300; *see also specific siderophores*

Signal peptidase II, specificity of for lipoproteins, 44

Signals, elimination of, 305-306

Signal transduction pathway
  in hypersensitive reaction, 70
  wound-induced proteinase inhibitors and,

170-171

SMV. *See* Soybean mosaic virus

Soft-rot *erwinias*, diseases caused by, 45

Soil, detection of pathogens in, 328-329

Solanum demissum, disease resistance genes from, 73

Soybean
  avirulence gene elicitation of hypersensitive response in, 69
  chitinases in, 140
  disease resistance genes in, 74
  DNA segments introduced into, 4
  functional cloning in, 80
  phytoalexins in, 196
  Pseudomonas identified from root washings of, 34
  resistance genes in, 68
  wildfire of, 120
  wound-induced expression of jasmonic acid in, 163

Soybean mosaic virus (SMV), coat protein-mediated resistance to, 98-99

Species-level resistance, avirulence genes in, 69

Specific elicitors, plant recognition of, 67-68

Spore germination inhibitors, as preformed antimicrobial defense chemicals, 194-195

Sporidiobolus pararoseus, killer phenomenon in, 262

Staphylococcus aureus, synthetic peptides against, 183

Stem and pith necrosis, *Pseudomonas corrugata* in, 32

Stem disease, potential losses in Australia due to, 22

Stilbenes, as induced antimicrobial defense chemicals, 196

Storage costs, reduction of, 10

Streptomyces hygroscopicus, bialophos produced by, 120-121

Streptomycin, bacterial resistance to, 25

Stress tolerance, of transgenic plants, 9-10

Striga spp., phenolics in, 191

Stripe rust, potential losses in Australia due to, 22

Strong resistance genes, effectiveness of, 74

Subtraction hybridization cloning, of disease resistance genes, 78-79

Sugarbeet
  coat protein-mediated resistance in, 92
  DNA segments introduced into, 4

Sugarcane, *Eldana saccharina* damage to,

282-284
Sunflower, transgenic, introduction of DNA
    segments in, 4
Sunn hemp mosaic virus (SHMV), coat
    protein-mediated resistance to, 93, 97
Surface-active peptides, in disease
    resistance, 176-187
Syringomycin, in plant diseases, 118
Systemic resistance, induction of, 303-305

Tabtoxin, structure of, 120
Tabtoxin resistance genes
    cloning of, 121-122
    construction of binary vector with,   129-
        130
    functions of, 122-126
    gene transfer into plants, 127-129
    introduction of into plants, 130-131,
        133-134
    Northern blotting of, 132
    nucleotide sequence of, 127
    in plant diseases, 118-119
    selection of tabtoxin-resistant plants,
        131-132
    self-resistance to toxic metabolites in
        microorganisms, 120-121
    tabtoxin-acetylating enzyme gene
        structure, 124-127
    wildfire disease resistance and, 131-133
Take-all disease
    biocontrol of, 15-36
    chitinase against, 221
    2,4-diacetylphloroglucinol against,
        297-298
    phenazine against, 296-297
    Pseudomonas fluorescens against,
        292-293
Tannins, as preformed antimicrobial defense
    chemicals, 195
tDNA. *See* Transfer DNA
Tentoxin, in plant diseases, 118
Terpenoids, as induced antimicrobial
    defense chemicals, 196
TEV. *See* Tobacco etch virus
Thalli, heterokaryotic, progeny from
    protoplast fusion in *Trichoderma* spp.
    and, 246
Thielaviopsis basicola, in black root rot,
    297, 304
Thiol-metalloproteinase inhibitors,
    accumulation of upon wounding, 169
Thiol proteinase inhibitor, wound-induced,
    166-168

Threonine dehydrase, wound-induced, 166,
    168
TMGMV. *See* Tobacco mild green mosaic
    virus
TMV. *See* Tobacco mosaic virus
TNV. *See* Tobacco necrosis virus
Tobacco
    agrocin 84 in, crown gall in, 281
    avirulence gene elicitation of
        hypersensitive response in, 69
    chitinases in, 140-141
    coat protein-mediated resistance in, 92
    DNA segments introduced into, 4-5, 7-9
    gene tagging in, 77
    hypersensitive response in, 57
    nuclear male sterility and, 10
    phenazine in, 297-298
    salicylic acid in, 200-202
    SB-37 gene cloned into, 179
    Shiva-1 gene cloned into, 179-182
Tobacco etch virus (TEV), coat protein-
    mediated resistance to, 94, 98-99
Tobacco golden mosaic geminivirus,
    expression of antisense RNAs against,
    90
Tobacco mild green mosaic virus
    (TMGMV), coat protein-mediated
    resistance to, 93
Tobacco mosaic virus (TMV)
    coat protein-mediated resistance to,
        92-93, 95, 97, 99-100
    lignins and, 193
    salicylic acid against, 200-202
    specific elicitor in, 68
    viral resistance genes in, 89-90
Tobacco necrosis virus (TNV)
    induced resistance to, 198
    salicylic acid against, 199-200, 203
Tobacco rattle virus (TRV), coat protein-
    mediated resistance to, 93, 95-96, 98
Tobacco streak virus (TSV), coat protein-
    mediated resistance to, 93
Tomato
    agrocin 84 in, 277-278, 281
    avirulence gene elicitation of
        hypersensitive response in, 69
    cell wall-degrading enzymes in, 142
    chitinases in, 140
    chromosome walking in, 75
    coat protein-mediated resistance in, 92,
        99-100
    DNA segments introduced into, 4-5
    gene tagging in, 77

Pseudomonas corrugata as minor
  pathogen of, 29, 32, 34
ripening and spoilage of, 11
Tomato mosaic virus (ToMV), coat protein-
  mediated resistance to, 93, 97-98
Tomato spotted wilt virus (TSWV), coat
  protein-mediated resistance to, 94
ToMV. *See* Tomato mosaic virus
Torulopsis glabrata, killer phenomenon in,
  262-263
tox gene, in lepidopterans, 282
Toxigenicity, in plant pathogens, 116-119
Transfer DNA (tDNA), in gene tagging, 78
Transposons
  in gene tagging, 77-78
  in subtraction hybridization cloning, 78
Trees, genetic transformation in, 74
Trichoderma harzianum
  in biocontrol activity, 212, 223, 225-231
  chitinase in, 219
  chromosomes in, 245
  endo-chitinases against, 214
  protoplast fusion and, 231, 245, 249,
    251-252
Trichoderma koningii, protoplast fusion and,
  249
Trichoderma reesei
  cellulase gene in, 213
  endo-chitinases against, 214
Trichoderma spp.
  chitinase in, 219-220
  improvement of as biocontrol agent, 223,
    225-231
  progeny from protoplast fusion in,
    239-253
    background information on, 237-238
    biological control efficacy of,
      252-253
    chromosome electrophoresis and,
      244-245, 250-251
    compatibility and, 246-247
    genetic analysis of progeny, 243
    genetic events following fusion,
      248-251
    haploidization agents and, 246
    heterokaryotic thalli and, 246
    hypothesis to explain postfusion
      genetic events, 251-252
    isozyme analysis and, 243-244, 250
    morphological and nutritional events
      following fusion, 248-249
    phenotypic markers and, 243
    protoplast fusion, 240-241

protoplast preparation, 239
resolution of heterokaryons, 243
RFLP analysis and, 245-246, 250-251
selection of progeny, 242
visualization of fused and
  regenerating protoplasts,
  240-242
Trichoderma viride
  cell wall-degrading enzyme levels after
    infection by, 142
  protoplast fusion and, 249
Trichosporon capitatum, killer phenomenon
  in, 262
Triticum aestivum L., host—pathogen
  interactions and, 342
Triticum durum Desf., host—pathogen
  interactions and, 342
Triticum spp., host-pathogen interactions
  and, 343
TRV. *See* Tobacco rattle virus
TSV. *See* Tobacco streak virus
TSWV. *See* Tomato spotted wilt virus
Tuber-specific proteins, in wound-induced
  gene activation, 166
Tuberonic acid, jasmonic acid structural
  similarity to, 165
Turnip, transgenic, introduction of DNA
  segments in, 4

Useful traits, addition of, 3
Ustilago avenae, killer phenomenon in,
  268-269
Ustilago hordei, killer phenomenon in,
  268-269
Ustilago maydis, killer phenomenon in,
  258-269
Ustilago nigra, killer phenomenon in, 268
Ustilago tritici, killer phenomenon in, 268

Variable number of tandem repeats
  (VNTRs), in DNA fingerprinting,
  343-344, 347-348, 351
Vegetative storage proteins, jasmonic acid in
  wound-induced expression of, 163
Venturia inequalis, phenolics against, 195
Verticillium albo-atrum, polymerase chain
  reaction analysis of, 334
Verticillium dahliae
  polymerase chain reaction analysis of,
    334
  in rhizosphere colonization, 292
Vibrio alginolyticus, sucrose operon of,
  283-284

Vibrio bioluminescent genes, in monitoring biocontrol bacteria in rhizosphere, 307

Vibrio cholerae, cholera toxin secretion by, 44, 53, 55

Victorin, *Pc-2* disease resistance gene interaction with, 79

Viral RNA, overcoming coat protein-mediated resistance by inoculation with, 96-97

VirB11 protein, of *Agrobacterium tumefaciens*, 49-51

Virulence, protein secretion pathways in, 39-58

VNTRs. *See* Variable number of tandem repeats

Volatile compounds, in biocontrol, 300-301, 304, 306; *see also specific compounds*

Walnut, transgenic, introduction of DNA segments in, 4

Water, detection of pathogens in, 328-329

Wheat
    Australian diseases of, 22
    chitinases in, 140
    fungal killer toxins in, 269
    hydroxamic acids in, 196
    Mycosphaerella graminicola in, 341-351
    Septoria tritici blotch in, 342, 347-351
    take-all disease of, 15-36, 221, 292-293, 296-298

Wighteone, as spore germination inhibitor, 195

Wildfire bacteria, pathogenic toxins by, 119-120

Wildfire disease, resistance to in transgenic plants, 131-133

World population, growth of, as reason for increased agricultural productivity, 2, 16

Wounding, proteinase inhibitors in potato response to, 157-171

Xanthomonas campestris pv. *alfalfae*, avirulence genes in, 69

Xanthomonas campestris pv. *campestris*
    cecropin B against, 178
    extracellular virulence proteins of, 45
    mutants of, with deficiencies in plant cell wall degrading enzymes, 39-40, 47, 49-50, 52, 56

Xanthomonas campestris pv. *glycines*, avirulence genes in, 69

Xanthomonas campestris pv. *holcicola*, avirulence genes in, 69

Xanthomonas campestris pv. *malvacearum*, avirulence genes in, 69

Xanthomonas campestris pv. *phaseoli*, avirulence genes in, 69

Xanthomonas campestris pv. *vesicatoria*, avirulence genes in, 69, 74

Xanthomonas campestris pv. *vignicola*, avirulence genes in, 69

xcpA gene, in *Pseudomonas aeruginosa*, 51

X-Gal chromogenic dye, *Pseudomonas* cleavage of, 27

Xylan, cloned gene product degradation of, 287

YAC libraries. *See* Yeast artificial chromosome libraries

Yeast, killer phenomenon in, 257-269

Yeast artificial chromosome (YAC) libraries, in chromosome walking, 76

Yellow spot, present losses in Australia due to, 22

Yersinia enterocolitica, Sec-dependent secretion pathway and, 50-51, 57

Yersinia spp., virulence proteins of, 58

Yop proteins, secretion of, 50-51, 58

yscC gene, of *Yersinia enterocolitica*, 50-51, 57